TIME SERIES ANALYSIS

PWS PUBLISHERS

Prindle, Weber & Schmidt • 🐿 • Duxbury Press • ♠ • PWS Engineering • ⚠ • Breton Publishers • ⚙
Statler Office Building • Boston, Massachusetts 02116

PWS Publishers is a division of Wadsworth, Inc.

Library of Congress Cataloging-in-Publication Data

Cryer, Jonathan D.
　　Time series analysis.
　　Bibliography: p
　　Includes index.
　　　1. Time-series analysis—Data processing.　2. Minitab
(Computer system)　I. Title.
QA280.C78 1985　　519.5'5'02855369　　83–13165

ISBN 0-87150-963-6

Printed in the United States of America

86　87　88　89　90　—　10　9　8　7　6　5　4　3　2　1

Editor: Michael Payne
Production Coordinator and Designer: S. London
Production: Bookman Productions
Typesetting: The Universities Press (Belfast) Ltd.
Cover Printing: New England Book Components
Printing and Binding: R.R. Donnelley & Sons Company

TIME SERIES ANALYSIS

JONATHAN D. CRYER
UNIVERSITY OF IOWA

 DUXBURY PRESS, BOSTON

To my family

PREFACE

The theory and practice of time series analysis have developed rapidly since the appearance in 1970 of the seminal work of George E. P. Box and Gwilym M. Jenkins, *Time Series Analysis: Forecasting and Control*. Since then many books on time series have been published. However, some of them give inadequate theoretical background for the models while others give too little practical application of the methods. This book presents both theory and applications at a level accessible to a wide variety of students and practitioners.

Our approach is to mix theory, application, and computer software throughout the book as they are naturally needed. In particular, we develop the applications using the general purpose Minitab* statistical system. Minitab is widely available for various micro-, mini-, and mainframe computers. A Minitab primer (Appendix J) is provided for those new to Minitab. Many other statistical packages containing time series analysis procedures, especially for microcomputers, are now on the market. With a little extra effort, most of them could also be used with the book.

Except for brief diversions, we deal only with time domain ARIMA (Box–Jenkins) models. One exception, Chapter 3, presents the special aspects of regression models that are relevant in a time series setting. In this chapter we also discuss the fitting of cosines as deterministic cyclical trends, but this is as close to the frequency domain as we get. The final chapter touches on special topics, including combined deterministic trend/ARIMA models, ARIMA models from a Kalman filter point of view, missing data, outliers, and nonnormal white noise ARIMA models.

The book was developed for a one-semester course usually attended by students in statistics, economics, business, engineering, and the quantitative social sciences. Calculus is assumed to the extent of minimizing sums of squares. Matrix algebra is needed only in one section of the special topics chapter. A calculus-based introduction to statistics is necessary for a thorough understanding of the theory. However, the required facts concerning expectation, variance, covariance, and correlation are reviewed in an appendix. Also, conditional expectation properties and minimum mean square error prediction are developed in appendices.

Actual time series data drawn from various disciplines are used throughout the book to illustrate the methodology. All the computer printout exhibits were produced from camera-ready copy of actual Minitab sessions, albeit printed on a Hewlett-Packard Laserjet printer. Minitab Release 5.1, running on a PRIME 850

---

* Minitab is a registered trademark of Minitab, Inc.

minicomputer, was used. Other computers and other releases may give slightly different numerical results.

The original idea for the book came from Brian Joiner. Thomas Ryan and Barbara Ryan of Minitab, Inc., provided help, encouragement, and prerelease Minitab software over the extended writing period.

A preliminary version of this book received a critical reading by many students. I would like, especially, to thank Norman Loomer and Jose M. Silva. A nearly final version was improved through the efforts of Perry Drake, Ken Johnson, Rhonda Knehans, Carlos Eduardo Mosconi, and Raja Selvam. Duxbury Press reviewers, at various stages of the manuscript, included Keh-Shin Lii (University of California, Riverside), Joseph Glaz (University of Connecticut), Jack Y. Narayan (Syracuse University), and Neil Polhemus (Statistical Graphics Corporation). William R. Bell (U.S. Bureau of the Census), Lynne Billard (University of Georgia), and J. Keith Ord (Pennsylvania State University) provided many helpful comments on the complete manuscript.

I would also like to express my appreciation to the staff and associates of Duxbury Press, especially Michael Payne, Susan London, Paul Monsour, and Hal Lockwood and Kim Kist. Karen Ransom coped with the idiosyncracies of our new Hewlett-Packard Laserjet printer to produce all the camera-ready Minitab printouts.

Lastly. I would like to thank my typist, who spent numerous evenings, weekends, and holidays turning my nearly illegible handwriting into handsome technical print.

Jonathan D. Cryer

CONTENTS

CHAPTER 5: MODELS FOR NONSTATIONARY SERIES, 83

CHAPTER 6: MODEL SPECIFICATION, 103

CHAPTER 7: PARAMETER ESTIMATION, 125

CHAPTER 8: MODEL DIAGNOSTICS, 144

CHAPTER 9: FORECASTING, 161

CHAPTER 10: SEASONAL MODELS, 196

CHAPTER 11 SPECIAL TOPICS, 231

CHAPTER 1 INTRODUCTION

Data that are obtained from observations of a phenomenon over time are extremely common. In business and economics we observe weekly interest rates, daily closing stock prices, monthly price indices, yearly sales figures, and so forth. In meteorology we observe daily highs and lows in temperature, annual precipitation and drought indices, and hourly wind speeds. In agriculture we record annual figures for crop and livestock production, soil erosion, and export sales. In the biological sciences we observe the electrical activity of the heart at millisecond intervals. The list of areas in which time series are observed and analyzed is endless. The purposes of time series analysis are generally two-fold: to understand or model the stochastic mechanism that gives rise to an observed series and to predict or forecast future values of a series based on the history of that series.

1.1 EXAMPLES OF TIME SERIES

Exhibit 1.1 shows a time plot of the quarterly U.S. unemployment rates (seasonally adjusted) from 1948-I (first quarter) through 1978-I. Notice that the variation from quarter to quarter is relatively smooth with no discernible patterns or trends. However, the data clearly do not arise from *independent* observations.

The average monthly temperatures recorded in Dubuque, Iowa, from January 1964 to December 1975 present a rather different time series (Exhibit 1.2). Here there are very clear patterns and very clear reasons for those patterns! However, there is still variation to be modeled. Not all Januaries are alike but all Januaries are similar, and we should exploit this in our model. This also applies to every other month. In addition, neighboring months, such as June and July, have similar average temperatures, and our model should account for this also.

Exhibit 1.3 presents a third type of time series. Here we have a time plot of the monthly milk production per cow from January 1962 to December 1975. In this case we have an overall upward trend, perhaps linear, but a seasonal pattern as well. Our model and forecasts should reflect these patterns.

EXHIBIT 1.1 Quarterly U.S. Unemployment Rates, 1948-I to 1978-I, Seasonally Adjusted (a) Graph produced with Minitab. (b) Graph produced with SAS/GRAPH and a ZETA plotter.

(a) `MTB > tsplot 4 'Unemp'`

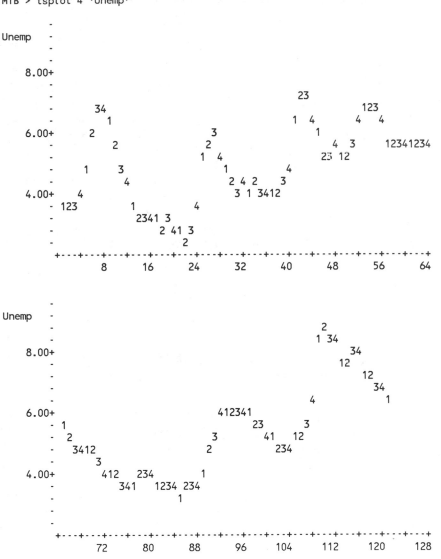

1.2 A MODEL-BUILDING STRATEGY

Finding appropriate models for time series is a nontrivial task. We will develop a multistep model-building strategy espoused so well by Box and Jenkins (1976). There are three main steps in the procedure, each of which may be used several times:

   1. model specification (or identification),

EXHIBIT 1.1 (continued)

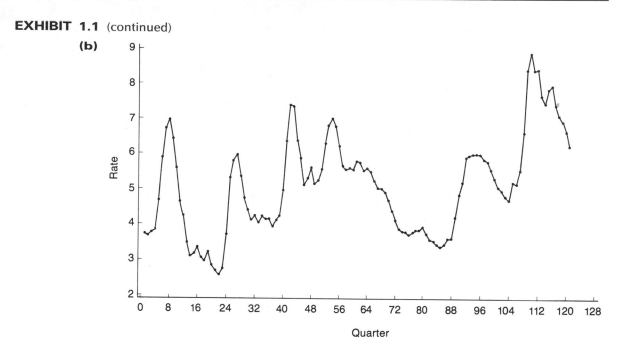

(b)

Rate (vertical axis), *Quarter* (horizontal axis)

2. **model fitting,** and

3. **model diagnostics.**

In model specification (or identification), the classes of time series models are selected that may be appropriate for a given observed series. In this step we look at the time plot of the series, compute many different statistics

EXHIBIT 1.2 Average Monthly Temperatures, Dubuque, Iowa, January 1964 to December 1975 (a) Graph produced with Minitab. (b) Graph produced with SAS/GRAPH and a ZETA plotter.

(a) `MTB > tsplot 12 'Temp'`

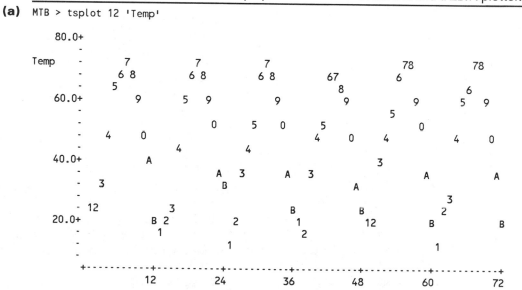

EXHIBIT 1.2 (continued)

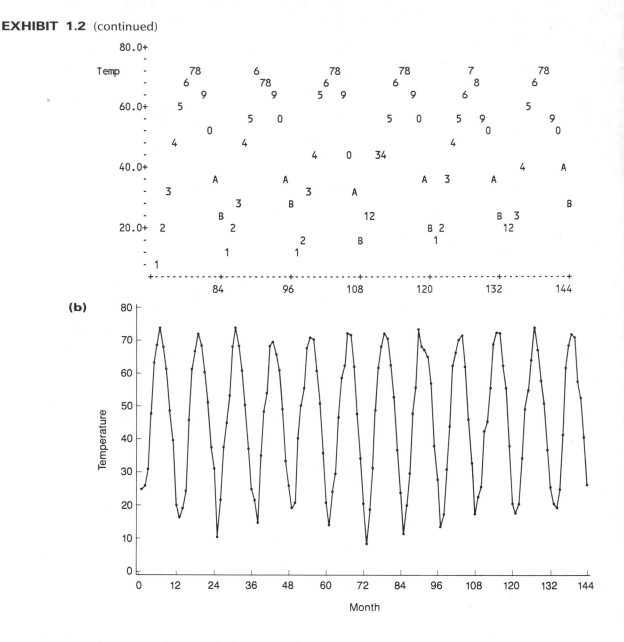

(b)

from the data, and also apply knowledge from the subject area in which the data arise, such as economics, physics, chemistry, or biology. It should be emphasized that the model chosen at this point is *tentative* and subject to revision later in the analysis.

In choosing a model, we shall attempt to adhere to the **principle of parsimony;** that is, the model used should require the smallest possible number of parameters that will adequately represent the data. Albert Einstein is quoted in Parzen (1982, p. 68) as remarking that "everything should be made as simple as possible but not simpler."

The model will inevitably involve one or more parameters whose values must be estimated from the observed series. Model fitting consists of finding the best possible estimates of those unknown parameters within a given model. We shall consider criteria such as least squares and maximum likelihood for estimation.

Model diagnostics is concerned with analyzing the quality of the model that we have specified and estimated. How well does the model fit the data? Are the assumptions of the model reasonably satisfied? If no inadequacies are found, the modeling may be assumed to be complete, and the model can be

EXHIBIT 1.3 Monthly Milk Production per Cow, January 1962 to December 1975 (a) Graph produced with Minitab. (b) Graph produced with SAS/GRAPH and a ZETA plotter.

(a)   MTB > tsplot 12 'Milkpr'

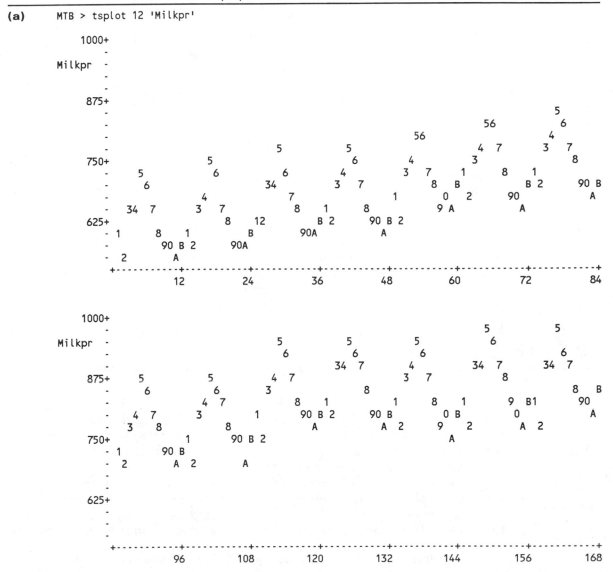

EXHIBIT 1.3 (continued)

(b)

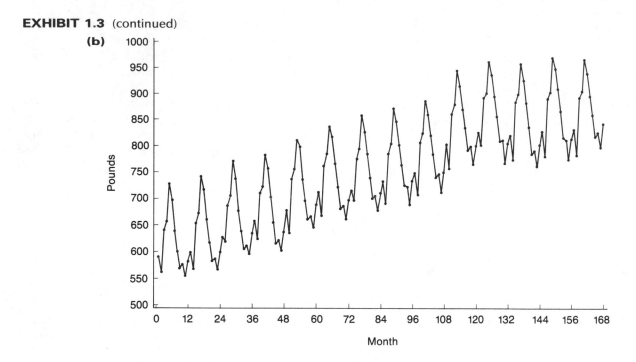

Month

used, for example, to forecast future series values. Otherwise, we choose another model in light of the inadequacies found; that is, we return to model specification. In this way we cycle through the three steps until, ideally, an acceptable model is found.

Because the computations required for each step in model building can be formidable, we shall rely on the readily available Minitab* statistical computing system, which will free us from much of the drudgery of computation. Readers unfamiliar with Minitab should consult Appendix J, "Minitab Primer," for a brief introduction and for a description of the presentation of Minitab commands in this book.

1.3 TIME SERIES PLOTS

According to Tufte (1983, p. 28), "The time-series plot is the most frequently used form of graphic design. With one dimension marching along to the regular rhythm of seconds, minutes, hours, days, weeks, months, years or millennia, the natural ordering of the time scale gives this design a strength and efficiency of interpretation found in no other graphic arrangement."

Figure 1.1 reproduces what appears to be the oldest known example of a time series plot, dating from the tenth (or possibly eleventh) century and showing the inclinations of the planetary orbits. Commenting on this artifact, Tufte says, "It appears as a mysterious and isolated wonder in the history of

* Minitab is a registered trademark of Minitab, Inc., State College, PA.

FIGURE 1.1   A Tenth-Century Time Series Plot

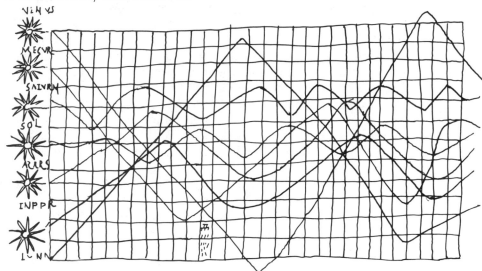

data graphics, since the next extant graphic of a plotted time-series shows up some 800 years later."

A plot of the observed values of a series versus time, such as in Exhibits 1.1 through 1.3, is the first step in any time series analysis. These plots are very easy to produce using the Minitab command TSPLOT. In its simplest version, the statement

TSPLOT of time series stored in column **C**

will plot the column of data on the vertical axis versus the integers 1, 2, . . . on the horizontal axis. (Here and throughout the book keystrokes that are mandatory are shown in boldface; other keystrokes are optional and do not affect the computation.)

For quarterly or monthly series, the command

TSPLOT [with period **K**] series in column **C**

is more useful. The monthly temperature data would be plotted using

TSPLOT with period **12** for temperature in **C1**

The TSPLOT command has four subcommands, TSTART, ORIGIN, INCREMENT, and START, which further control the plotting. Appendix J explains their use.

We also show in Exhibits 1.1 through 1.3 high-resolution plots produced with SAS/GRAPH* and a ZETA plotter. These plots could be included in a final report or a publication.

---

* SAS/GRAPH ™ is a registered trademark of SAS Institute, Inc., Cary, NC, USA.

1.4 OVERVIEW OF THE BOOK

Chapter 2 develops the basic ideas of mean, covariance, and correlation functions and ends with the important concept of stationarity. Chapter 3 discusses trend analysis and investigates how to estimate and check common deterministic trend models, such as those for linear and seasonal trends.

Chapter 4 begins the development of parametric models for stationary time series, namely the so-called autoregressive moving average (ARMA) models (also known as Box–Jenkins models). These models are then generalized in Chapter 5 to encompass certain types of stochastic nonstationary cases; these are known as ARIMA models.

Chapters 6, 7, and 8 form the heart of the model-building strategy for ARIMA modeling. Techniques are presented for tentatively specifying models (Chapter 6), efficiently estimating the parameters of those models using least squares and maximum likelihood (Chapter 7), and determining how well the models fit the data (Chapter 8).

Chapter 9 thoroughly develops the theory and methods of minimum mean square error forecasting for ARIMA models. Chapter 10 extends the ideas of Chapters 4 through 9 to stochastic seasonal models. Finally, Chapter 11 presents several special topics in time series, including Kalman filtering and missing data.

CHAPTER 2 FUNDAMENTAL CONCEPTS

This chapter describes the fundamental concepts in the theory of time series models. In particular, we introduce the concepts of stochastic process, mean and covariance function, stationary process, and autocorrelation function.

2.1 TIME SERIES AND STOCHASTIC PROCESSES

Examples of observed series were given in Chapter 1. We now need to consider the probabilistic structure that underlies these observations. We write Z_t for the observation made at time t. The units of time will vary with the application; they could be years, quarters, months, days, or even microseconds, depending on the situation to be modeled. We do assume that the observations are *equally spaced in time*.

In order to model the uncertainty in our observations, we assume that for each time point t, Z_t is a *random variable*. Thus the behavior of Z_t will be determined by a probability distribution. However, the most important feature of the time series models is that we assume that the observations made at different time points are statistically dependent. *It is precisely this dependence that we wish to investigate*. So, for two time points t and s, the joint behavior of Z_t and Z_s will be determined from a joint bivariate distribution; more generally, the probabilistic nature of the collection of random observations $Z_1, Z_2, \ldots, Z_n$ will be reflected in their multivariate joint distribution. The finite set of observations $Z_1, Z_2, \ldots, Z_n$ will be considered as a portion of a much longer sequence going indefinitely into the future and possibly into the past.

The sequence of random variables $\{Z_1, Z_2, \ldots\}$ or $\{\ldots, Z_{-1}, Z_0, Z_1, Z_2, \ldots\}$ is called a **stochastic process.** It is known that the complete probabilistic structure of such a process is determined by the set of distributions of all finite collections of Z's. Fortunately, we shall not have to deal explicitly with these multivariate distributions. Much of the information in these joint distributions can be described in terms of means, variances, and covariances. Consequently, we shall concentrate our efforts on these first and second moments. (If the joint distributions of the Z's are multivariate *normal* distributions, then the first and second moments completely determine the distributions.)

2.2 MEANS AND COVARIANCES

For a stochastic process $\{Z_t : t = 0, \pm 1, \pm 2, \ldots\}$ the **mean function** (or mean sequence) is defined by

$$\mu_t = E(Z_t) \quad \text{for } t = 0, \pm 1, \pm 2, \ldots$$

that is, μ_t is just the expected value of the process at time t. In general, μ_t can be different at each time t.

The **autocovariance function** is defined as

$$\gamma_{t,s} = \text{Cov}(Z_t, Z_s) \quad \text{for } t, s = 0, \pm 1, \pm 2, \ldots$$

where $\text{Cov}(Z_t, Z_s) = E[(Z_t - \mu_t)(Z_s - \mu_s)] = E(Z_t Z_s) - \mu_t \mu_s$

Finally, the **autocorrelation function** is given by

$$\rho_{t,s} = \text{Corr}(Z_t, Z_s) \quad \text{for } t, s = 0, \pm 1, \pm 2, \ldots$$

where

$$\text{Corr}(Z_t, Z_s) = \frac{\text{Cov}(Z_t, Z_s)}{[\text{Var}(Z_t)\,\text{Var}(Z_s)]^{1/2}} = \frac{\gamma_{t,s}}{(\gamma_{t,t} \cdot \gamma_{s,s})^{1/2}}$$

Appendix A reviews the basic properties of expectation, covariance, and correlation.

We recall that both covariance and correlation are measures of (linear) dependence between random variables but that the unitless correlation is somewhat easier to interpret. The following properties follow from known results and our definitions:

$$
\begin{array}{lll}
\textbf{1.} \; \gamma_{t,t} = \text{Var}(Z_t), & \rho_{t,t} = 1 & \\
\textbf{2.} \; \gamma_{t,s} = \gamma_{s,t}, & \rho_{t,s} = \rho_{s,t} & \text{(2-1)}\\
\textbf{3.} \; |\gamma_{t,s}| \leq \sqrt{(\gamma_{t,t}\gamma_{s,s})}, & |\rho_{t,s}| \leq 1 &
\end{array}
$$

Values of $\rho_{t,s}$ near ± 1 indicate strong (linear) dependence, whereas values near zero indicate weak or no (linear) dependence. If $\rho_{t,s} = 0$, we say that Z_t and Z_s are uncorrelated. If Z_t and Z_s are independent, then $\rho_{t,s} = 0$. These facts are illustrated in Exhibit 4.2, which shows a scatter of pairs with a correlation of about $\frac{1}{2}$, and in Exhibit 4.3, where the correlation is zero.

To investigate the covariance properties of various time series models, the following result will be used repeatedly: If $c_1, c_2, \ldots, c_m$ and $d_1, d_2, \ldots, d_n$ are constants and $t_1, t_2, \ldots, t_m$ and $s_1, s_2, \ldots, s_n$ are time points, then

$$\text{Cov}\left[\sum_{i=1}^{m} c_i Z(t_i), \sum_{j=1}^{n} d_j Z(s_j)\right] = \sum_{i=1}^{m} \sum_{j=1}^{n} c_i d_j \, \text{Cov}[Z(t_i), Z(s_j)] \qquad \text{(2-2)}$$

In words, the covariance between two linear combinations is the sum of all covariances between terms of those linear combinations. (In this and the following result, we temporarily write $Z(t) = Z_t$ to avoid subscripts on subscripts.)

The proof of Equation (2-2), though somewhat tedious, is a straightforward application of the linear properties of expectation. As a special case, we obtain the well-known result

$$\text{Var}\left[\sum_{i=1}^{n} c_i Z(t_i)\right] = \sum_{i=1}^{n} c_i^2 \text{Var}\left[Z(t_i)\right] + 2 \sum_{i=2}^{n} \sum_{j=1}^{i-1} c_i c_j \text{Cov}\left[Z(t_i), Z(t_j)\right] \qquad \textbf{(2-3)}$$

THE RANDOM WALK

Let $a_1, a_2, \ldots$ be independent, identically distributed random variables, each with zero mean and variance σ_a^2. The time series that we observe, $\{Z_t\}$, is then constructed as follows:

$$
\begin{aligned}
Z_1 &= a_1 \\
Z_2 &= a_1 + a_2 \\
&\vdots \\
Z_t &= a_1 + a_2 + \cdots + a_t
\end{aligned}
\qquad \textbf{(2-4)}
$$

Alternatively, we can write

$$Z_t = Z_{t-1} + a_t \qquad \textbf{(2-5)}$$

If the a's are interpreted as the size of the "steps" taken forward or backward at time t, then Z_t is the position of a "random walker" at time t. From (2-4) we obtain the mean function:

$$
\begin{aligned}
\mu_t = E(Z_t) &= E(a_1 + a_2 + \cdots + a_t) \\
&= E(a_1) + E(a_2) + \cdots + E(a_t)
\end{aligned}
$$

Thus, since $E(a_t) = 0$, we have

$$\mu_t = 0 \quad \text{for all } t \qquad \textbf{(2-6)}$$

Also,

$$
\begin{aligned}
\text{Var}(Z_t) &= \text{Var}(a_1 + a_2 + \cdots + a_t) \\
&= \text{Var}(a_1) + \text{Var}(a_2) + \cdots + \text{Var}(a_t) \\
&\qquad \text{(since the } a\text{'s are independent)} \\
&= \sigma_a^2 + \sigma_a^2 + \cdots + \sigma_a^2 \quad (t \text{ terms})
\end{aligned}
$$

or

$$\text{Var}(Z_t) = t\sigma_a^2 \qquad \textbf{(2-7)}$$

Notice that the process variance increases linearly with time.

Suppose now that $1 \le t \le s$. Then we have

$$
\begin{aligned}
\gamma_{t,s} &= \text{Cov}(Z_t, Z_s) \\
&= \text{Cov}(a_1 + a_2 + \cdots + a_t, a_1 + a_2 + \cdots + a_t + a_{t+1} + \cdots + a_s)
\end{aligned}
$$

Equation (2-2) says that we need to add all covariances between terms of $a_1 + a_2 + \cdots + a_t$ and terms of $a_1 + a_2 + \cdots + a_t + \cdots + a_s$. Since the a's are assumed to be independent, the nonzero covariances will occur only when a subscript in the first sum matches a subscript in the second sum. Thus we get

$$\gamma_{t,s} = \text{Cov}\,(a_1, a_1) + \text{Cov}\,(a_2, a_2) + \cdots + \text{Cov}\,(a_t, a_t)$$
$$= \sigma_a^2 + \sigma_a^2 + \cdots + \sigma_a^2 \quad (t \text{ terms})$$
$$= t\sigma_a^2$$

and we can write

$$\gamma_{t,s} = t\sigma_a^2 \quad \text{for } 1 \le t \le s \tag{2-8}$$

Since $\gamma_{t,s} = \gamma_{s,t}$, this specifies the autocovariance function for all time points t and s. The autocorrelation function is then simply

$$\rho_{t,s} = \sqrt{\frac{t}{s}} \quad \text{for } 1 \le t \le s \tag{2-9}$$

For example,

$$\rho_{1,2} = \sqrt{\frac{1}{2}} = 0.707, \quad \rho_{8,9} = \sqrt{\frac{8}{9}} = 0.943,$$

$$\rho_{5,1} = \sqrt{\frac{1}{5}} = 0.447, \quad \text{and} \quad \rho_{1,50} = \sqrt{\frac{1}{50}} = 0.014$$

The values of Z at neighboring time points are more and more strongly correlated as time goes by. The values of Z at distant time points are nearly uncorrelated. A computer simulation of a random walk is illustrated in Exhibit 2.1, where the a's were selected from a standard normal distribution. Note that

EXHIBIT 2.1 A Random Walk

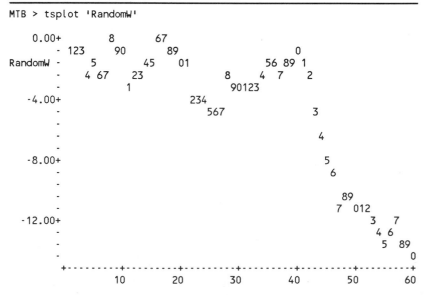

even though the theoretical mean function is zero for all time points, the facts that the variance $\text{Var}(Z_t) = t\sigma_a^2$ increases over time and that the correlation of neighboring Z-values gets closer to 1 indicate that we should expect long excursions of the process away from the mean level of zero.

The simple random walk process provides a good model for such diverse phenomena as the movement of common stock prices and the position of small particles suspended in a fluid (so-called Brownian motion).

A MOVING AVERAGE

As a second example, suppose that Z_t is given by

$$Z_t = a_t - \frac{1}{2}a_{t-1}$$

where (as always throughout this book), the a's are assumed to be independent and identically distributed with zero mean and variance σ_a^2. Here

$$\mu_t = E(Z_t)$$

$$= E(a_t) - \frac{1}{2}E(a_{t-1})$$

$$= 0$$

and

$$\text{Var}(Z_t) = \text{Var}\left(a_t - \frac{1}{2}a_{t-1}\right)$$

$$= \sigma_a^2 + \left(\frac{1}{2}\right)^2 \sigma_a^2$$

$$= 1.25\sigma_a^2$$

Also,

$$\text{Cov}(Z_t, Z_{t-1}) = \text{Cov}\left(a_t - \frac{1}{2}a_{t-1}, a_{t-1} - \frac{1}{2}a_{t-2}\right)$$

$$= \text{Cov}(a_t, a_{t-1}) - \frac{1}{2}\text{Cov}(a_t, a_{t-2}) - \frac{1}{2}\text{Cov}(a_{t-1}, a_{t-1})$$

$$+ \frac{1}{4}\text{Cov}(a_{t-1}, a_{t-2})$$

$$= -\frac{1}{2}\text{Cov}(a_{t-1}, a_{t-1}) \quad \text{(since the } a\text{'s are independent)}$$

or

$$\gamma_{t,t-1} = -\frac{1}{2}\sigma_a^2 \quad \text{for all } t$$

Furthermore,

$$\text{Cov}(Z_t, Z_{t-2}) = \text{Cov}\left(a_t - \frac{1}{2}a_{t-1}, a_{t-2} - \frac{1}{2}a_{t-3}\right)$$

$$= 0 \quad \text{(since the } a\text{'s are independent)}$$

Similarly, Cov $(Z_t, Z_{t-k}) = 0$ for $k \geq 2$, so we may write

$$
\gamma_{t,s} = \begin{cases} -\dfrac{1}{2}\sigma_a^2 & \text{if } |t - s| = 1 \\ 0 & \text{if } |t - s| > 1 \end{cases}
$$

For $\rho_{t,s}$ we then have

$$
\rho_{t,s} = \begin{cases} -0.4 & \text{if } |t - s| = 1 \\ 0 & \text{if } |t - s| > 1 \end{cases}
$$

Notice that $\rho_{2,1} = \rho_{3,2} = \rho_{4,3} = \rho_{9,8} = -0.4$. Values of Z precisely 1 time unit apart have exactly the same correlation no matter where they occur in time. Furthermore, $\rho_{3,1} = \rho_{4,2} = \rho_{t,t-2}$; more generally, $\rho_{t,t-k}$ is the same for all t. This leads us to the important concept of stationarity.

2.3 STATIONARITY

In order to make statistical inferences about the structure of a stochastic process on the basis of a finite observed record of that process, we must make some simplifying (and presumably reasonable) assumptions about that structure. The most important such assumption is that of **stationarity.** The basic idea of stationarity is that the probability laws governing the process do not change with time—that is, the process is in statistical equilibrium. Specifically, a stochastic process $\{Z_t\}$ is said to be **strictly stationary** if the joint distribution of $Z(t_1), Z(t_2), \ldots, Z(t_n)$ is the same as the joint distribution of $Z(t_1 - k)$, $Z(t_2 - k), \ldots, Z(t_n - k)$ for all choices of time points $t_1, t_2, \ldots, t_n$ and all choices of time lag k.

Thus, when $n = 1$ the univariate distribution of Z_t is the same as that of Z_{t-k} for any k; in other words, the Z's are (marginally) identically distributed. It then follows that $E(Z_t) = E(Z_{t-k})$ for all t and k, so that the mean function μ_t is constant for all time. Additionally, Var $(Z_t) =$ Var (Z_{t-k}) must be constant for all time. When $n = 2$ we see that the bivariate distribution of (Z_t, Z_s) must be the same as that of (Z_{t-k}, Z_{s-k}), from which we have Cov $(Z_t, Z_s) =$ Cov (Z_{t-k}, Z_{s-k}) for all t, s, and k. Putting $k = s$ first and then $k = t$, we obtain

$$
\begin{aligned}
\gamma_{t,s} = \text{Cov}(Z_t, Z_s) &= \text{Cov}(Z_{t-s}, Z_0) \\
&= \text{Cov}(Z_0, Z_{s-t}) \\
&= \text{Cov}(Z_0, Z_{|t-s|}) \\
&= \gamma_{0,|t-s|}
\end{aligned}
$$

that is, the covariance between Z_t and Z_s depends on time *only through the time difference* $|t - s|$ and not on the actual times t and s. Thus, for a stationary process we can simplify our notation and write

$$
\gamma_k = \text{Cov}(Z_t, Z_{t-k}) \quad \text{and} \quad \rho_k = \text{Corr}(Z_t, Z_{t-k}) \tag{2-10}
$$

Note also that
$$\rho_k = \gamma_k/\gamma_0$$

The general properties given in Equations (2-1) now become

1. $\gamma_0 = \text{Var}(Z_t),$ $\rho_0 = 1$
2. $\gamma_k = \gamma_{-k},$ $\rho_k = \rho_{-k}$ **(2-11)**
3. $|\gamma_k| \le \gamma_0,$ $|\rho_k| \le 1$

If a process is strictly stationary and has finite variance, then the covariance function must depend only on the time lag.

A definition that is similar to that of strict stationarity but is mathematically weaker is the following: A stochastic process $\{Z_t\}$ is said to be **weakly** (or **second-order**) **stationary** if

1. the mean function is constant over time, and
2. $\gamma_{t,t-k} = \gamma_{0,k}$ for all time t and lag k.

In this book the term *stationary* when used alone will always mean weakly stationary. However, if the joint distributions of a process are all multivariate normal, then it can readily be shown that the two definitions coincide.

WHITE NOISE

A very important example of a stationary process is the so-called **white noise** process, which is defined as a sequence of independent, identically distributed random variables $\{a_t\}$. Its importance stems not from the fact that it is an interesting model itself but from the fact that many useful processes can be constructed from white noise. The fact that $\{a_t\}$ is strictly stationary is easy to see since

$$\Pr[a(t_1) \le x_1, a(t_2) \le x_2, \ldots, a(t_n) \le x_n]$$
$$= \Pr[a(t_1) \le x_1]\Pr[a(t_2) \le x_2] \cdots \Pr[a(t_n) \le x_n] \quad \text{(by independence)}$$
$$= \Pr[a(t_1 - k) \le x_1]\Pr[a(t_2 - k) \le x_2] \cdots \Pr[a(t_n - k) \le x_n)$$
$$\text{(identical distribution)}$$
$$= \Pr[a(t_1 - k) \le x_1, a(t_2 - k) \le x_2, \ldots, a(t_n - k) \le x_n]$$

as required. Also $\mu_t = E(a_t)$ is constant and

$$\gamma_{t,s} = \begin{cases} \text{Var}(a_t) & \text{if } t = s \\ 0 & \text{if } t \ne s \end{cases}$$

Alternatively, we can write

$$\rho_k = \begin{cases} 1 & \text{for } k = 0 \\ 0 & \text{for } k \ne 0 \end{cases} \qquad \text{(2-12)}$$

The term *white noise* arises from the fact that a frequency analysis of the model (not considered in this book) shows that, in analogy with white light, all frequencies enter equally. We shall usually assume that white noise has zero mean and denote Var (a_t) by σ_a^2.

The moving average example in Section 2.2, where $Z_t = a_t - \frac{1}{2}a_{t-1}$, is an instance of a nontrivial, stationary stochastic process constructed from white noise. In our new notation, we have for the moving average process that

$$\rho_k = \begin{cases} 1 & \text{for } k = 0 \\ -0.4 & \text{for } k = \pm 1 \\ 0 & \text{for } |k| \geq 2 \end{cases}$$

RANDOM COSINE WAVE

As a somewhat different example,* consider the stochastic process defined as follows:

$$Z_t = \cos\left[2\pi\left(\frac{t}{12} + \Phi\right)\right], \quad t = 0, \pm 1, \pm 2, \ldots$$

where Φ is selected from a uniform distribution on the interval from 0 to 1. Notice that Φ is selected once for the whole process, not at each time point. A sample from such a process will appear highly deterministic, since Z_t will repeat itself identically every 12 time units and look like a perfect (discrete time) cosine curve. However, its maximum will not occur at $t = 0$ but will be determined by the random phase Φ. Still, the statistical properties of this process can be computed as follows:

$$E(Z_t) = E\left\{\cos\left[2\pi\left(\frac{t}{12} + \Phi\right)\right]\right\}$$

$$= \int_0^1 \cos\left[2\pi\left(\frac{t}{12} + \phi\right)\right] d\phi$$

$$= \frac{1}{2\pi} \sin\left[2\pi\left(\frac{t}{12} + \phi\right)\right]\Big|_0^1$$

$$= \frac{1}{2\pi}\left[\sin\left(\frac{2\pi t}{12} + 2\pi\right) - \sin\left(\frac{2\pi t}{12}\right)\right]$$

$$= 0 \quad \text{(since the sines must agree)}$$

So $\mu_t = 0$ for all t. Also,

$$\gamma_{t,s} = E(Z_t Z_s)$$

$$= E\left\{\cos\left[2\pi\left(\frac{t}{12} + \Phi\right)\right]\cos\left[2\pi\left(\frac{s}{12} + \Phi\right)\right]\right\}$$

$$= \int_0^1 \cos\left[2\pi\left(\frac{t}{12} + \phi\right)\right]\cos\left[2\pi\left(\frac{s}{12} + \phi\right)\right] d\phi$$

$$= \frac{1}{2}\int_0^1 \left\{\cos\left[2\pi\left(\frac{t-s}{12}\right)\right] + \cos\left[2\pi\left(\frac{t+s}{12} + 2\phi\right)\right]\right\} d\phi$$

$$= \frac{1}{2}\left\{\cos\left[2\pi\frac{t-s}{12}\right] + \frac{1}{4\pi}\sin\left[2\pi\left(\frac{t+s}{12} + 2\phi\right)\right]\Big|_0^1\right\}$$

$$= \frac{1}{2}\cos\left(2\pi\frac{|t-s|}{12}\right)$$

*This example contains optional material that is not needed in order to understand the remainder of this book.

So the process is *stationary* with autocorrelation function

$$\rho_k = \cos\left(\frac{2\pi k}{12}\right), \qquad k = 0, \pm 1, \pm 2, \ldots \tag{2-13}$$

This example suggests that it is difficult to assess whether a stationary model is appropriate for a given series on the basis of the time plot of the series.

The random walk of Section 2.2, where $Z_t = a_1 + a_2 + \cdots + a_t$, is also constructed from white noise but is *not* stationary. For example, Var $(Z_t) = t\sigma_a^2$ is *not* constant; furthermore, the covariance function $\gamma_{t,s} = t\sigma_a^2$ for $0 \le t \le s$ does *not* depend only on time lag. However, suppose that instead of analyzing $\{Z_t\}$ directly, we consider the differences of successive Z-values. Then $Z_t - Z_{t-1} = a_t$, so the *differenced series* $\{Z_t - Z_{t-1}\}$ is stationary. This represents a simple example of a technique found to be extremely useful in applications. Clearly, many real time series cannot be reasonably modeled by stationary processes, since they are not in statistical equilibrium but are evolving over time. However, we can frequently transform nonstationary series into stationary series by simple techniques such as differencing. Such techniques will be vigorously pursued in the remaining chapters. ∎

CHAPTER 2 EXERCISES

2.1. Suppose $E(X) = 2$,   Var $(X) = 9$, $E(Y) = 0$,   Var $(Y) = 4$,   and Corr $(X, Y) = \frac{1}{4}$. Find:
   a. Var $(X + Y)$
   b. Cov $(X, X + Y)$
   c. Corr $(X + Y, X - Y)$

2.2. If X and Y are dependent but Var $(X) =$ Var (Y), find

$$\text{Cov} (X + Y, X - Y).$$

2.3. Let X have a distribution with mean μ and variance σ^2, and let $Z_t = X$ for all t.
   a. Show that Z_t is strictly and weakly stationary.
   b. Find the autocovariance function for Z_t.
   c. Draw a "typical" time plot of Z_t.

2.4. Let $\{a_t\}$ be zero-mean white noise. Find the autocorrelation function for the following two processes:

   a. $Z_t = a_t + \dfrac{1}{3} a_{t-1}$

   b. $Z_t = a_t + 3a_{t-1}$
   c. You should have discovered that both series are stationary and have the same autocorrelation functions. Do you think that these models could be distinguished on the basis of observations of $\{Z_t\}$?

2.5. Suppose $Z_t = 5 + 2t + X_t$, where $\{X_t\}$ is a zero-mean stationary series with autocovariance function γ_k.

a. Find the mean function for $\{Z_t\}$.

b. Find the autocovariance function for $\{Z_t\}$.

c. Is $\{Z_t\}$ stationary? (Why or why not?)

2.6. Suppose $\{Z_t\}$ is stationary with autocovariance function γ_k.

a. Show that $W_t = \nabla Z_t = Z_t - Z_{t-1}$ is stationary by finding the mean and autocovariance function for W_t.

b. Show that $U_t = \nabla W_t = \nabla^2 Z_t = Z_t - 2Z_{t-1} + Z_{t-2}$ is stationary. (You need not find the autocovariance function.)

c. In general, show that

$$W_t = \sum_{j=0}^{n} c_j Z_{t-j}$$

is stationary for any "weights" c_j.

2.7. Suppose $Z_t = \beta_0 + \beta_1 t + X_t$ where $\{X_t\}$ is stationary.

a. Show that $\{Z_t\}$ is not stationary but that $\nabla Z_t = Z_t - Z_{t-1}$ is stationary.

b. In general, show that if $Z_t = \mu_t + X_t$ where μ_t is a polynomial in t of degree d and $\{X_t\}$ is stationary, then $\nabla^m Z_t$ is stationary for $m \geq d$ and nonstationary for $0 \leq m < d$.

We will see in Chapter 5 that these results may be further generalized to certain nonstationary $\{X_t\}$ processes.

2.8. Let $\{X_t\}$ be a zero-mean, unit-variance, stationary process with autocorrelation function ρ_k. Suppose that μ_t is a nonconstant function and that σ_t is a positive-valued nonconstant function. The observed series is formed as

$$Z_t = \mu_t + \sigma_t \cdot X_t$$

a. For the Z-process, find the mean, variance, and autocovariance functions.

b. Show that the autocorrelation function for the Z-process depends only on lag. Is the Z-process stationary?

c. Is it possible to have a series with a constant mean and Corr (Z_t, Z_{t-k}) free of t but with $\{Z_t\}$ not stationary?

2.9. Suppose Cov $(X_t, X_{t-k}) = \gamma_k$ is free of t but that $E(X_t) = 3t$.

a. Is $\{X_t\}$ stationary?

b. Let $Z_t = 7 - 3t + X_t$. Is $\{Z_t\}$ stationary?

2.10. Suppose that $Z_t = (a_t + a_{t-12})/2$. Show that $\{Z_t\}$ is stationary and that for $k > 0$, ρ_k is nonzero only for $k = 12$.

2.11. Suppose $Z_t = A + X_t$ where $\{X_t\}$ is stationary and A is random but independent of $\{X_t\}$. Find the mean and autocorrelation function for $\{Z_t\}$ in terms of the mean and variance of A and the mean and autocovariance function of $\{X_t\}$.

2.12. Let $\{Z_t\}$ be stationary with autocovariance function γ_k. Let

$$\bar{Z} = \frac{1}{n} \sum_{t=1}^{n} Z_t$$

Show that

$$\text{Var}(\bar{Z}) = \frac{1}{n} \sum_{k=-n+1}^{n-1} \left(1 - \frac{|k|}{n}\right)\gamma_k$$

$$= \frac{\gamma_0}{n} + \frac{2}{n} \sum_{k=1}^{n-1} \left(1 - \frac{k}{n}\right)\gamma_k$$

2.13. Let $\{a_t\}$ be white noise but with nonzero mean $E(a_t) = \mu_a \neq 0$. Let $\{Z_t\}$ be the random walk formed from $\{a_t\}$; that is,

$$Z_t = a_1 + a_2 + \cdots + a_t \quad \text{for } t = 1, 2, \ldots$$

The process $\{Z_t\}$ is called a **random walk with drift.**
  a. Find the mean function for $\{Z_t\}$.
  b. Find the autocovariance function for $\{Z_t\}$.

2.14. Consider the random walk defined by:

$$Z_t = Z_{t-1} + a_t, \quad t > 1$$

with $Z_1 = a_1$.
  a. Use the above representation to show that

$$\mu_t = \mu_{t-1}, \quad t > 1$$

with initial condition

$$\mu_1 = E(a_1) = 0$$

Hence show that $\mu_t = 0$ for $t \geq 0$.
  b. Similarly, show that

$$\text{Var}(Z_t) = \text{Var}(Z_{t-1}) + \sigma_a^2, \quad t > 1$$

with $\text{Var}(Z_1) = \sigma_a^2$, and hence

$$\text{Var}(Z_t) = t \cdot \sigma_a^2$$

  c. For $1 \leq t \leq s$ use

$$Z_s = Z_t + a_{t+1} + a_{t+2} + \cdots + a_s$$

to show that

$$\text{Cov}(Z_t, Z_s) = \text{Var}(Z_t)$$

and hence that

$$\text{Cov}(Z_t, Z_s) = \min(t, s) \cdot \sigma_a^2$$

(This exercise gives alternative derivations of Equations (2-6), (2-7), and (2-8)).

2.15. Let $\{a_t\}$ be white noise and let c be a constant with $|c| < 1$. The series $\{Z_t\}$ is then constructed as follows:

$$Z_1 = a_1$$
$$Z_t = cZ_{t-1} + a_t \quad \text{for } t > 1$$

  a. Show that $E(Z_t) = 0$ but that $\{Z_t\}$ is not stationary, since, for example,

$$\text{Var}(Z_t) = \sigma_a^2(1 + c^2 + c^4 + \cdots + c^{2t-2})$$

b. Show that

$$\text{Corr}\,(Z_t, Z_{t-1}) = c\left[\frac{\text{Var}\,(Z_{t-1})}{\text{Var}\,(Z_t)}\right]^{1/2}$$

and, in general,

$$\text{Corr}\,(Z_t, Z_{t-k}) = c^k\left[\frac{\text{Var}\,(Z_{t-k})}{\text{Var}\,(Z_t)}\right]^{1/2} \quad \text{for } k > 0$$

c. For large t, argue that

$$\text{Var}\,(Z_t) \approx \frac{\sigma_a^2}{1 - c^2} \quad \text{and} \quad \text{Corr}\,(Z_t, Z_{t-k}) \approx c^k \quad \text{for } k > 0$$

so that $\{Z_t\}$ could be called asymptotically stationary.

d. Suppose now that we alter the definition of $\{Z_t\}$ by putting

$$Z_1 = \frac{a_1}{\sqrt{1 - c^2}}$$

and then let $\{Z_t\}$ develop as before; that is,

$$Z_t = cZ_{t-1} + a_t \quad \text{for } t > 1$$

Show that $\{Z_t\}$ is now stationary.

2.16. Two processes $\{Z_t\}$ and $\{Y_t\}$ are said to be **independent** if for any time points $t_1, t_2, \ldots, t_n$ and $s_1, s_2, \ldots, s_m$, the random variables $\{Z(t_1), Z(t_2), \ldots, Z(t_n)\}$ are independent of the random variables $\{Y(s_1), Y(s_2), \ldots, Y(s_m)\}$. Show that if $\{Z_t\}$ and $\{Y_t\}$ are independent stationary processes, then $W_t = Z_t + Y_t$ is stationary.

2.17. *Signal plus noise*. Let $\{X_t\}$ be a time series in which we are interested. However, because the measurement process itself is subject to error in some circumstances, we actually observe $Z_t = X_t + a_t$. We assume that $\{X_t\}$ and $\{a_t\}$ are independent processes and that $\{a_t\}$ is white noise. We call $\{X_t\}$ the **signal** and $\{a_t\}$ the **measurement noise,** or **error.**

If $\{X_t\}$ is stationary with autocorrelation function ρ_k, show that $\{Z_t\}$ is also stationary with

$$\text{Corr}\,(Z_t, Z_{t-k}) = \frac{\rho_k}{1 + \sigma_a^2/\sigma_X^2}, \quad k \geq 1$$

We call σ_X^2/σ_a^2 the **signal-to-noise ratio,** or **SNR.** Note that the larger the value of the SNR, the closer the autocorrelation function of the observed series $\{Z_t\}$ is to that of the desired signal $\{X_t\}$.

2.18. Evaluate the mean and autocovariance function for each of the following processes. In each case determine whether the process is stationary.

a. $Z_t = \theta_0 + ta_t$.

b. $W_t = \nabla Z_t$ where Z_t is as in (a).

c. $Z_t = a_t \cdot a_{t-1}$.

2.19. (Mathematical statistics required). Suppose that

$$Z_t = A \cos\,[2\pi(ft + \Phi)]$$

where A and Φ are independent random variables and f is a constant frequency. The phase Φ is assumed to be uniformly distributed on the interval $(0, 1)$, and the amplitude A has a *Rayleigh* distribution with p.d.f.

$$f(a) = a \exp\left(-\frac{a^2}{2}\right) \quad \text{for } a > 0$$

a. Show that $\{Z_t\}$ is stationary.

b. Show that for each time point t, Z_t has a normal distribution. (It can also be shown that all finite dimensional distributions are multivariate normal so that the process is also strictly stationary.) (*Hint*: Let $X = A \cos [2\pi(ft + \Phi)]$ and $Y = A \sin [2\pi(ft + \Phi)]$.)

APPENDIX A EXPECTATION, VARIANCE, AND COVARIANCE

We shall assume throughout this book that all random variables are continuous with probability density functions (p.d.f.'s).

Let X have p.d.f. $f(x)$. The **expected value** of X is defined as

$$E(X) = \int_{-\infty}^{\infty} xf(x)\,dx$$

(if $\int_{-\infty}^{\infty} |x|\,f(x)\,dx < \infty$; otherwise $E(X)$ is undefined). $E(X)$ is also called the **expectation** of X or the **mean** of X and is often denoted μ_X.

PROPERTIES OF EXPECTATION

E1. If $h(x)$ is a function such that $\int_{-\infty}^{\infty} |h(x)|\,f(x)\,dx < \infty$, then $E[h(X)]$ $= \int_{-\infty}^{\infty} h(x)f(x)\,dx$.

E2. More generally, if X and Y have joint p.d.f. $f(x,y)$ and $\int_{-\infty}^{\infty} \int_{-\infty}^{\infty} |h(x,y)|\,f(x,y)\,dx\,dy < \infty$, then

$$E[h(X, Y)] = \int_{-\infty}^{\infty} \int_{-\infty}^{\infty} h(x,y)f(x,y)\,dx\,dy$$

E3. As a corollary to E2 we have the important result

$$E(aX + bY + c) = aE(X) + bE(Y) + c$$

E4. Also

$$E(XY) = \int_{-\infty}^{\infty} \int_{-\infty}^{\infty} xyf(x,y)\,dx\,dy$$

The **variance** of a random variable X is defined as

$$\text{Var}(X) = E[X - E(X)]^2$$

(provided $E(X)$ is defined and $\int_{-\infty}^{\infty} x^2 f(x)\,dx < \infty$). The variance of X is also denoted σ_X^2.

PROPERTIES OF VARIANCE

V1. $\text{Var}(X) \geq 0$

V2. $\text{Var}(a + bX) = b^2\,\text{Var}(X)$

V3. If X and Y are independent, then

$$\text{Var}(X + Y) = \text{Var}(X) + \text{Var}(Y)$$

V4. $\text{Var}(X) = E(X^2) - [E(X)]^2$

The positive square root of the variance of X is called the **standard deviation** of X and is often denoted σ_X. The random variable $(X - \mu_X)/\sigma_X$ is called the **standardized** version of X. The mean and standard deviation of a standardized random variable are always zero and one, respectively.

The **covariance** of X and Y is defined as

$$\text{Cov}(X, Y) = E[(X - \mu_X)(Y - \mu_Y)]$$

PROPERTIES OF COVARIANCE

CV1. If X and Y are independent, then $\text{Cov}(X, Y) = 0$.

CV2. $\text{Cov}(a + bX, c + dY) = bd\,\text{Cov}(X, Y)$

CV3. $\text{Cov}(X, Y) = E(XY) - E(X)E(Y)$

CV4. $\text{Var}(X + Y) = \text{Var}(X) + \text{Var}(Y) + 2\,\text{Cov}(X, Y)$

CV5. $\text{Cov}(X + Y, Z) = \text{Cov}(X, Z) + \text{Cov}(Y, Z)$

CV6. $\text{Cov}(X, X) = \text{Var}(X)$

CV7. $\text{Cov}(X, Y) = \text{Cov}(Y, X)$

The **correlation coefficient** of X and Y, denoted $\text{Corr}(X, Y)$ or ρ, is defined as

$$\text{Corr}(X, Y) = \frac{\text{Cov}(X, Y)}{\sqrt{\text{Var}(X)\,\text{Var}(Y)}}$$

PROPERTIES OF CORRELATION

CR1. $-1 \leq \text{Corr}(X, Y) \leq 1$

CR2. $\text{Corr}(a + bX, c + dY) = \text{sign}(bd)\,\text{Corr}(X, Y)$ where

$$\text{sign}(bd) = \begin{cases} 1 & \text{if } bd > 0 \\ 0 & \text{if } bd = 0 \\ -1 & \text{if } bd < 0 \end{cases}$$

CR3. $\text{Corr}(X, Y) = \text{Cov}\left(\dfrac{X - \mu_X}{\sigma_X}, \dfrac{Y - \mu_Y}{\sigma_Y}\right)$

CR4. $\text{Corr}(X, Y) = \pm 1$ if and only if there are constants a and b such that $P(Y = a + bX) = 1$.

EXAMPLE 1: Suppose X and Y are random variables with $\mu_X = 0$, $\sigma_X^2 = 1$, $\mu_Y = 10$, $\sigma_Y^2 = 9$, and $\rho = \frac{1}{2}$. Then

1. From E3, $E(2X + Y) = 2E(X) + E(Y) = 2(0) + 10 = 10$.

2. Using CV4 and the definition of ρ, we have

$$\text{Var}(X - Y) = \text{Var}(X) + \text{Var}(-Y) + 2\,\text{Cov}(X, -Y)$$
$$= \text{Var}(X) + (-1)^2\,\text{Var}(Y) - 2\,\text{Cov}(X, Y)$$
$$= 1 + 9 - 2\left(\frac{1}{2}\right)(1)(3) = 7$$

3. From CV2, CV5, and CV6,

$$\text{Cov}(X - Y, Y) = \text{Cov}(X, Y) + \text{Cov}(-Y, Y)$$
$$= \text{Cov}(X, Y) - \text{Cov}(Y, Y)$$
$$= \text{Cov}(X, Y) - \text{Var}(Y)$$
$$= \left(\frac{1}{2}\right)(1)(3) - 9 = \frac{3 - 18}{2} = -\frac{15}{2}$$

4. Using the results in steps 2 and 3 we have

$$\text{Corr}(X - Y, Y) = \frac{-15/2}{\sqrt{(7)(9)}} = -\frac{5}{2\sqrt{7}}$$ ∎

EXAMPLE 2:   Suppose X, Y, and Z are random variables with means of 10 and variances of 4. Suppose also that $\text{Cov}(X, Y) = \text{Cov}(Y, Z) = 2$ and $\text{Cov}(X, Z) = 1$. Then

1. Using CV5 twice and CV6, we obtain

$$\text{Cov}(X + Y, Y + Z) = \text{Cov}(X, Y + Z) + \text{Cov}(Y, Y + Z)$$
$$= \text{Cov}(X, Y) + \text{Cov}(X, Z) + \text{Cov}(Y, Y) + \text{Cov}(Y, Z)$$
$$= 2 + 1 + 4 + 2$$
$$= 9$$

2. From V2, CV4, and CV5,

$$\text{Var}\left(\frac{X + Y + Z}{3}\right) = \frac{1}{9}\,\text{Var}(X + Y + Z)$$

$$= \frac{1}{9}\,[\text{Var}(X) + \text{Var}(Y + Z) + 2\,\text{Cov}(X, Y + Z)]$$

$$= \frac{1}{9}\,[\text{Var}(X) + \text{Var}(Y) + \text{Var}(Z) + 2\,\text{Cov}(X, Y)$$
$$+ 2\,\text{Cov}(Y, Z) + 2\,\text{Cov}(X, Z)]$$

$$= \frac{3(4) + 2(2 + 2) + 2(1)}{9} = \frac{(12 + 8 + 2)}{9}$$

$$= \frac{22}{9}$$ ∎

CHAPTER 3 TRENDS

In a general time series, the mean function is a totally arbitrary function of time. On the other hand, as we have seen, in a stationary time series the mean function must be constant in time. Frequently we need to model series that exhibit behavior contrary to that which is expected when there is a constant mean but have a fairly simple trend to them. This chapter presents some methods for modeling these trends.

3.1 DETERMINISTIC VERSUS STOCHASTIC TRENDS

The modeling of a trend in times series is not necessarily an easy task. The time plot of a series can be viewed quite differently by different observers. Consider again Exhibit 2.1. One might say that the plot exhibits a general downward trend with some occasional upturns. However, recall that this is a time plot from a simulation of a random walk with a mean function that is exactly zero. The perceived trend is really just an artifact of the strong positive correlation between nearby time points and the increasing variance in the process as time increases. Another simulation of exactly the same model could show quite different trends. Some authors have described such trends as **stochastic trends** (see Box and Jenkins, 1976, for example), although there is no generally accepted definition of a stochastic trend.

The average monthly temperature series plotted in Exhibit 1.2 shows a cyclical or seasonal trend, but here the reason for the trend is quite clear—the Northern Hemisphere's changing inclination toward the sun. In this case, a possible model for the temperature might be $Z_t = \mu_t + X_t$, where μ_t is a deterministic mean function that is periodic with period 12; that is,

$$\mu_t = \mu_{t+12} \quad \text{for all } t$$

We might then assume that X_t, the unobserved variation around μ_t, has zero mean for all t so that indeed μ_t is the mean function for the observed series Z_t. We could describe this model as having a **deterministic trend** as opposed to the stochastic trend considered earlier. In other situations we might hypothesize a linear deterministic trend, that is, $\mu_t = \beta_0 + \beta_1 t$, or perhaps a quadratic trend, $\mu_t = \beta_0 + \beta_1 t + \beta_2 t^2$. Note that an implication of the model $Z_t = \mu_t + X_t$ with $E(X_t) = 0$ for all t is that the deterministic trend μ_t applies for all time. Thus if $\mu_t = \beta_0 + \beta_1 t$, we are assuming that the *same* linear trend applies forever. We should therefore have good reasons for assuming such a model—not just because the series looks somewhat linear over the time period observed.

In this chapter we shall consider methods for modeling only deterministic trends. Stochastic trends will be discussed again in Chapter 5, and stochastic seasonal models will be discussed in Chapter 10. Many authors use the word *trend* only for a slowly changing mean function, such as a linear or quadratic function, and use the term *seasonal component* for a mean function that varies cyclically. We do not find it useful to make such a distinction here.

3.2 ESTIMATION OF A CONSTANT MEAN

We first consider the simple situation where a constant mean function is assumed. Our model may then be written as

$$Z_t = \mu + X_t \tag{3-1}$$

where $E(X_t) = 0$ for all t. We wish to estimate μ with our observed time series $Z_1, Z_2, \ldots, Z_n$.

The most common estimate of μ is the sample mean or average:

$$\bar{Z} = \frac{1}{n} \sum_{t=1}^{n} Z_t \tag{3-2}$$

Under the minimal assumptions of Equation (3-1) we see that $E(\bar{Z}) = \mu$; therefore $\bar{Z}$ is an unbiased estimate of μ. To investigate the precision of $\bar{Z}$ as an estimate of μ, we need to make further assumptions concerning X_t.

Suppose $\{X_t\}$ is a stationary time series with autocorrelation function ρ_k. Then the same autocorrelation function applies to $\{Z_t\}$, and by Exercise 2.12 we have: If $\{Z_t\}$ is stationary

$$\begin{aligned}
\text{Var}\,(\bar{Z}) &= \frac{\gamma_0}{n} \sum_{k=-n+1}^{n-1} \left(1 - \frac{|k|}{n}\right) \rho_k \\
&= \frac{\gamma_0}{n} \left[1 + 2 \sum_{k=1}^{n-1} \left(1 - \frac{k}{n}\right) \rho_k\right]
\end{aligned} \tag{3-3}$$

Notice that the first factor, γ_0/n, is the population variance over sample size—a concept with which we are familiar in simpler contexts. If the $\{X_t\}$ series is in fact just white noise, then $\rho_k = 0$ for $k \geq 1$ and $\text{Var}\,(\bar{Z})$ reduces to simply γ_0/n.

In the stationary moving average model $Z_t = a_t - \frac{1}{2} a_{t-1}$ considered in Chapter 2, we found that $\rho_1 = -0.4$ and $\rho_k = 0$ for $k > 1$. In this case we have

$$\begin{aligned}
\text{Var}\,(\bar{Z}) &= \frac{\gamma_0}{n} \left[1 + 2\left(1 - \frac{1}{n}\right)(-0.4)\right] \\
&= \frac{\gamma_0}{n} \left[1 - (0.8)\frac{n-1}{n}\right]
\end{aligned}$$

For values of n usually occurring in time series ($n > 50$, say), the factor

$(n - 1)/n$ will be close to 1, so that we have

$$\text{Var}(\bar{Z}) \approx 0.2 \frac{\gamma_0}{n}$$

We see that the negative correlation at lag 1 has improved the estimation of the mean compared with the estimation obtained in a white noise situation. Because the series tends to oscillate across the mean, the sample mean obtained is more precise.

On the other hand, if $\rho_k \geq 0$ for all $k \geq 1$, we see from Equation (3-3) that $\text{Var}(\bar{Z})$ will be larger than γ_0/n. Here the positive correlations make estimation of the mean *more* difficult than in a white noise series. In general, some correlations will be positive and some negative, and Equation (3-3) must be used to assess the total effect.

For many stationary processes, the correlation function decays rapidly with increasing lag; thus

$$\sum_{k=0}^{\infty} \rho_k < \infty \tag{3-4}$$

(The random cosine wave of Chapter 2 is an exception.)

Under assumption (3-4) and given a large sample size n, the following useful approximation follows from Equation (3-3) (see, for example, Anderson, 1971, p. 459):

$$\text{Var}(\bar{Z}) \approx \frac{\gamma_0}{n} \sum_{k=-\infty}^{\infty} \rho_k \quad \text{for large } n \tag{3-5}$$

Notice that the variance is essentially inversely proportional to the sample size n.

As an example, suppose that $\rho_k = \phi^{|k|}$ where $|\phi| < 1$. Summing a geometric series yields

$$\text{Var}(\bar{Z}) \approx \left(\frac{1 + \phi}{1 - \phi}\right)\left(\frac{\gamma_0}{n}\right) \tag{3-6}$$

For a nonstationary $\{X_t\}$ process (but with a zero-mean function), the precision of $\bar{Z}$ as an estimate of μ is quite different. As an example, suppose $\{X_t\}$ is a zero-mean random walk as described in Chapter 2 so that $\gamma_{t,s} = t\sigma_a^2$ for $1 \leq t \leq s$. Then

$$\text{Var}(\bar{Z}) = \frac{1}{n^2} \text{Var}\left(\sum_{t=1}^{n} Z_t\right)$$

$$= \frac{1}{n^2}\left[\sum_{t=1}^{n} \text{Var}(Z_t) + 2\sum_{s=2}^{n}\sum_{t=1}^{s-1} \text{Cov}(Z_t, Z_s)\right]$$

$$= \frac{1}{n^2}\left[\sum_{t=1}^{n} t\sigma_a^2 + 2\sum_{s=2}^{n}\sum_{t=1}^{s-1} t\sigma_a^2\right]$$

Using known results for the sum of n consecutive integers or squares of integers, these sums may be worked out with the following result:

$$\text{Var}(\bar{Z}) = \sigma_a^2(2n + 1)\frac{n + 1}{6n}$$

(3-7)

Notice that for this case the variance of our estimate actually increases as the record length n increases. Clearly this is unacceptable, and we need to consider other techniques for nonstationary series.

3.3 REGRESSION METHODS

The classical statistical method of regression analysis may be readily used to estimate the parameters of common, nonconstant trend models. We shall consider the most useful ones: linear, quadratic, seasonal means, and cosine trends.

LINEAR AND QUADRATIC TRENDS

Consider the linear deterministic trend

$$\mu_t = \beta_0 + \beta_1 t$$

(3-8)

where the **slope** and **intercept,** β_1 and β_0, respectively, are unknown parameters. The classical least-squares (or regression) method is to choose as estimates of β_0 and β_1 values that minimize

$$Q(\beta_0, \beta_1) = \sum_{t=1}^{n} (Z_t - \beta_0 - \beta_1 t)^2$$

The solution may be obtained by computing the partial derivatives with respect to β_0 and β_1, setting them both to zero, and solving the resulting linear equations in β_0 and β_1. Denoting the solutions by b_0 and b_1, we find that

and

$$b_1 = \frac{\sum_{t=1}^{n} (Z_t - \bar{Z})(t - \bar{t})}{\sum_{t=1}^{n} (t - \bar{t})^2}$$

(3-9)

$$b_0 = \bar{Z} - b_1\bar{t}$$

where $\bar{t} = (n + 1)/2$, the average of $1, 2, \ldots, n$. These formulae can be simplified somewhat, and various formulae for computing b_1 are well-known. However, computations will be done by Minitab using the REGRESS command. In its simplest case the command has the form:

REGRESS the series in **C** on **1** predictor in **C**

where the predictor in C should be the values of t, namely $1, 2, \ldots, n$. These predictor values can be placed directly into a column using the SET command with patterned data:

SET into column **C**

the integers **1 : K**

END

The 1 : K is an abbreviation for the list $1, 2, \ldots, K$, and K would be replaced by the number n. (See Appendix J, "Minitab Primer.")

      For example, suppose we wish to estimate a deterministic time trend for the milk production data shown in Exhibit 1.3. This series contains 168 observations that we have previously stored in a file named milkpr. Exhibit 3.1 shows the Minitab commands and output. (The BRIEF command controls the amount of output from the regression commands that follow. Use HELP BRIEF, or see the *Minitab Reference Manual* for details on the different levels of output that can be obtained.)

      From the output, the regression equation shows us that:

$$b_0 = 612 \quad \text{and} \quad b_1 = 1.69$$

(If desired, more digits can be obtained from the coefficients column in the table. The rest of the output will be interpreted in Section 3.5.) Exhibit 3.2 shows the milk production series onto which the estimated linear time trend has been superimposed.

      Quadratic time trends can be estimated in a similar manner. We first need to set up a column of t^2 values. If C2 contains t, that is, $1, 2, \ldots, n$, then we can use

LET C3 = C2 ∗ C2

EXHIBIT 3.1 Time Trend Estimation for Milk Production Series

```
MTB > set 'milkpr' into c1
C1
    589    561    640    656    .    .    .

MTB > set c2
DATA> 1:168
DATA> end
MTB > brief
MTB > regress c1 on 1 predictor in c2

The regression equation is
C1 = 612 + 1.69 C2

Predictor      Coef       Stdev     t-ratio
Constant     611.682      9.414      64.97
C2           1.69262     0.09663     17.52

s = 60.74      R-sq = 64.892    R-sq(adj) = 64.680

Analysis of Variance

SOURCE        DF         SS          MS
Regression     1      1132003     1132003
Error        166       612439        3689
Total        167      1744442
```

EXHIBIT 3.2 Estimated Time Trend Superimposed on Milk Production Series

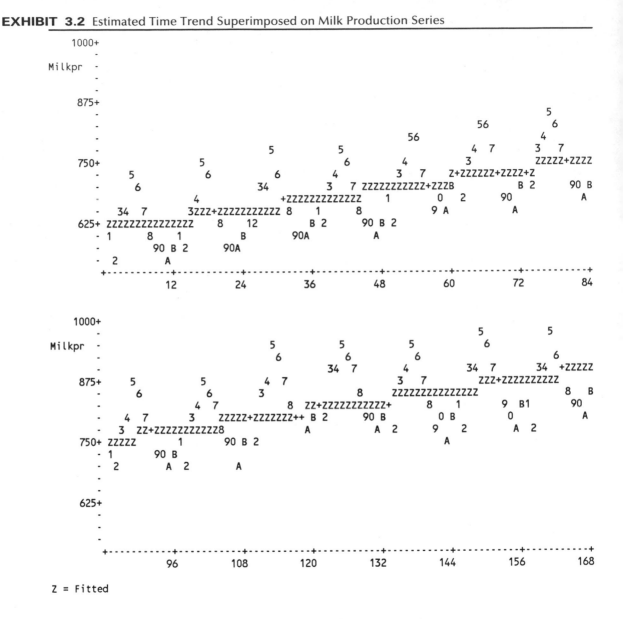

Z = Fitted

to get t^2 into C3. With the series in C1, we then use the regression command
with two predictors, t and t^2:

REGRESS **C1** on **2** predictors in **C2** and **C3**

CYCLICAL OR SEASONAL TRENDS

Consider now modeling and estimating seasonal trends, such as for the average
monthly temperature data in Exhibit 1.2. Here we assume that the observed

series can be represented as

$$Z_t = \mu_t + X_t$$

where

$$\mu_t = \mu_{t+12} \quad \text{for all } t$$

and $E(X_t) = 0$ for all t.

The most general assumption for μ_t is that there are 12 constants (parameters) $\beta_1, \beta_2, \ldots, \beta_{12}$, giving the expected average monthly temperatures for each of the twelve months. We may write

$$\mu_t = \begin{cases} \beta_1 & \text{for } t = 1, 13, 25, \ldots \\ \beta_2 & \text{for } t = 2, 14, 26, \ldots \\ \;\;\vdots \\ \beta_{12} & \text{for } t = 12, 24, 36, \ldots \end{cases} \tag{3-10}$$

This is sometimes called the **seasonal means** model.

For the regression command to fit such a model, we must first set up 12 columns of dummy, or indicator, variables. The first contains a 1 for each January and zero otherwise, the second contains a 1 for each February and zero otherwise, and so on. Minitab accomplishes this most easily with two commands: SET for patterned data and INDICATOR. We first use

SET C2

12(1 : 12)

END

to get a column of codes for the various months, that is, 1 for January, 2 for February, and so on. The 1 : 12 is a Minitab abbreviation for the list $1, 2, \ldots, 12$, and the first 12 is a repetition factor that repeats this list once for each of the 12 years. We then use:

INDICATORS for C2, put into C3–C14

to obtain the required ones and zeroes in columns C3 through C14.

Since we have used an indicator for each of the 12 months, our model does not also include an overall constant term, or β_0. Thus we must use the NOCONSTANT subcommand with the reGRESS command.

Because we are using a large number of columns, it is convenient to introduce the NAME command for keeping track of the columns. This command has the format:

NAME for C is 'Name' [, for C is 'Name', . . .]

A name may be up to eight characters long and contain letters, numbers, and periods, but it must start with a letter. Either the column number or its name may be used in any succeeding command. Anytime a name is used, it must be enclosed in single quotes (apostrophes). Names will also be used on all outputs involving that column.

For our temperature example, we could use:

NAME C3 'Jan' C4 'Feb' C5 'Mar' C6 'Apr' C7 'May' C8 'Jun'

NAME C9 'Jul' C10 'Aug' C11 'Sep' C12 'Oct' C13 'Nov' C14 'Dec'

EXHIBIT 3.3 Seasonal Means Estimates for Monthly Temperature Series

```
The regression equation is
Temp = 16.6 Jan + 20.6 Feb + 32.5 Mar + 46.5 Apr
          + 58.1 May + 67.5 Jun + 71.7 Jul
          + 69.3 Aug + 61.0 Sep + 51.0 Oct
          + 36.6 Nov + 23.6 Dec
```

Predictor	Coef	Stdev	t-ratio
Noconstant			
Jan	16.6083	0.9870	16.83
Feb	20.6500	0.9870	20.92
Mar	32.4750	0.9870	32.90
Apr	46.5250	0.9870	47.14
May	58.0917	0.9870	58.86
Jun	67.5000	0.9870	68.39
Jul	71.7166	0.9870	72.66
Aug	69.3333	0.9870	70.25
Sep	61.0250	0.9870	61.83
Oct	50.9750	0.9870	51.65
Nov	36.6500	0.9870	37.13
Dec	23.6417	0.9870	23.95

```
s = 3.419
Analysis of Variance
```

SOURCE	DF	SS	MS
Regression	12	360329	30027
Error	132	1543	12
Total	144	361872	

With the data in C1, our Minitab program would now simply be

REGRESS **C1** on **12** predictors in **C3–C14;**

NOCONSTANT.

The output is shown in Exhibit 3.3 and shows, for example, that the estimate of β_1 is 16.6, of β_2 is 20.6, and so forth.

In general, a seasonal means model can be estimated for a monthly series of K years stored in C1 by using the following sequence of Minitab commands:

SET C2

K(1 : 12)

END

INDI C2 C3–C14

REGR C1 12 C3–C14;

NOCO.

For K years of **quarterly** data, the squence would be

SET C2

K(1 : 4)

END

INDI C2 C3–C6

REGR C1 4 C3–C6;

NOCO.

EXHIBIT 3.4 Linear Trend and Seasonal Means Estimates for Milk Production Series

```
The regression equation is
Milkpr = 1.73 Time + 591 Jan + 551 Feb + 644 Mar
              + 658 Apr + 719 May + 691 Jun
              + 641 Jul + 599 Aug + 556 Sep
              + 560 Oct + 529 Nov + 566 Dec
```

Predictor	Coef	Stdev	t-ratio
Noconstant			
Time	1.72776	0.02575	67.11
Jan	590.578	4.770	123.82
Feb	551.065	4.781	115.27
Mar	643.551	4.792	134.30
Apr	658.395	4.803	137.07
May	719.381	4.815	149.42
Jun	691.011	4.826	143.18
Jul	641.212	4.838	132.55
Aug	598.912	4.849	123.50
Sep	556.328	4.861	114.44
Oct	559.814	4.873	114.88
Nov	528.801	4.885	108.25
Dec	565.501	4.897	115.47

```
s = 16.14
Analysis of Variance
```

SOURCE	DF	SS	MS
Regression	13	97394272	7491867
Error	155	40392	261
Total	168	97434656	

The reader who has studied regression methods will probably recognize that, in the absence of any other predictor variables, the seasonal means estimate of β_j will be just the average of all data at "season" j; that is, for monthly data the estimate of β_1 is the average of all Januaries, of β_2 the average of all Februaries, and so forth. However, if we simultaneously estimate a linear trend and seasonal means, this will not be the case. The milk production series illustrates this well. Exhibit 3.4 shows the output from such a fit—the predictor variables are time and indicator variables for each of the 12 months. Note that the slope estimate changes slightly from 1.69 to 1.73 when we add the seasonal means portion of the model to the linear time trend model considered earlier (Exhibit 3.1).

COSINE TRENDS

The seasonal means model for monthly data contains 12 independent parameters and does not take the shape of the seasonal trend into account at all. It does not, for example, use the fact that Januaries and Februaries are similar and that Junes and Julys are similar. In some cases seasonal trends can be modeled economically with cosine curves that incorporate the smooth change expected from one time period to the next while still preserving the seasonality.

Consider the cosine curve

$$\mu_t = \beta \cos (2\pi ft + \phi) \tag{3-11}$$

We call β the **amplitude,** f the **frequency,** and ϕ the **phase** of the curve. As t varies, the curve oscillates between a maximum of β and a minimum of $-\beta$. Since the curve repeats itself exactly every $1/f$ time units, $1/f$ is the **period.** As noted in Chapter 2, the phase ϕ serves to set the origin on the t-axis. For monthly data, the most important frequency is $f = \frac{1}{12}$, because such a cosine curve will repeat itself every 12 months.

Equation (3-11) is inconvenient for estimation because the parameters β and ϕ do not both enter the expression linearly. Fortunately, a trigonometric identity is available that reparameterizes (3-11) more conveniently, namely,

$$\beta \cos (2\pi ft + \phi) = \beta_1 \cos (2\pi ft) + \beta_2 \sin (2\pi ft) \tag{3-12}$$

where

$$\beta = \sqrt{(\beta_1^2 + \beta_2^2)}, \qquad \phi = \text{Arctan} \, (-\beta_2/\beta_1) \tag{3-13}$$

and conversely,

$$\beta_1 = \beta \cos \phi, \qquad \beta_2 = -\beta \sin \phi \tag{3-14}$$

To estimate the parameters β_1 and β_2 with the regression command, we first need to set up a column containing $\cos (2\pi ft)$ and a second containing $\sin (2\pi ft)$. We shall illustrate the necessary Minitab commands with the average monthly temperature series and the fundamental frequency $f = \frac{1}{12}$. We shall also need a constant term to reflect the overall level of the series. Our model for the complete trend is then

$$\mu_t = \beta_0 + \beta_1 \cos \left(\frac{2\pi t}{12}\right) + \beta_2 \sin \left(\frac{2\pi t}{12}\right)$$

which we estimate by using the following commands:

```
LET     K1 = 2 * 3.14159/12
SET     C2
        1:144
        END
LET     C3 = COS(K1 * C2)
LET     C4 = SIN(K1 * C2)
NAME    C1 'Temp' C3 'Cos' C4 'Sin'
REGR    C1 2 C3 C4
```

The output from these commands applied to the temperature series is given in Exhibit 3.5, where we see that the estimate of β_1 is -22.0 and of β_2 is -15.2. Notice that we are estimating only three parameters here, whereas in the seasonal means model 12 parameters were needed.

Additional cosine functions at other frequencies will frequently be used to model cyclical trends. For monthly series, the higher harmonic frequencies, such as $\frac{2}{12}$ and $\frac{3}{12}$ are especially pertinent and will improve the fit at the expense of adding more parameters to the model. In fact, it may be shown that any

EXHIBIT 3.5 Cosine Trend Estimation for Monthly Temperature Series

```
The regression equation is
Temp = 46.3 - 22.0 Cos - 15.2 Sin

Predictor      Coef       Stdev     t-ratio
Constant     46.2659     0.3088     149.82
Cos         -22.0457     0.4367     -50.48
Sin         -15.2318     0.4367     -34.88

s = 3.706      R-sq = 96.390    R-sq(adj) = 96.339

Analysis of Variance

SOURCE       DF         SS          MS
Regression    2        51698       25849
Error       141         1936          14
Total       143        53634
```

periodic trend with period 12 may be expressed *exactly* by the sum of six pairs of cosine-sine functions. These ideas are discussed in detail in Fourier analysis or spectral analysis and will not be pursued here. The interested reader may consult Anderson (1971), Bloomfield (1976), and Fuller (1976).

3.4 RELIABILITY AND EFFICIENCY OF REGRESSION ESTIMATES

We assume that the observed series is represented as $Z_t = \mu_t + X_t$, where μ_t is a deterministic trend of the kind considered above and $\{X_t\}$ is a zero-mean stationary process with covariance and correlation functions γ_k and ρ_k, respectively. The regression command always estimates parameters according to the criterion of *least squares* regardless of whether we are fitting linear trends, seasonal means, cosine curves, or whatever.

We first consider the easiest case—the seasonal means. As mentioned earlier, the seasonal means estimates are just the seasonal averages; thus, if we have N (full) years of monthly data, we can write the estimate for the jth season as

$$\hat{\beta}_j - \frac{1}{N} \sum_{i=0}^{N-1} Z_{j+12i}$$

Since $\hat{\beta}_j$ is an average like $\bar{Z}$ but uses only every 12th observation, Equation (3-3) can easily be modified to give Var $(\hat{\beta}_j)$. We replace n by N (years) and ρ_k by ρ_{12k} to get

$$\text{Var}(\hat{\beta}_j) = \frac{\gamma_0}{N}\left[1 + 2\sum_{k=1}^{N-1}\left(1 - \frac{k}{N}\right)\rho_{12k}\right] \quad \text{for } j - 1, 2, \ldots, 12 \qquad \text{(3-15)}$$

We notice that if $\{X_t\}$ is white noise, then Var $(\hat{\beta}_j)$ is γ_0/N. Furthermore, if several ρ_k are nonzero but $\rho_{12k} = 0$, then we still have Var $(\hat{\beta}_j) = \gamma_0/N$. In any case, only the seasonal correlations $\rho_{12}, \rho_{24}, \rho_{36}, \ldots$ enter into Equation (3-15). Since

N will rarely be very large (except perhaps for quarterly data), approximations like Equation (3-5) will usually not be useful.

We turn now to the cosine trends. For any frequency of the form $f = m/n$ for $1 \leq m < n/2$, explicit expressions are available for the estimates $\hat{\beta}_1$ and $\hat{\beta}_2$, the amplitudes of the cosine and sine, respectively:

$$\hat{\beta}_1 = \frac{2}{n} \sum_{t=1}^{n} \cos\left(\frac{2\pi mt}{n}\right) Z_t \quad \text{and} \quad \hat{\beta}_2 = \frac{2}{n} \sum_{t=1}^{n} \sin\left(\frac{2\pi mt}{n}\right) Z_t \tag{3-16}$$

Because these are linear functions of the Z's, we may evaluate their variances using Equation (2-3). We find that

$$\text{Var}(\hat{\beta}_1) = \frac{2\gamma_0}{n}\left[1 + \frac{4}{n} \sum_{s=2}^{n} \sum_{t=1}^{s-1} \cos\left(\frac{2\pi mt}{n}\right) \cos\left(\frac{2\pi ms}{n}\right) \rho_{s-t}\right] \tag{3-17}$$

where we have used the fact that $\sum_{t=1}^{n} \cos^2(2\pi mt/n) = n/2$. However, the double sum generally does not reduce further. A similar expression holds for $\text{Var}(\hat{\beta}_2)$ if we replace the cosines by sines.

If $\{X_t\}$ is white noise, we get just $2\gamma_0/n$. If $\rho_1 \neq 0$, $\rho_k = 0$ for $k > 1$, and $m/n = 1/12$, then the variance reduces to

$$\text{Var}(\hat{\beta}_1) = \frac{2\gamma_0}{n}\left\{1 + \frac{4\rho_1}{n} \sum_{t=1}^{n-1} \cos\left(\frac{\pi t}{6}\right) \cos\left[\frac{\pi(t+1)}{6}\right]\right\} \tag{3-18}$$

To illustrate the effect of the cosine terms, we have calculated some representative values:

n	$\text{Var}(\hat{\beta}_1)$
25	$\left(\frac{2\gamma_0}{n}\right)(1 + 1.71\rho_1)$
50	$\left(\frac{2\gamma_0}{n}\right)(1 + 1.75\rho_1)$
100	$\left(\frac{2\gamma_0}{n}\right)(1 + 1.71\rho_1)$
500	$\left(\frac{2\gamma_0}{n}\right)(1 + 1.73\rho_1)$
∞	$\left(\frac{2\gamma_0}{n}\right)\left(1 + 2\rho_1 \cos\left(\frac{\pi}{6}\right)\right) = \left(\frac{2\gamma_0}{n}\right)(1 + 1.732\rho_1)$

$$\tag{3-19}$$

If $\rho_1 = -0.4$, then this large sample value is $1 + 1.732(-0.4) = 0.307$. The variance is thus reduced by about 70% when compared with the white noise case.

In some circumstances, seasonal means and cosine trends could both be considered as competing models for a cyclical trend. But if a cosine trend is an adequate model, how much do we lose if we use the seasonal means model containing considerably more parameters? To approach this problem, we must first consider how to compare the models. The parameters themselves are not directly comparable, but we can compare the estimates of the trend at comparable time points. Consider the two estimates for the trend in January, that is, μ_1. With seasonal means, this estimate is just the January average, which has variance

$$\frac{\gamma_0}{N}\left[1 + 2 \sum_{k=1}^{N-1} \left(1-\frac{k}{N}\right)\rho_{12k}\right]$$

With the cosine trend model, the corresponding estimate is

$$\hat{\mu}_1 = \hat{\beta}_0 + \hat{\beta}_1 \cos\left(\frac{2\pi}{12}\right) + \hat{\beta}_2 \sin\left(\frac{2\pi}{12}\right)$$

To compute the variance of this estimate, we need one more fact: With this model, the estimates $\hat{\beta}_0$, $\hat{\beta}_1$, and $\hat{\beta}_2$ are uncorrelated. This follows from the orthogonality relationships of the cosines and sines involved. See Bloomfield (1976) or Fuller (1976) for more details. For the cosine model, then, we have

$$\text{Var}(\hat{\mu}_1) = \text{Var}(\hat{\beta}_0) + \cos^2\left(\frac{\pi}{6}\right)\text{Var}(\hat{\beta}_1) + \sin^2\left(\frac{\pi}{6}\right)\text{Var}(\hat{\beta}_2)$$

For our first comparison, assume that the stochastic component is white noise. Then the variance of our estimate in the seasonal means model is just γ_0/N. For the cosine model we use Equation (3-17) twice to get

$$\text{Var}(\hat{\mu}_1) = \frac{\gamma_0}{n}\left[1 + 2\cos^2\left(\frac{\pi}{6}\right) + 2\sin^2\left(\frac{\pi}{6}\right)\right]$$

$$= \frac{3\gamma_0}{n}$$

since $\cos^2\theta + \sin^2\theta = 1$. Thus the ratio of the standard deviation in the cosine model to that in the seasonal means model is

$$\left(\frac{3\gamma_0/n}{\gamma_0/N}\right)^{1/2} = \left(\frac{3N}{n}\right)^{1/2}$$

In particular, for the monthly temperature series we have $n = 144$ and $N = 12$; thus, the ratio is

$$\left(\frac{3(12)}{144}\right)^{1/2} = 0.5$$

Thus in the cosine model we estimate the trend for January with a standard deviation that is only half as large as it would be with a seasonal means model—a substantial gain.

Suppose now that the stochastic component is such that $\rho_1 \neq 0$ but $\rho_k = 0$ for $k > 1$. With a seasonal means model, the variance of the January trend will be unchanged (see Equation (3-15)). For the cosine trend model, if we have a reasonably large sample size, we may use Equation (3-19), an identical expression for Var $(\hat{\beta}_2)$, and Equation (3-3) for Var $(\hat{\beta}_0)$ to obtain

$$\text{Var}\,(\hat{\mu}_1) = \frac{\gamma_0}{n}\left\{1 + 2\rho_1 + 2\left[1 + 2\rho_1 \cos\left(\frac{2\pi}{12}\right)\right]\right\}$$

$$= \frac{\gamma_0}{n}\left\{3 + 2\rho_1\left[1 + 2\cos\left(\frac{\pi}{6}\right)\right]\right\}$$

If $\rho_1 = -0.4$, then we have $0.814\gamma_0/n$, and the ratio of the standard deviation in the cosine case to the standard deviation in the seasonal means case is

$$\left(\frac{0.814\gamma_0/n}{\gamma_0/N}\right)^{1/2} = \left(\frac{0.814N}{n}\right)^{1/2}$$

If we take $n = 144$ and $N = 12$, the ratio is

$$\left(\frac{0.814}{12}\right)^{1/2} = 0.26$$

—a very substantial reduction indeed!

We now turn again to linear trends. For linear trends an alternative formula to Equation (3-9) for b_1 is more convenient:

$$b_1 = \frac{\displaystyle\sum_{t=1}^{n}(t-\bar{t})Z_t}{\displaystyle\sum_{t=1}^{n}(t-\bar{t})^2} \tag{3-20}$$

Since b_1 is a linear combination of Z-values, some progress can be made in computing Var (b_1). We have

$$\text{Var}\,(b_1) = \frac{12\gamma_0}{n(n^2-1)}\left[1 + \frac{24}{n(n^2-1)}\sum_{s=2}^{n}\sum_{t=1}^{s-1}(t-\bar{t})(s-\bar{t})\rho_{s-t}\right] \tag{3-21}$$

where we have used $\sum_{t=1}^{n}(t-\bar{t})^2 = n(n^2-1)/12$. Again, the double sum does not in general reduce.

To illustrate the effect of Equation (3-21), consider again the case where $\rho_1 \neq 0$ but $\rho_k = 0$ for $k > 1$. Then after some algebraic manipulation, again involving the sum of consecutive integers and their squares, Equation (3-21) can be reduced to

$$\text{Var}\,(b_1) = \frac{12\gamma_0}{n(n^2-1)}\left[1 + 2\rho_1\left(1 - \frac{3}{n}\right)\right]$$

For large n we can neglect the $3/n$ term and use

$$\text{Var}\,(b_1) \approx \frac{12\gamma_0}{n(n^2-1)}(1 + 2\rho_1) \tag{3-22}$$

If $\rho_1 = -0.4$, then $1 + 2\rho_1 = 0.2$, and the variance of b_1 is only 20% of what it would be if $\{X_t\}$ were white noise. Of course, if $\rho_1 > 0$, then the variance would be larger than for the white noise case.

We turn now to comparing the least-squares estimates with the so-called **best linear unbiased estimates** (BLUE) or the **generalized least squares** (GLS) estimates.

If the stochastic component $\{X_t\}$ is not white noise, estimates of the unknown parameters in the trend function may be made that are linear functions of the data, are unbiased, and have the smallest variance among all such estimates—the so-called BLUE or GLS estimates. These estimates and their variances can be expressed fairly explicitly by using certain matrices and their inverses. (Details may be found in Draper and Smith, 1981.) However, constructing these estimates requires complete knowledge of the covariance function of the stochastic component, a function that is unknown in virtually all real applications. It is possible to iteratively estimate the covariance function for $\{X_t\}$ based on a preliminary estimate of the trend. The trend is then estimated again using the estimated covariance function for $\{X_t\}$ and thus iterated to an approximate BLUE for the trend. This method will not be pursued here, however.

Fortunately, there are some results based on large sample sizes that support the use of the simpler least-squares estimates for the types of trends that we have considered. In particular, we have the following result (see Fuller (1976), pp. 388–393, for more details): We assume that the trend is either a polynomial in time, a trigonometric polynomial, seasonal means, or a linear combination of these. Then for a very general stationary stochastic component $\{X_t\}$, the least-squares estimates for trend have the *same* variance as the best linear unbiased estimates for large sample sizes.

Although the simple least-squares estimates may be asymptotically efficient, *it does not follow* that the estimated standard deviations of the coefficients as printed out by all regression routines, including Minitab, are correct. We shall elaborate on this point in the next section.

We also caution the reader that the result above is restricted to certain kinds of trends and cannot, in general, be extended to regression on arbitrary predictor variables, such as other time series. For example, Fuller (1976, pp. 420–422) shows that if $Z_t = \beta Y_t + X_t$, where $\{X_t\}$ has a simple stochastic structure but $\{Y_t\}$ is also a stationary series, then the least-squares estimate of β can be very inefficient and biased even for large samples.

3.5 INTERPRETING THE REGRESSION OUTPUT

We have already noted that the standard regression routines calculate least-squares estimates of the unknown regression coefficients—the betas. As such, the estimates are reasonable under minimal assumptions on the stochastic component $\{X_t\}$. However, some of the regression output depends heavily on the usual regression assumption that $\{X_t\}$ is white noise, and some depends on the further assumption that $\{X_t\}$ is approximately normally distributed. We begin with the items that depend least on the assumptions.

Consider the regression output under the BRIEF command, as in Exhibit 3.1. We shall write $\hat{\mu}_t$ for the estimated trend regardless of the assumed parametric form for μ_t. For example, for the linear time trend we have $\hat{\mu}_t = b_0 + b_1 t$. For each t the unobserved stochastic component X_t can be estimated (predicted) by $Z_t - \hat{\mu}_t$. If the $\{X_t\}$ process has constant variance, then we can estimate the standard deviation of X_t, namely $\sqrt{\gamma_0}$, by

$$S = \left[\frac{1}{n - p} \sum_{t=1}^{n} (Z_t - \hat{\mu}_t)^2 \right]^{1/2}$$

where p is the number of parameters estimated in μ_t and $n - p$ is the so-called degrees of freedom for S. The value of S gives an absolute measure of the goodness-of-fit of the estimated trend—the smaller the value of S, the better the fit. However, a value of S of, say, 60.74 is somewhat difficult to interpret.

A unitless measure of the goodness-of-fit of the trend is the value of R^2, also called the **coefficient of determination.** One interpretation of R^2 is that it is the square of the sample correlation coefficient between the observed series and the estimated trend. It is also the fraction of the variation in the series that is explained by the estimated trend. According to Exhibit 3.5, about 96% of the variation in the monthly temperature series is explained by the cosine trend at frequency $\frac{1}{12}$. The R-sq(adj) value is a small adjustment to R^2 that yields an approximately unbiased estimate based on the number of parameters estimated in the trend. Various formulas for computing R^2 may be found in any book on regression, such as Draper and Smith (1981).

The standard deviations of the coefficients labeled Stdev on the output need to be interpreted carefully. They are appropriate *only* when the stochastic component is white noise—the usual regression assumption. For example, in Exhibit 3.1 the value 0.09663 is obtained from the square root of the value given by Equation (3-21) when $\rho_k = 0$ for $k > 0$ and with γ_0 estimated by S^2, that is,

$$0.09663 = \left[\frac{12(60.74)^2}{168(168^2 - 1)} \right]^{1/2}$$

Similarly, in Exhibit 3.3 the value 0.9870 is just $S/\sqrt{12}$ (12 years) from Equation (3-15), and in Exhibit 3.5 the value 0.4367 is $S\sqrt{(2/144)}$ from Equation (3-17). The important point is that these standard deviations assume a white noise stochastic component that will rarely be true for time series.

The t-ratios, then, are just the estimated regression coefficients, each divided by their respective estimated standard deviations. If the stochastic component is *normally distributed white noise*, then these ratios provide test statistics for checking the significance of the regression coefficients. The significance levels are given by the t-distribution with $n - p$ degrees of freedom.

3.6 RESIDUAL ANALYSIS

As we have already noted, the unobserved stochastic component $\{X_t\}$ can be estimated, or predicted, by

$$\hat{X}_t = Z_t - \hat{\mu}_t$$

Predicted is really a better term. We reserve the term *estimate* for an unknown parameter, and the term *predictor* for an estimate of an unobserved random variable. We call $\hat{X}_t$ the tth **residual.** If the trend model is reasonably correct, then the residuals should behave roughly like the true stochastic component, and various assumptions about the stochastic component can be assessed by looking at the residuals. If the stochastic component is white noise, then the residuals should behave like independent (normal) random variables with zero mean and standard deviation S. Since a least-squares fit of any trend containing a constant term will automatically produce residuals with a zero mean, we might consider the **standardized residuals** $\hat{X}_t/S$ and check their independence and possible normality. In Minitab, an extension of the REGRESS command will automatically store the standardized residuals (and the estimated trend $\hat{\mu}_t$) into specified columns. The format is as follows:

REGRESS **C** on **K** pred. **C**, . . . , **C** [put stand. res. in **C** [trend in **C**]]

(Minitab actually uses a slightly different standardization. Each residual is divided by the estimated standard deviation of *that* residual; see the *Minitab Reference Manual*. Unstandardized residuals may be obtained with the sub-command RESIDUAL.)

With the residuals in hand, the next step is to examine various residual plots. We first look at the plot of the residuals over time. The command TSPLOT applied to the residuals will of course accomplish this. If the data are seasonal, we should specify the seasonal period in TSPLOT so that residuals associated with the same season can be identified easily. Exhibit 3.6 shows such a plot for the residuals of the monthly temperature data fitted by seasonal means. If the stochastic component is white noise and the trend is adequately modeled, we would expect such a plot to suggest a rectangular scatter with no discernible trends whatsoever. There are no striking departures from randomness apparent in Exhibit 3.6.

Next we look at the residuals versus the corresponding trend estimate, or fitted value, as in Exhibit 3.7. Once more we are looking for patterns. Are small residuals associated with small trend values, and large residuals with large trend values? Exhibit 3.7 certainly does not indicate any such pattern.

Gross nonnormality can be assessed by plotting a histogram of the residuals using the HISTOGRAM command:

HISTOGRAM of the data in **C**

Normality can be checked more carefully by plotting the so-called **normal scores** of the residuals. The Minitab command is

NSCORES of the data in **C**, put into **C**

followed by the PLOT command:

PLOT the data in **C** versus normal scores in **C**

The tth normal score is defined to be the $(t - \frac{3}{8})/(n + \frac{1}{4})$ percentage point of the standard normal distribution. With normally distributed data, a plot of the tth ordered data value versus the corresponding normal score should fall approximately on a straight line.

EXHIBIT 3.6 Residual Time Plot for Temperature Series with Seasonal Means Model

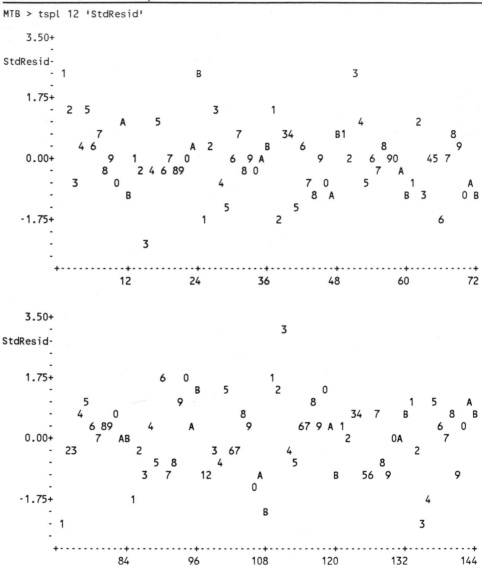

A powerful test of normality can be based on the sample correlation coefficient between the data (residuals) and the corresponding normal scores. The Minitab command

CORRELATION between **C** and **C**

will do the calculation after the normal scores have been obtained. Critical values for this correlation are given in Table 3.1. For a given sample size and significance level, we would reject normality if the observed correlation falls *below* the appropriate critical value in the table.

EXHIBIT 3.7 Residuals versus Fitted Values for Temperature Series with Seasonal Means Model

```
MTB > plot 'StdResid' vs. 'Fitted'

      3.20+
         -                         *
  StdResid-        *
         -                 *       *
         -        *                              *
      1.60+       *    2          *                       *          *
         -        *       *               *        *      2    *
         -             *                  *        2           2              *
         -        *       3       2                2    *              3    3
         -        *   *   *               4        2    *         4       5  2  *
     -0.00+       *    2   *              2        *    3    *    4       2     4
         -            3            *      *        2    *         *       2  3  2
         -        *   *            2      *        *    2    2             2  *
         -        *   *   4        2      *        *    *    2    *        *
         -                                *                 *    *     *     *
     -1.60+       *                        *    *           *         *
         -        *   *   *
         -
         -        *                2
      -------+---------+---------+---------+---------+---------+------Fitted
            20.0      30.0      40.0      50.0      60.0      70.0
```

Randomness (independence) in the residuals can be tested in several ways. The **runs test** examines the residuals in sequence to look for patterns. For example, patterns will occur if the residuals are correlated or if the fitted trend missed the actual trend in the series in a systematic way—say, too low for a stretch, then too high, then too low again, and so on.

The Minitab command for the runs test is

RUNS above and below **K** for the data in **C**

Since we are analyzing residuals, the usual value for *K* will be zero. The output includes the number of runs, the expected number of runs, and the standard

TABLE 3.1   Critical Values for the Normal-Scores Correlation Test for Normality

Sample Size	Significance Level		
n	0.01	0.05	0.10
10	0.880	0.918	0.935
15	0.911	0.930	0.951
20	0.929	0.950	0.960
25	0.941	0.958	0.966
30	0.949	0.964	0.971
40	0.960	0.972	0.977
50	0.966	0.976	0.981
60	0.971	0.980	0.984
75	0.976	0.984	0.987
100	0.981	0.986	0.989
150	0.987	0.991	0.992
200	0.990	0.993	0.994

EXHIBIT 3.8 Histogram of Standardized Residuals for Temperature Series with Seasonal Means Model

```
MTB > histogram of 'StdResid'

Histogram of StdResid   N = 144

Midpoint   Count
   -2.5       3   ***
   -2.0       3   ***
   -1.5       5   *****
   -1.0      23   ***********************
   -0.5      24   ************************
   -0.0      28   ****************************
    0.5      29   *****************************
    1.0      15   ***************
    1.5       8   ********
    2.0       2   **
    2.5       3   ***
    3.0       1   *
```

EXHIBIT 3.9 Normal-Scores Plot for Temperature Series with Seasonal Means Model

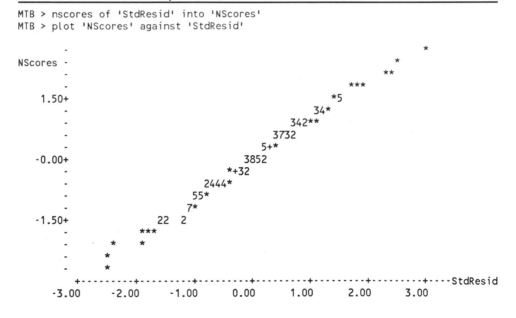

```
MTB > nscores of 'StdResid' into 'NScores'
MTB > plot 'NScores' against 'StdResid'
```

EXHIBIT 3.10 Runs Test for Temperature Series with Seasonal Means Model

```
MTB > runs test around 0 for 'StdResid'

    StdResid

    K =     0.0000


    THE OBSERVED NO. OF RUNS =  65
    THE EXPECTED NO. OF RUNS =  72.8750
    75 OBSERVATIONS ABOVE K    69 BELOW
            THE TEST IS SIGNIFICANT AT  0.1873
            CANNOT REJECT AT ALPHA = 0.05
```

deviation under the assumption of independence. The observed significance level is also given.

Exhibits 3.8, 3.9, and 3.10 give the histogram, normal-scores plot, and runs test, respectively, for the temperature series residuals with a seasonal means trend. The histogram is reasonably symmetrical but does contain a few more extreme values than we would expect. The normal-scores plot is reasonably straight, again showing minor bumps at the extremes. The sample correlation coefficient between the residuals and the corresponding normal scores was calculated to be 0.997. Here the sample size is 144, and according to Table 3.1 we would not reject normality at any of the given significance levels. Furthermore, the results of the runs test do not suggest a lack of independence in the stochastic component.

THE SAMPLE AUTOCORRELATION FUNCTION

Another very important diagnostic tool for examining dependence is the sample autocorrelation function. Consider any sequence of data $Y_1, Y_2, \ldots, Y_n$— whether residuals, original data, or some transformation of the original data, such as differenced data. Tentatively assuming stationarity, we would like to estimate the autocorrelation function ρ_k for a variety of lags $k = 1, 2, \ldots$. The obvious way to do this is to compute the sample correlation for the pairs (Y_1, Y_{1+k}), $(Y_2, Y_{2+k}), \ldots, (Y_{n-k}, Y_n)$. Except for two minor modifications, this leads to our definition.

The **sample autocorrelation function** r_k is defined by

$$r_k = \frac{\sum\limits_{t-k+1}^{n} (Y_t - \bar{Y})(Y_{t-k} - \bar{Y})}{\sum\limits_{t=1}^{n} (Y_t - \bar{Y})^2} \quad \text{for } k = 0, 1, 2, \ldots \qquad (3\text{-}23)$$

Notice that we have used the "grand" sample mean $\bar{Y}$ in all places and have also divided by the "grand sum of squares" rather than the product of the two separate standard deviations used in the ordinary correlation coefficient. However, since we are assuming a stationary sequence, these modifications seem appropriate. We also note that the numerator contains $n - k$ cross products although the denominator always contains n. For a variety of reasons, this has become the standard definition for the sample autocorrelation function. A plot of r_k versus k is sometimes called a **correlogram.**

Here we are interested in discovering dependence in the stochastic component; therefore the sample autocorrelation function for the residuals is of interest. The Minitab command ACF does the required computations and has the form

ACF [up to lag **K**] for the series in **C** [store in **C**]

If the maximum lag K is not specified, the ACF is calculated out to lag $\sqrt{n} + 10$.

For the residuals in the seasonal means fit of the temperature data, Exhibit 3.11 shows the resulting sample autocorrelation function. As we would suspect from our earlier results, no strong correlations show up at any of the lags out to 36.

The values of r_k are, of course, *estimates* of ρ_k; consequently, they have their own sampling distributions, standard errors, and other properties. For now we shall use r_k as a descriptive tool and defer discussion of those topics until Chapters 6 and 8.

Let us now consider these residual analyses for the milk production series. Suppose that we fit a combined linear trend and seasonal means trend to this series. Exhibit 3.12 shows a time plot of the residuals from such an

EXHIBIT 3.11 Sample Autocorrelation Function of Residuals for Temperature Series with Seasonal Means Model.

```
ACF of StdResid

           -1.0 -0.8 -0.6 -0.4 -0.2  0.0  0.2  0.4  0.6  0.8  1.0
           +----+----+----+----+----+----+----+----+----+----+
    1   0.094                         XXX
    2  -0.099                         XXX
    3  -0.009                          X
    4  -0.110                        XXXX
    5  -0.048                         XX
    6   0.138                          XXXX
    7  -0.004                          X
    8  -0.128                        XXXX
    9  -0.076                         XXX
   10  -0.006                          X
   11  -0.133                        XXXX
   12  -0.055                         XX
   13   0.095                          XXX
   14  -0.064                         XXX
   15   0.071                          XXX
   16   0.057                          XX
   17  -0.017                          X
   18  -0.066                         XXX
   19  -0.005                          X
   20  -0.065                         XXX
   21  -0.016                          X
   22   0.050                          XX
   23   0.008                          X
   24  -0.194                       XXXXXX
   25   0.018                          X
   26   0.051                          XX
   27  -0.004                          X
   28   0.211                          XXXXXX
   29   0.096                          XXX
   30  -0.061                         XXX
   31  -0.021                         XX
   32   0.094                          XXX
   33  -0.002                          X
   34   0.048                          XX
   35   0.035                          XX
   36  -0.170                        XXXXX
```

EXHIBIT 3.12 Residual Time Plot for Milk Production Series with Linear Trend and Seasonal Means
Model

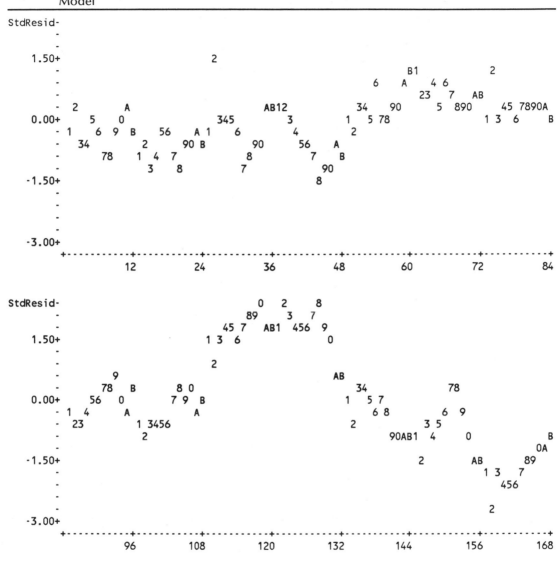

undertaking. Here it is clear that we have *not* achieved randomness in the residuals—they change much too smoothly for white noise. Exhibit 3.13 shows a tendency for larger scatter in the residuals associated with larger fitted values.

Exhibits 3.14 and 3.15 illustrate a lack of normality,* which is confirmed by referring the normal scores versus residuals correlation of 0.988 to Table 3.1.

---

* The normal-scores test assumes independence of the observations. The tests of independence should be performed before testing for normality.

EXHIBIT 3.13 Residuals versus Fitted Values for Milk Production Series with Linear Trend and Seasonal Means Model

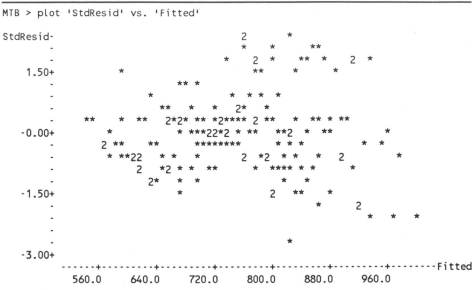

```
MTB > plot 'StdResid' vs. 'Fitted'

StdResid-                                   2        *
       -                                   *     *      **
       -                            *    2  *    ** *    2  *
  1.50+           *                     **       *       *
       -                   **  *        *
       -               *              *   *  *  *
       -            **    *   *   2*      *
       -      **    *  **   2*2* ** *2**** 2 **   *  ** * **
 -0.00+        *          * ***22*2 * **   **2  *  **       *
       -    2 **    **      ********   * * *        *  *
       -    * **22   * *    *      2  *2 * *  *  2       *
       -        2   *2 * *  **      * **** * *     *
       -       2*   * *          *   *
 -1.50+         *          2  **   *
       -                   *        2
       -                         *  *  *
       -
       -                 *
 -3.00+
       -----+---------+---------+---------+---------+---------+--------Fitted
          560.0     640.0     720.0     800.0     880.0     960.0
```

With our sample size of 168, we would reject normality at level 0.05 and nearly reject it at level 0.01.

Exhibit 3.16 indicates a lack of independence, as does the sample autocorrelation function in Exhibit 3.17. Notice in particular how the smooth change in the residuals is reflected in the large positive values of r_k for the small lags. Clearly there is much left to be modeled in the milk production series, which we shall return to in Chapters 10 and 11.

EXHIBIT 3.14 Histogram of Residuals for Milk Production Series with Linear Trend and Seasonal Means Model

```
MTB > histogram of 'StdResid'

Histogram of StdResid   N = 168

Midpoint    Count
    -2.5        1    *
    -2.0        4    ****
    -1.5       10    **********
    -1.0       22    **********************
    -0.5       32    ********************************
    -0.0       35    ***********************************
     0.5       34    **********************************
     1.0        8    ********
     1.5       10    **********
     2.0        9    *********
     2.5        3    ***
```

EXHIBIT 3.15 Normal-Scores Plot for Milk Production Series with Linear Trend and Seasonal Means Model

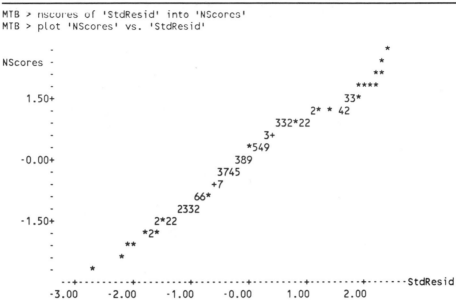

```
MTB > nscores of 'StdResid' into 'NScores'
MTB > plot 'NScores' vs. 'StdResid'
```

EXHIBIT 3.16 Runs Test for Milk Production Series with Linear Trend and Seasonal Means Model

```
MTB > runs test around 0 using 'StdResid'

StdResid

K =      0.0000


THE OBSERVED NO. OF RUNS =   29
THE EXPECTED NO. OF RUNS =   84.8095
80 OBSERVATIONS ABOVE K    88 BELOW
            THE TEST IS SIGNIFICANT AT   0.0000
```

CHAPTER 3 EXERCISES

3.1. Verify Equations (3-9) for the least-squares estimates when
$$Z_t = \beta_0 + \beta_1 t + X_t.$$

3.2. Suppose $Z_t = \mu + a_t - a_{t-1}$. Find Var $(\bar{Z})$ where

$$\bar{Z} = \frac{1}{n} \sum_{t=1}^{n} Z_t$$

Note any unusual results.

3.3. Verify Equation (3-6).

3.4. If $Z_t = \mu + a_t + a_{t-1}$, find Var $(\bar{Z})$.

3.5. Write a Minitab program that will estimate a model combining a

EXHIBIT 3.17 Sample Autocorrelation Function of Residuals for Milk Production Series with Linear Trend and Seasonal Means Model

```
MTB > acf to lag 36 for 'StdResid'

ACF of StdResid
```

		-1.0 -0.8 -0.6 -0.4 -0.2 0.0 0.2 0.4 0.6 0.8 1.0
		+----+----+----+----+----+----+----+----+----+----+
1	0.883	XXXXXXXXXXXXXXXXXXXXXXX
2	0.824	XXXXXXXXXXXXXXXXXXXXXX
3	0.768	XXXXXXXXXXXXXXXXXXXX
4	0.687	XXXXXXXXXXXXXXXXXX
5	0.627	XXXXXXXXXXXXXXXX
6	0.555	XXXXXXXXXXXXXX
7	0.528	XXXXXXXXXXXXXX
8	0.477	XXXXXXXXXXXX
9	0.442	XXXXXXXXXXX
10	0.395	XXXXXXXXXX
11	0.342	XXXXXXXXX
12	0.334	XXXXXXXXX
13	0.268	XXXXXXX
14	0.210	XXXXX
15	0.166	XXXXX
16	0.122	XXXX
17	0.088	XXX
18	0.028	XX
19	-0.001	X
20	-0.028	XX
21	-0.048	XX
22	-0.078	XXX
23	-0.094	XXX
24	-0.101	XXXX
25	-0.142	XXXXX
26	-0.180	XXXXX
27	-0.219	XXXXX
28	-0.250	XXXXXX
29	-0.275	XXXXXXX
30	-0.293	XXXXXXX
31	-0.298	XXXXXXX
32	-0.307	XXXXXXXX
33	-0.302	XXXXXXXX
34	-0.306	XXXXXXXX
35	-0.305	XXXXXXXX
36	-0.307	XXXXXXXX

linear time trend and seasonal means for a quarterly series of length 10 years. Assume that the series has already been stored in column 1.

3.6. Consider fitting a cosine seasonal trend at the fundamental frequency $\frac{1}{4}$ to quarterly data. Assuming the series is of length 43 and already stored in column 1, write a Minitab program to estimate β_0, β_1, and β_2.

3.7. Suppose we want to fit the monthly Dubuque temperature series with a sine–cosine pair at the harmonic frequency $\frac{2}{12}$ in addition to the sine–cosine pair at the fundamental frequency $\frac{1}{12}$. Estimate such a mean function, and compare the results to those in Exhibit 3.5.

3.8. Time series with deterministic trends as discussed in this chapter are easily simulated in Minitab. We first set 'time' into a column, then calculate a linear time trend, say, and finally add an error term using the RANDOM command. For example:

SET C1
    1 : 100
    END
LET C2 = 10 + 2 ∗ C1
RANDOM 100 std normal errors into **C3**
LET C4 = C2 + C3

At this point C2 contains the trend, C3 contains the errors, and C4 contains the observed series.

   Use Minitab to simulate a variety of trend models. Estimate the (known) parameters in those series with the REGRESSION command and compare your estimates with the known values.

   3.9. Verify Equation (3-7).

   3.10. Verify Equation (3-21).

CHAPTER 4 MODELS FOR STATIONARY TIME SERIES

This chapter discusses the basic concepts of the broad parametric family of autoregressive moving average (ARMA) time series models, which have assumed great importance in modeling real-world situations.

4.1 GENERAL LINEAR PROCESSES

We shall always let $\{Z_t\}$ denote the observed time series; from here on, $\{a_t\}$ will represent an unobserved white noise series, that is, a sequence of identically distributed, zero-mean, independent random variables. (For much of our work, the assumption of independence can be replaced by the weaker assumption that the $\{a_t\}$ are uncorrelated random variables, but we shall not pursue that slight generality.)

A **general linear process** $\{Z_t\}$ is one that can be represented as a weighted linear combination of the present and past terms of a white noise process:

$$Z_t = a_t + \psi_1 a_{t-1} + \psi_2 a_{t-2} + \cdots \tag{4-1}$$

If the right-hand side of this expression is truly an infinite series, then certain restrictions must be placed on the ψ's for the right-hand side to be mathematically meaningful. For our purposes, it suffices to assume that

$$\sum_{i=1}^{\infty} \psi_i^2 < \infty \tag{4-2}$$

We should note that since $\{a_t\}$ is unobservable, there is no loss in the generality of Equation (4-1) if we assume that the coefficient of a_t is 1, effectively $\psi_0 = 1$.

An important, nontrivial example to which we shall return many times is the case where the ψ's form an exponentially (geometrically) decaying sequence:

$$\psi_j = \phi^j$$

where ϕ is a number between -1 and $+1$. Then

$$Z_t = a_t + \phi a_{t-1} + \phi^2 a_{t-2} + \cdots$$

For this example

$$E(Z_t) = E(a_t + \phi a_{t-1} + \cdots) = E(a_t) + \phi E(a_{t-1}) + \cdots$$
$$= 0$$

so that $\{Z_t\}$ has a constant mean of zero. Also,

$$\text{Var}(Z_t) = \text{Var}(a_t + \phi a_{t-1} + \phi^2 a_{t-2} + \cdots)$$
$$= \text{Var}(a_t) + \phi^2 \text{Var}(a_{t-1}) + \phi^4 \text{Var}(a_{t-2}) + \cdots \quad \text{(since the}$$
$$a\text{'s are independent)}$$
$$= \sigma_a^2(1 + \phi^2 + \phi^4 + \cdots)$$
$$= \sigma_a^2 \frac{1}{1 - \phi^2}$$

by assuming the geometric series in the penultimate line. Furthermore,

$$\text{Cov}(Z_t, Z_{t-1}) = \text{Cov}(a_t + \phi a_{t-1} + \phi^2 a_{t-2} + \cdots, a_{t-1} + \phi a_{t-2} + \cdots)$$
$$= \text{Cov}(\phi a_{t-1}, a_{t-1}) + \text{Cov}(\phi^2 a_{t-2}, \phi a_{t-2}) + \cdots \quad \text{(all}$$
$$\text{other terms being zero)}$$
$$= \phi \sigma_a^2 + \phi^3 \sigma_a^2 + \phi^5 \sigma_a^2 + \cdots$$
$$= \phi \sigma_a^2(1 + \phi^2 + \phi^4 + \cdots)$$
$$= \frac{\phi \sigma_a^2}{1 - \phi^2} \quad \text{(again summing a geometric series)}$$

Thus,

$$\text{Corr}(Z_t, Z_{t-1}) = \frac{\dfrac{\phi \sigma_a^2}{1 - \phi^2}}{\dfrac{\sigma_a^2}{1 - \phi^2}} = \phi$$

In a similar manner we can calculate

$$\text{Cov}(Z_t, Z_{t-k}) = \frac{\phi^k \sigma_a^2}{1 - \phi^2}$$

and

$$\boxed{\text{Corr}(Z_t, Z_{t-k}) = \phi^k \quad \text{for } k = 0, 1, 2, \ldots} \tag{4-3}$$

It is important to note that the processes defined in this way will be stationary—the autocovariance structure depends only on lagged time and not on absolute time. For a general linear process $Z_t = a_t + \psi_1 a_{t-1} + \cdots$, calculations similar to those done above yield the following results:

$$\boxed{E(Z_t) = 0 \qquad \gamma_k = \text{Cov}(Z_t, Z_{t-k}) = \sigma_a^2 \sum_{i=0}^{\infty} \psi_i \psi_{i+k}, \qquad k \geq 0} \tag{4-4}$$

with $\psi_0 = 1$. A process with a nonzero mean μ may be obtained by adding μ to the right-hand side of Equation (4-1). Since the mean does not affect the covariance properties of a process, we shall assume a zero mean until we begin fitting models to data.

4.2 MOVING AVERAGE PROCESSES

In the case where only a finite number of ψ's are nonzero, we have what is called a **moving average process.** In this case we change notation somewhat and write

$$Z_t = a_t - \theta_1 a_{t-1} - \theta_2 a_{t-2} - \cdots - \theta_q a_{t-q} \tag{4-5}$$

We call such a series a **moving average of order q** and abbreviate the name as MA(q). (The confusing change of notation from ψ's to $-\theta$'s will actually be convenient later on.) The terminology *moving* average arises from the fact that Z_t is obtained by applying the weights $1, -\theta_1, -\theta_2, \ldots, -\theta_q$ to the variables $a_t, a_{t-1}, a_{t-2}, \ldots, a_{t-q}$ and then *moving* the same weights 1 unit of time forward and applying them to $a_{t+1}, a_t, a_{t-1}, \ldots, a_{t-q+1}$ to obtain Z_{t+1}. Moving average models were first considered by Slutsky (1927) and Wold (1938).

THE FIRST-ORDER MOVING AVERAGE PROCESS

We shall consider in detail the simple, but nevertheless very important, moving average process of order 1, that is, the MA(1) series. Rather than specialize the formulas in (4-4), it is instructive to rederive the results. The model is $Z_t = a_t - \theta a_{t-1}$. (Since only one θ is involved, we drop the redundant subscript 1.) Clearly $E(Z_t) = 0$ and $\text{Var}(Z_t) = \sigma_a^2(1 + \theta^2)$. Now

$$\text{Cov}(Z_t, Z_{t-1}) = \text{Cov}(a_t - \theta a_{t-1}, a_{t-1} - \theta a_{t-2})$$
$$= \text{Cov}(-\theta a_{t-1}, a_{t-1}) = -\theta \sigma_a^2$$

and

$$\text{Cov}(Z_t, Z_{t-2}) = \text{Cov}(a_t - \theta a_{t-1}, a_{t-2} - \theta a_{t-3})$$
$$= 0$$

since there are no a's with subscripts in common between Z_t and Z_{t-2}. Similarly, $\text{Cov}(Z_t, Z_{t-k}) = 0$ whenever $k \geq 2$; that is, *the process has no correlation beyond lag 1.* This fact will be important later when we need to choose suitable models for real data.

In summary: for an MA(1) series $Z_t = a_t - \theta a_{t-1}$,

$$E(Z_t) = 0$$
$$\gamma_0 = \text{Var}(Z_t) = \sigma_a^2(1 + \theta^2)$$
$$\gamma_1 = -\theta \sigma_a^2$$
$$\rho_1 = \frac{-\theta}{1 + \theta^2} \tag{4-6}$$

and

$$\gamma_k = \rho_k = 0 \quad \text{for } k \geq 2$$

Some numerical values for ρ_1 versus θ in Table 4.1 help illustrate the possibilities. Note that ρ_1 values for θ negative can be obtained by simply negating the value given for the corresponding positive θ-value.

TABLE 4.1   Lag 1 Autocorrelation for MA(1) Processes

θ	$\rho_1 = -\theta/(1 + \theta^2)$
0.0	−0.0000
0.1	−0.0099
0.2	−0.1923
0.3	−0.2752
0.4	−0.3448
0.5	−0.4000
0.6	−0.4412
0.7	−0.4698
0.8	−0.4878
0.9	−0.4972
1.0	−0.5000

A simple argument using calculus will show that the largest value that ρ_1 can attain is 0.5, which occurs when $\theta = -1$, and the smallest value is −0.5, which occurs when $\theta = +1$. In addition, if θ is replaced by $1/\theta$ in ρ_1, we obtain the *same* numerical value for ρ_1. So, for example, ρ_1 is the same for $\theta = 0.5$ as it is for $\theta = 1/0.5 = 2$. If we knew that an MA(1) process had $\rho_1 = 0.4$, we still could not tell the precise value for θ. This somewhat bothersome point will be discussed further when we define **invertibility** in Section 4.5.

Exhibit 4.1 shows a time plot of a computer simulation of an MA(1) series with $\theta = -0.9$ and normally distributed white noise. Appendix B gives a Minitab program for doing such simulations. Recall from Table 4.1 that for this model $\rho_1 = 0.4972$; thus there is a moderately strong positive correlation at lag 1. This correlation is quite evident in the plot of the series, since consecutive observations tend to be closely related. If an observation is above the mean level of zero, then the adjacent observation also tends to be above the mean. The plot is relatively *smooth* over time with only occasional large fluctuations. The lag 1 correlation is also apparent when the scatter of the pairs (Z_t, Z_{t-1}) is shown, as in Exhibit 4.2. The lack of correlation at lag 2 is shown clearly in Exhibit 4.3, which plots the pairs (Z_t, Z_{t-2}).

Similar plots are shown in Exhibits 4.4, 4.5, and 4.6 for MA(1) simulations with $\theta = 0.9$ so that $\rho_1 = -0.4972$. The negative correlation at lag 1 is now evident in both the time plot and the scatter plot of (Z_t, Z_{t-1}). Note the *jagged* shape of the time plot (Exhibit 4.4) compared with the smooth plot in Exhibit 4.1.

THE SECOND-ORDER MOVING AVERAGE PROCESS

Consider now the moving average process of order 2:

$$Z_t = a_t - \theta_1 a_{t-1} - \theta_2 a_{t-2}$$

Here

$$\gamma_0 = \text{Var}(Z_t) = \text{Var}(a_t - \theta_1 a_{t-1} - \theta_2 a_{t-2}) = (1 + \theta_1^2 + \theta_2^2)\sigma_a^2$$

$$\gamma_1 = \text{Cov}(Z_t, Z_{t-1}) = \text{Cov}(a_t - \theta_1 a_{t-1} - \theta_2 a_{t-2}, a_{t-1} - \theta_1 a_{t-2} - \theta_2 a_{t-3})$$

$$= \text{Cov}(-\theta_1 a_{t-1}, a_{t-1}) + \text{Cov}(-\theta_2 a_{t-2}, -\theta_1 a_{t-2})$$

$$= [-\theta_1 + (-\theta_2)(-\theta_1)]\sigma_a^2$$

$$= (-\theta_1 + \theta_1 \theta_2)\sigma_a^2$$

EXHIBIT 4.1 Time Plot of an MA(1) Process with $\theta = -0.9$

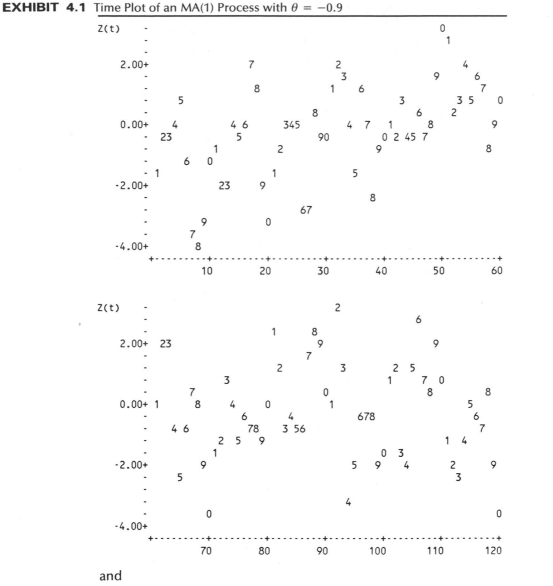

and

$$\gamma_2 = \text{Cov}\,(Z_t, Z_{t-2}) = \text{Cov}\,(a_t - \theta_1 a_{t-1} - \theta_2 a_{t-2}, a_{t-2} - \theta_1 a_{t-3} - \theta_2 a_{t-4})$$
$$= \text{Cov}\,(-\theta_2 a_{t-2}, a_{t-2}) = -\theta_2 \sigma_a^2$$

Thus for an MA(2) process,

$$\rho_1 = \frac{-\theta_1 + \theta_1 \theta_2}{1 + \theta_1^2 + \theta_2^2}$$

$$\rho_2 = \frac{-\theta_2}{1 + \theta_1^2 + \theta_2^2}$$

(4-7)

$$\rho_k = 0 \quad \text{for } k \geq 3$$

and, as before,

EXHIBIT 4.2 Scatter Plot of Pairs (Z_t, Z_{t-1}) from an MA(1) Series with $\theta = -0.9$: Sample Correlation = 0.528, Theoretical Correlation = 0.497

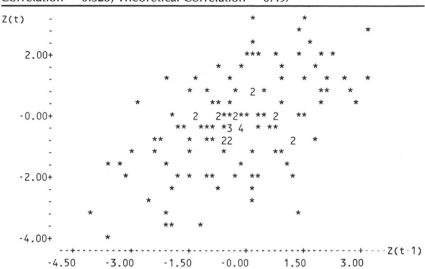

EXHIBIT 4.3 Scatter Plot of Pairs (Z_t, Z_{t-2}) from an MA(1) Series with $\theta = -0.9$: Sample Correlation = 0.007, Theoretical Correlation = 0

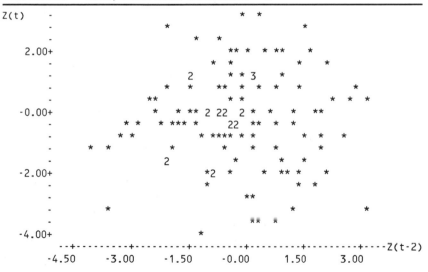

For the specific case $Z_t = a_t - a_{t-1} + 0.6a_{t-2}$, we have

$$\rho_1 = \frac{-1 + (1)(-0.6)}{1 + (1)^2 + (0.6)^2} = \frac{-1.6}{2.36} = -0.678$$

and

$$\rho_2 = \frac{0.6}{2.36} = 0.254$$

EXHIBIT 4.4 Time Plot of an MA(1) Series with $\theta = 0.9$

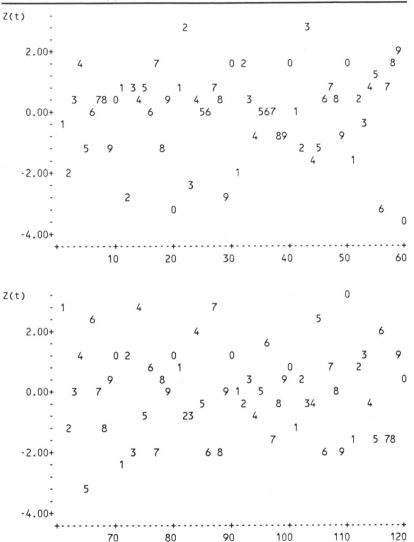

A computer simulation of this process is shown in Exhibit 4.7. Scatter plots in Exhibits 4.8, 4.9, and 4.10 exhibit the lagged correlation structure of the series.

Formulas for the autocorrelation function of a general MA(q) process can now be obtained easily:

$$\rho_k = \begin{cases} \dfrac{-\theta_k + \theta_1\theta_{k+1} + \theta_2\theta_{k+2} + \cdots + \theta_{q-k}\theta_q}{1 + \theta_1^2 + \theta_2^2 + \cdots + \theta_q^2} & \text{for } k = 1, 2, \ldots, q \\[2ex] 0 & \text{for } k \geq q + 1 \end{cases} \tag{4-8}$$

where the numerator of ρ_q is just $-\theta_q$. It is important to emphasize that the autocorrelation is zero after q lags and can follow almost any pattern for lags 1 through q.

EXHIBIT 4.5 Scatter Plot of Pairs (Z_t, Z_{t-1}) from an MA(1) Series with $\theta = 0.9$: Sample Correlation = -0.519, Theoretical Correlation = -0.497

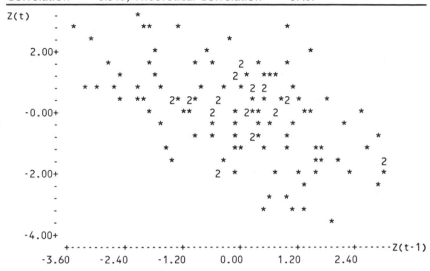

EXHIBIT 4.6 Scatter Plot of Pairs (Z_t, Z_{t-2}) from an MA(1) Series with $\theta = 0.9$: Sample Correlation = -0.044, Theoretical Correlation = 0

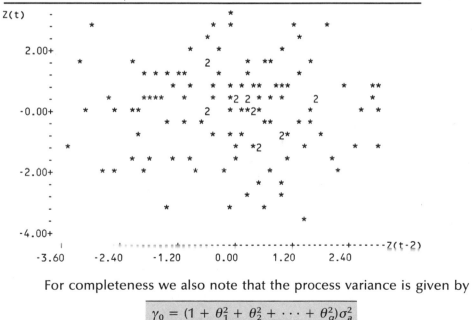

For completeness we also note that the process variance is given by

$$\gamma_0 = (1 + \theta_1^2 + \theta_2^2 + \cdots + \theta_q^2)\sigma_a^2 \qquad \text{(4-9)}$$

4.3 AUTOREGRESSIVE PROCESSES

Autoregressive processes are as their name implies—regressions on themselves. Specifically, a pth-order **autoregressive process** $\{Z_t\}$ satisfies the

EXHIBIT 4.7 Time Plot of an MA(2) Series with $\theta_1 = 1$ and $\theta_2 = -0.6$

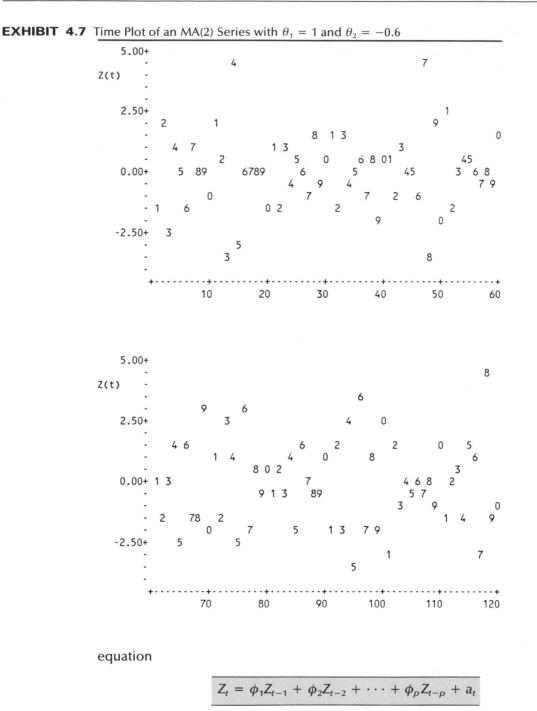

equation

$$Z_t = \phi_1 Z_{t-1} + \phi_2 Z_{t-2} + \cdots + \phi_p Z_{t-p} + a_t \qquad \text{(4-10)}$$

The current value of the series Z_t is a linear combination of the p most recent past values of itself plus an "innovation" term a_t, which incorporates everything new in the series at time t that is not explained by the past values. Thus we *assume* a_t is independent of $Z_{t-1}, Z_{t-2}, \ldots$. Yule (1927) carried out the original work on autoregressive processes.

EXHIBIT 4.8 Scatter Plot of Pairs (Z_t, Z_{t-1}) from an MA(2) Series with $\theta_1 = 1$ and $\theta_2 = -0.6$: Sample Correlation $= -0.735$, Theoretical Correlation $= -0.678$

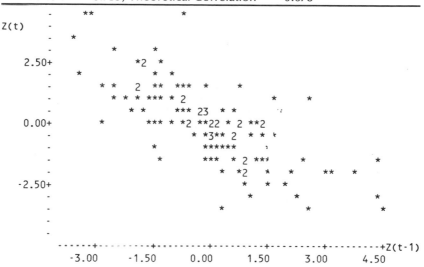

EXHIBIT 4.9 Scatter Plot of Pairs (Z_t, Z_{t-2}) from an MA(2) Series with $\theta_1 = 1$ and $\theta_2 = -0.6$: Sample Correlation $= 0.355$, Theoretical Correlation $= 0.254$

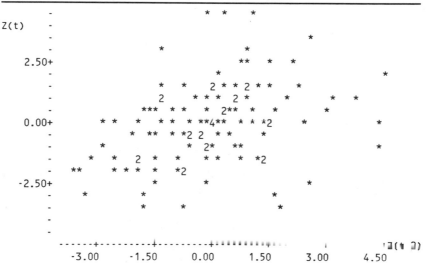

THE FIRST-ORDER AUTOREGRESSIVE PROCESS

Again, it is instructive to consider the first-order model, abbreviated AR(1), in detail. Assume that the series is stationary and satisfies

$$Z_t = \phi Z_{t-1} + a_t \tag{4-11}$$

Conditions for stationarity will be considered later. As usual, we assume that the mean of the series has been subtracted out so that Z_t has a zero mean. We

EXHIBIT 4.10 Scatter Plot of Pairs (Z_t, Z_{t-3}) from an MA(2) Process with $\theta_1 = 1$ and $\theta_2 = -0.6$: Sample Correlation = -0.117, Theoretical Correlation = 0

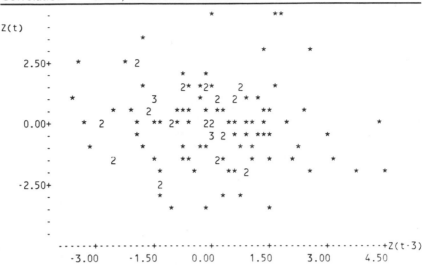

first take variances of both sides of Equation (4-11) and obtain

$$\gamma_0 = \phi^2 \gamma_0 + \sigma_a^2$$

Solving for γ_0 yields

$$\gamma_0 = \frac{\sigma_a^2}{1 - \phi^2} \tag{4-12}$$

Notice the immediate implication that $\phi^2 < 1$ or $|\phi| < 1$. Now take Equation (4-11), multiply both sides by Z_{t-k} ($k = 1, 2, \ldots$) and take expected values:

$$E(Z_t Z_{t-k}) = \phi E(Z_{t-1} Z_{t-k}) + E(a_t Z_{t-k})$$

Since the series is assumed to be stationary and since a_t and Z_{t-k} are independent, we obtain

$$\gamma_k = \phi \gamma_{k-1} \quad \text{for } k = 1, 2, \ldots \tag{4-13}$$

Setting $k = 1$, we get $\gamma_1 = \phi \gamma_0 = \phi \sigma_a^2 / (1 - \phi^2)$. With $k = 2$ we obtain $\gamma_2 = \phi \gamma_1 = \phi^2 \sigma_a^2 / (1 - \phi^2)$. Now it is easy to see that in general

$$\gamma_k = \frac{\phi^k \sigma_a^2}{1 - \phi^2}$$

and thus

$$\rho_k = \frac{\gamma_k}{\gamma_0} = \phi^k \quad \text{for } k = 0, 1, 2, \ldots \tag{4-14}$$

Since $|\phi| < 1$, the autocorrelation function is an exponentially decreasing curve as the number of lags k increases. If $0 < \phi < 1$, all correlations are positive; if $-1 < \phi < 0$, the lag 1 autocorrelation is negative ($\rho_1 = \phi$) and the signs of

FIGURE 4.1 Autocorrelation Functions for Several AR(1) Models (a) $\phi = 0.9$. (b) $\phi = 0.4$. (c) $\phi = -0.7$

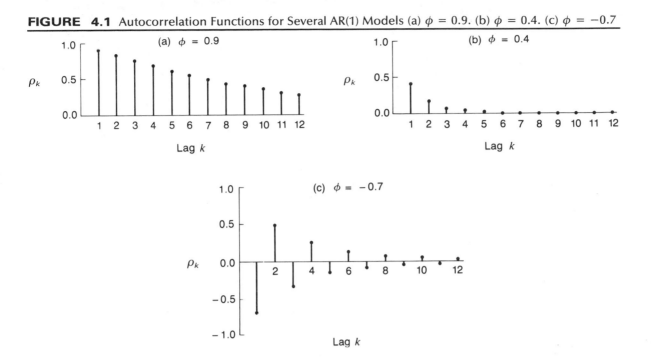

successive autocorrelations alternate from positive to negative with their magnitudes decreasing exponentially. Several autocorrelation functions are graphed in Figure 4.1.

Notice that for ϕ near ± 1, the exponential decay is quite slow ($[0.9]^6 = 0.53$), but for smaller ϕ the decay is rapid ($[0.4]^6 = 0.0041$). With ϕ near ± 1, the strong correlation will extend over many lags and produce a relatively smooth series if ϕ is positive and a very jagged series if ϕ is negative. Computer simulations of such series illustrate these facts in Exhibits 4.11, 4.12, and 4.13. Appendix B shows how to use Minitab to produce simulations from autoregressive models.

The perceptive reader will have noticed that the basic defining equation (4-10) or (4-11) for an autoregressive process is *not* in the form of a general linear process of Equation (4-1). Consider in particular the AR(1) model $Z_t = \phi Z_{t-1} + a_t$ that is valid for all t. If we use this equation with $t - 1$ replacing t, we get $Z_{t-1} = \phi Z_{t-2} + a_{t-1}$. Substituting this expression into the original equation gives us

$$Z_t = \phi(\phi Z_{t-2} + a_{t-1}) + a_t$$
$$= a_t + \phi a_{t-1} + \phi^2 Z_{t-2}$$

If we repeat this substitution process, say $k - 1$ times, we get

$$Z_t = a_t + \phi a_{t-1} + \phi^2 a_{t-2} + \cdots + \phi^{k-1} a_{t-(k-1)} + \phi^k Z_{t-k} \qquad \text{(4-15)}$$

Assuming $|\phi| < 1$ and letting k increase without bound, it seems reasonable (this is almost a rigorous proof) that we should obtain the infinite series

EXHIBIT 4.11 Time Plot of an AR(1) Series with $\phi = 0.9$

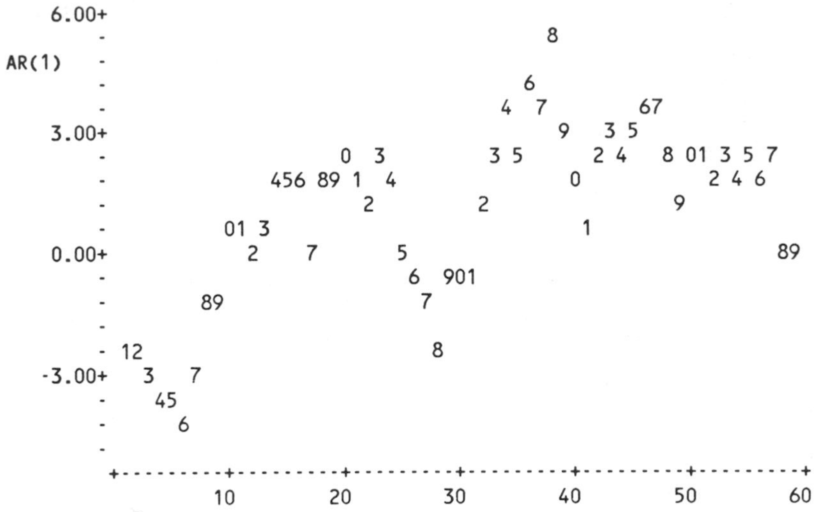

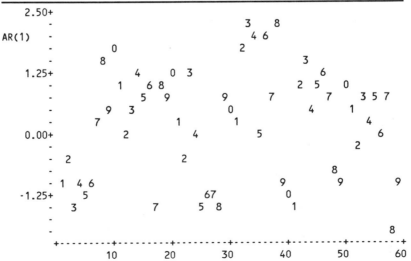

representation

$$Z_t = a_t + \phi a_{t-1} + \phi^2 a_{t-2} + \cdots \qquad (4\text{-}16)$$

This, of course, is in the form of a general linear process with $\psi_j = \phi^j$, which we already investigated in Section 4.1. This representation also reemphasizes the need for the restriction $|\phi| < 1$.

It can be shown that, subject to the restriction that a_t be independent of $Z_{t-1}, Z_{t-2}, \ldots$ and that $\sigma_a^2 > 0$, the solution of $Z_t = \phi Z_{t-1} + a_t$ will be stationary if and only if $|\phi| < 1$. The requirement $|\phi| < 1$ is usually called the **stationarity**

EXHIBIT 4.12 Time Plot of an AR(1) Series with $\phi = 0.4$

EXHIBIT 4.13 Time Plot of an AR(1) Series with $\phi = -0.7$

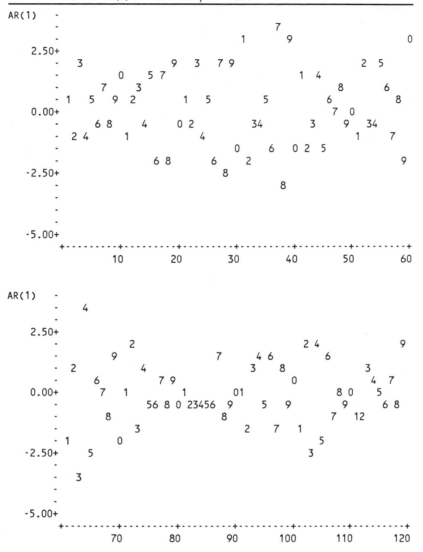

condition (see Box and Jenkins, 1976, p. 53; Nelson, 1973, p. 39) even though more than stationarity is involved. (See also Exercise 4.11.)

At this point we should note that the autocorrelation function for the AR(1) process has been derived in two different ways. The first method used the general linear process representation leading up to Equation (4-3). The second method used the defining equation $Z_t = \phi Z_{t-1} + a_t$ and the development of Equations (4-12), (4-13), and (4-14). A third derivation is obtained by multiplying both sides of Equation (4-15) by Z_{t-k}, taking the expected values of both sides, and using the fact that $a_t, a_{t-1}, \ldots, a_{t-(k-1)}$ are independent of Z_{t-k}. The second method should be especially noted, since it will generalize nicely to higher-order processes.

THE SECOND-ORDER AUTOREGRESSIVE PROCESS

Consider now the series satisfying

$$Z_t = \phi_1 Z_{t-1} + \phi_2 Z_{t-2} + a_t \qquad (4\text{-}17)$$

where, as usual, we assume that a_t is independent of $Z_{t-1}, Z_{t-2}, \ldots$. To discuss stationarity of second- and higher-order AR processes, we introduce the **AR characteristic polynomial**

$$\phi(x) = 1 - \phi_1 x - \phi_2 x^2$$

and the corresponding **AR characteristic equation**

$$1 - \phi_1 x - \phi_2 x^2 = 0$$

We recall that a quadratic equation always has two roots (possibly complex). It may be shown that, subject to the assumption that a_t be independent of $Z_{t-1}, Z_{t-2}, \ldots$, a stationary solution to Equation (4-17) exists if and only if the roots of the AR characteristic equation exceed unity in absolute value (modulus). This statement will generalize to the pth-order case without change. It also applies to the first-order case, where the AR characteristic equation is just $1 - \phi x = 0$ with root $1/\phi$, which exceeds 1 in absolute value if and only if $|\phi| < 1$.

In the second-order case, the two roots can easily be found to be

$$\frac{\phi_1 \pm \sqrt{\phi_1^2 + 4\phi_2}}{-2\phi_2}$$

After some algebraic manipulation, we find that these roots will exceed 1 in modulus if and only if simultaneously

$$\phi_1 + \phi_2 < 1, \quad \phi_2 - \phi_1 < 1, \quad \text{and} \quad |\phi_2| < 1 \qquad (4\text{-}18)$$

As with the AR(1) model, we call these the **stationarity conditions** for the AR(2) case.

To derive the autocorrelation function for the AR(2) case, we take defining Equation (4-17), multiply both sides by $Z_{t-k}, k = 1, 2, \ldots$, and take expectations. Assuming stationarity, zero means, and that a_t is independent of Z_{t-k}, we get

$$\gamma_k = \phi_1 \gamma_{k-1} + \phi_2 \gamma_{k-2}, \qquad k = 1, 2, \ldots \qquad (4\text{-}19)$$

or, dividing through by γ_0,

$$\rho_k = \phi_1 \rho_{k-1} + \phi_2 \rho_{k-2}, \qquad k = 1, 2, \ldots \qquad (4\text{-}20)$$

Equations (4-19) and/or (4-20) are called the **Yule–Walker equations.**

Setting $k = 1$ and using $\rho_0 = 1$ and $\rho_{-1} = \rho_1$, we get

$$\rho_1 = \phi_1 + \phi_2 \rho_1$$

or

$$\rho_1 = \frac{\phi_1}{1 - \phi_2} \tag{4-21}$$

Using the known values for ρ_0 and ρ_1, Equation (4-20) can now be used to get ρ_2:

$$\begin{aligned} \rho_2 &= \phi_1\rho_1 + \phi_2\rho_0 \\ &= \frac{\phi_2(1 - \phi_2) + \phi_1^2}{(1 - \phi_2)} \end{aligned} \tag{4-22}$$

Successive values of ρ_k may be easily calculated from the recursive relationship of Equation (4-20).

Although Equation (4-20) is very efficient for numerically calculating the autocorrelation function from given values of ϕ_1 and ϕ_2, it is also desirable for other purposes to have a more explicit formula for ρ_k. The form of the explicit solution critically depends on the roots of the AR characteristic equation $1 - \phi_1 x - \phi_2 x^2 = 0$. Denoting the reciprocals of these roots by G_1 and G_2, we have

$$G_1 = \frac{-2\phi_2}{\phi_1 + \sqrt{\phi_1^2 + 4\phi_2}} \quad \text{and} \quad G_2 = \frac{-2\phi_2}{\phi_1 - \sqrt{\phi_1^2 + 4\phi_2}}$$

For the case $G_1 \neq G_2$, it can be shown that

$$\rho_k = \frac{(1 - G_2^2)G_1^{k+1} - (1 - G_1^2)G_2^{k+1}}{(G_1 - G_2)(1 + G_1G_2)}, \qquad k = 0, 1, \ldots \tag{4-23}$$

If the roots are complex, that is, if $\phi_1^2 + 4\phi_2 < 0$, then ρ_k may be rewritten as

$$\rho_k = R^k \frac{\sin(\Theta k + \Phi)}{\sin \Phi}, \qquad k = 0, 1, 2, \ldots \tag{4-24}$$

where $R = \sqrt{-\phi_2}$, $\cos \Theta = \phi_1/(2\sqrt{-\phi_2})$, and $\tan \Phi = [(1 - \phi_2)/(1 + \phi_2)] \tan \Theta$. For completeness we note that if the roots are equal, $(\phi_1^2 + 4\phi_2 = 0)$, then we have

$$\rho_k = \left(1 + \frac{1 - \phi_2}{1 - \phi_2}k\right)\left(\frac{\phi_1}{2}\right)^k, \qquad k = 0, 1, 2, \ldots \tag{4-25}$$

A good discussion of the derivations of these formulae can be found in Fuller (1976, Sections 2.4–2.5).

The specific details of these expressions are of little importance to us. We need only note that the autocorrelation function can assume a wide variety of shapes. In all cases, ρ_k dies out exponentially fast as the lag k increases. In the case of complex roots, ρ_k displays a *damped sine wave* behavior with damping factor $R, 0 \leq R < 1$, frequency Θ, and phase Φ. Illustrations of the possible shapes are given in Figure 4.2. Exhibit 4.14 shows a computer simulation of a series with $\phi_1 = 1.5$ and $\phi_2 = -0.75$. The periodic behavior of ρ_k shown in

FIGURE 4.2 Autocorrelation Functions for Several AR(2) Models

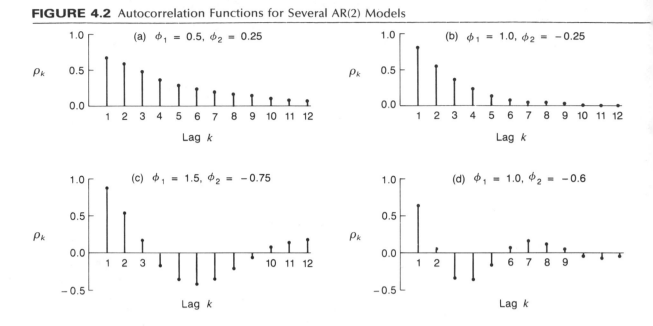

Figure 4.2(c) is clearly reflected in the nearly periodic behavior of the series at the same period of $\frac{360}{30} = 12$ time units.

The process variance γ_0 can be expressed in terms of the model parameters ϕ_1, ϕ_2, and σ_a^2 as follows: Taking the variance of both sides of Equation (4-17) yields

$$\gamma_0 = (\phi_1^2 + \phi_2^2)\gamma_0 + 2\phi_1\phi_2\gamma_1 + \sigma_a^2 \tag{4-26}$$

Setting $k = 1$ in Equation (4-19) gives a second linear equation for γ_0 and γ_1, $\gamma_1 = \phi_1\gamma_0 + \phi_2\gamma_1$, which can be solved simultaneously with Equation (4-26) to obtain

$$\begin{aligned}
\gamma_0 &= \frac{(1 - \phi_2)\sigma_a^2}{(1 - \phi_2)(1 - \phi_1^2 - \phi_2^2) - 2\phi_2\phi_1^2} \\
&= \left(\frac{1 - \phi_2}{1 + \phi_2}\right)\frac{\sigma_a^2}{(1 - \phi_2)^2 - \phi_1^2}
\end{aligned} \tag{4-27}$$

The coefficients ψ_j in the linear process representation (4-1) for an AR(2) series are more complex than for the AR(1) case. However, we can substitute the representation (4-1) into the AR(2) defining equation (4-17) for Z_t, Z_{t-1}, and Z_{t-2} and then equate coefficients of Z_j. Doing so gives us the relationships

$$\begin{aligned}
\psi_0 &= 1 \\
\psi_1 - \phi_1\psi_0 &= 0
\end{aligned} \tag{4-28}$$

and

$$\psi_j - \phi_1\psi_{j-1} - \phi_2\psi_{j-2} = 0 \quad \text{for } j = 2, 3, \ldots$$

EXHIBIT 4.14 Time Plot of an AR(2) Series with $\phi_1 = 1.5$ and $\phi_2 = -0.75$

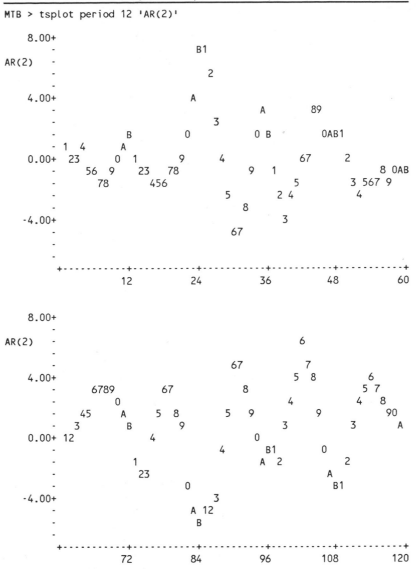

```
MTB > tsplot period 12 'AR(2)'
```

These may be solved recursively to obtain $\psi_0 = 1$, $\psi_1 = \phi_1$, $\psi_2 = \phi_1^2 + \phi_2$, and so on. One can show that an explicit solution is

$$\psi_j = \frac{G_1^{j+1} - G_2^{j+1}}{G_1 - G_2}$$

where, as before, G_1 and G_2 are the reciprocals of the roots of the AR characteristic equation and $G_1 \neq G_2$. If the roots are complex, one can write ψ_j as a damped sine wave with the same damping factor R and frequency Θ, as in Equation (4-24).

THE GENERAL AUTOREGRESSIVE PROCESS

Consider now the pth-order autoregressive model

$$Z_t = \phi_1 Z_{t-1} + \phi_2 Z_{t-2} + \cdots + \phi_p Z_{t-p} + a_t \qquad \text{(4-29)}$$

with AR characteristic polynomial

$$\phi(x) = 1 - \phi_1 x - \phi_2 x^2 - \cdots - \phi_p x^p \qquad \text{(4-30)}$$

and corresponding AR characteristic equation

$$1 - \phi_1 x - \phi_2 x^2 - \cdots - \phi_p x^p = 0 \qquad \text{(4-31)}$$

As noted earlier, assuming a_t is independent of $Z_{t-1}, Z_{t-2}, \ldots$, a stationary solution $\{Z_t\}$ to Equation (4-29) exists if and only if the p roots of the AR characteristic equation all exceed unity in modulus. Numerically finding the roots of a pth-degree polynomial is a nontrivial task, but a simple algorithm based on Schur's theorem can be used to check on the stationarity condition. See Appendix C.

Assuming stationarity (and a zero mean, as usual), we may multiply Equation (4-29) by Z_{t-k}, take expectations, and obtain the important recursive relationship

$$\rho_k = \phi_1 \rho_{k-1} + \phi_2 \rho_{k-2} + \cdots + \phi_p \rho_{k-p}, \qquad k \geq 1 \qquad \text{(4-32)}$$

Putting $k = 1, 2, \ldots, p$ into Equation (4-32) and using $\rho_0 = 1$ and $\rho_{-k} = \rho_k$, we get the **Yule–Walker** equations

$$
\begin{aligned}
\rho_1 &= \phi_1 && + \phi_2 \rho_1 && + \cdots + \phi_p \rho_{p-1} \\
\rho_2 &= \phi_1 \rho_1 && + \phi_2 && + \cdots + \phi_p \rho_{p-2} \\
&\,\,\vdots \\
\rho_p &= \phi_1 \rho_{p-1} + \phi_2 \rho_{p-2} + \cdots + \phi_p
\end{aligned}
\qquad \text{(4-33)}
$$

Given values for $\phi_1, \phi_2, \ldots, \phi_p$, these linear equations can be solved for $\rho_1, \rho_2, \ldots, \rho_p$. Then Equation (4-32) can be used to obtain ρ_k at higher lags.

Noting that

$$
\begin{aligned}
E(a_t Z_t) &= E[a_t(\phi_1 Z_{t-1} + \phi_2 Z_{t-2} + \cdots + \phi_p Z_{t-p} + a_t)] \\
&= E(a_t^2) \\
&= \sigma_a^2
\end{aligned}
$$

we may multiply Equation (4-29) by Z_t, take expectations, and find

$$\gamma_0 = \phi_1 \gamma_1 + \phi_2 \gamma_2 + \cdots + \phi_p \gamma_p + \sigma_a^2$$

which, using $\rho_k = \gamma_k/\gamma_0$, can be written as

$$\gamma_0 = \frac{\sigma_a^2}{1 - \phi_1 \rho_1 - \phi_2 \rho_2 - \cdots - \phi_p \rho_p} \qquad \text{(4-34)}$$

and express the process variance γ_0 in terms of the parameters $\sigma_a^2, \phi_1, \phi_2, \ldots, \phi_p$, and the now-known values $\rho_1, \rho_2, \ldots, \rho_p$.

Of course, explicit solutions for ρ_k are essentially impossible in this generality, but we can say that ρ_k will be a linear combination of exponentially decaying terms (corresponding to the *real* roots of the characteristic equation) and damped sine wave terms (corresponding to *complex* roots).

Assuming stationarity, the process can also be expressed in the general linear form of Equation (4-1), but the ψ-weights are complicated functions of the parameters $\phi_1, \phi_2, \ldots, \phi_p$. The weights can be found numerically; see Equation (4-41).

4.4 THE MIXED AUTOREGRESSIVE–MOVING AVERAGE MODEL

If we assume that the series is partly autoregressive and partly moving average, we obtain a quite general time series model. In general, if

$$Z_t = \phi_1 Z_{t-1} + \phi_2 Z_{t-2} + \cdots + \phi_p Z_{t-p} + a_t - \theta_1 a_{t-1}$$
$$- \theta_2 a_{t-2} - \cdots - \theta_q a_{t-q} \qquad \text{(4-35)}$$

we say that $\{Z_t\}$ is a mixed **autoregressive moving average** process of orders p and q, respectively; we abbreviate the name to ARMA(p, q). As usual, we discuss an important special case first.

THE ARMA(1, 1) MODEL

The defining equation can be written

$$Z_t = \phi Z_{t-1} + a_t - \theta a_{t-1} \qquad \text{(4-36)}$$

To derive Yule–Walker type equations, we first note that

$$E(a_t Z_t) = E[a_t(\phi Z_{t-1} + a_t - \theta a_{t-1})]$$
$$= \sigma_a^2$$

and

$$E(a_{t-1} Z_t) = E[a_{t-1}(\phi Z_{t-1} + a_t - \theta a_{t-1})]$$
$$= \phi \sigma_a^2 - \theta \sigma_a^2$$
$$= (\phi - \theta)\sigma_a^2$$

If we multiply Equation (4-36) by Z_{t-k} and take expectations, we have

$$\gamma_0 = \phi \gamma_1 + [1 - \theta(\phi - \theta)]\sigma_a^2 \quad \text{for } k = 0$$
$$\gamma_1 = \phi \gamma_0 - \theta \sigma_a^2 \qquad\qquad \text{for } k = 1$$

and

$$\gamma_k = \phi \gamma_{k-1} \qquad\qquad\qquad \text{for } k \geq 2$$

(4-37)

Solving the first two equations yields

$$\gamma_0 = \frac{(1 - 2\theta\phi + \theta^2)}{1 - \phi^2} \sigma_a^2 \tag{4-38}$$

and then solving the simple recursion gives

$$\gamma_k = \frac{(1 - \theta\phi)(\phi - \theta)}{1 - \phi^2} \phi^{k-1} \sigma_a^2 \text{ for } k \geq 1$$

or

$$\rho_k = \frac{(1 - \theta\phi)(\phi - \theta)}{1 - 2\theta\phi + \theta^2} \phi^{k-1} \quad \text{for } k \geq 1 \tag{4-39}$$

Note that this autocorrelation function decays exponentially as the lag k increases. The damping factor is ϕ, but the decay starts from the initial value ρ_1, which also depends on θ. This is in contrast to the AR(1) autocorrelation, which also decays with damping factor ϕ but always from the initial value $\rho_0 = 1$. For example, if $\phi = 0.8$ and $\theta = 0.4$, then $\rho_1 = 0.418$, $\rho_2 = 0.335$, $\rho_3 = 0.268$, and so on. Several shapes for ρ_k are possible, depending on the sign of ρ_1 and the sign of ϕ.

The general linear form of the model can be obtained in the same manner that led to Equation (4-16). We find

$$Z_t = a_t + (\phi - \theta) \sum_{j=1}^{\infty} \phi^{j-1} a_{t-j} \tag{4-40}$$

that is,

$$\psi_j = (\phi - \theta)\phi^{j-1} \quad \text{for } j \geq 1$$

We should now mention the obvious stationarity condition $|\phi| < 1$, or equivalently, the root of the AR characteristic equation $1 - \phi x = 0$ must exceed unity in absolute value.

For the general ARMA(p, q) model, we state the following facts without proof: Subject to a_t being independent of $Z_{t-1}, Z_{t-2}, \ldots$, a stationary solution $\{Z_t\}$ satisfying Equation (4-35) exists if and only if the roots of the AR characteristic equation $\phi(x) = 0$ all exceed unity in modulus.

If the stationarity conditions are satisfied, then the model can be written as a general linear process with weights ψ_j determined from

$$\begin{aligned}
\psi_0 &= 1 \\
\psi_1 &= -\theta_1 + \phi_1 \\
\psi_2 &= -\theta_2 + \phi_2 + \phi_1\psi_1 \\
&\vdots \\
\psi_j &= -\theta_j + \phi_p\psi_{j-p} + \cdots + \phi_1\psi_{j-1}
\end{aligned} \tag{4-41}$$

where we take $\psi_j = 0$ for $j < 0$ and $\theta_j = 0$ for $j > q$. (See Box and Jenkins, 1976, pp. 95–96; or Fuller, 1976, pp. 56–61.)

Again assuming stationarity, the autocorrelation function can easily be shown to satisfy

$$\rho_k = \phi_1\rho_{k-1} + \phi_2\rho_{k-2} + \cdots + \phi_p\rho_{k-p} \quad \text{for } k > q \qquad \text{(4-42)}$$

Similar equations can be developed for $k = 0, 1, \ldots, q$ that involve $\theta_1, \theta_2, \ldots, \theta_q$. An algorithm suitable for numerical computation of the complete autocorrelation function is given in Appendix D.

4.5 INVERTIBILITY

We have seen that for the MA(1) process we get exactly the same autocorrelation function if θ is replaced by $1/\theta$. In the exercises we find a similar problem for an MA(2) model. This lack of uniqueness of MA models given their autocorrelation functions must be rectified before we try to infer the values of the parameters from an observed time series. It turns out that this nonuniqueness is related to the seemingly unrelated question stated next.

   We have seen that an autoregressive process can always be reexpressed as an infinite-order moving average process through the ψ-weights. However, for some purposes the autoregressive representations are also convenient. Can a moving average process be expressed as an autoregression?

   To fix ideas, consider an MA(1) model:

$$Z_t = a_t - \theta a_{t-1} \qquad \text{(4-43)}$$

First rewriting as $a_t = Z_t + \theta a_{t-1}$ and then replacing t by $t - 1$ and substituting for a_{t-1}, we get

$$a_t = Z_t + \theta(Z_{t-1} + \theta a_{t-2})$$
$$= Z_t + \theta Z_{t-1} + \theta^2 a_{t-2}$$

If $|\theta| < 1$, we may continue this substitution into the past and obtain [compare with Equations (4-15) and (4-16)]

$$a_t = Z_t + \theta Z_{t-1} + \theta^2 Z_{t-2} + \cdots \qquad \text{(4-44)}$$

or

$$Z_t = (-\theta Z_{t-1} - \theta^2 Z_{t-2} - \cdots) + a_t \qquad \text{(4-45)}$$

If $|\theta| < 1$, we see that the MA(1) can be "inverted" into an infinite-order autoregressive process. We say that the MA(1) model is *invertible*.

   For a general MA(q) model, we define the **MA characteristic polynomial** as

$$\theta(x) = 1 - \theta_1 x - \theta_2 x^2 - \cdots - \theta_q x^q \qquad \text{(4-46)}$$

and the corresponding **MA characteristic equation**

$$1 - \theta_1 x - \theta_2 x^2 - \cdots - \theta_q x^q = 0 \qquad \text{(4-47)}$$

It can then be shown that the MA(q) model is **invertible,** that is, there are constants π_j such that

$$Z_t = \sum_{j=1}^{\infty} \pi_j Z_{t-j} + a_t \qquad \text{(4-48)}$$

if and only if the roots of the MA characteristic equation exceed 1 in absolute value. (Compare with Equation (4-30) and the stationarity condition for AR models.)

Because it can be shown that there is only one invertible MA(q) model with a given autocorrelation function, the uniqueness problem is also solved if we restrict attention to the physically sensible class of invertible models.

For a general ARMA(p, q) model we shall require both stationarity and invertibility.

CHAPTER 4 EXERCISES

4.1. Sketch the autocorrelation functions for each of the following ARMA models:

a. AR(2) with $\phi_1 = 1.2$ and $\phi_2 = -0.7$
b. AR(2) with $\phi_1 = -1$ and $\phi_2 = -0.6$
c. MA(2) with $\theta_1 = 1.2$ and $\theta_2 = -0.7$
d. MA(2) with $\theta_1 = -1$ and $\theta_2 = -0.6$
e. ARMA(1, 1) with $\phi = 0.7$ and $\theta = 0.4$
f. ARMA(1, 1) with $\phi = 0.7$ and $\theta = -0.4$

4.2. Suppose $\{Z_t\}$ is an AR(1) process with $-1 < \phi < 1$.
a. Find the covariance function for $W_t = Z_t - Z_{t-1}$ in terms of ϕ and σ_a^2.
b. In particular, show that Var $(W_t) = 2\sigma_a^2/(1 + \phi)$.

4.3. Find the autocorrelation function for the process defined by

$$Z_t = 5 + a_t - \frac{1}{2} a_{t-1} + \frac{1}{4} a_{t-2}$$

4.4. Consider an MA(6) model with $\theta_1 = 0.5$, $\theta_2 = -0.25$, $\theta_3 = 0.125$, $\theta_4 = -0.0625$, $\theta_5 = 0.03125$, and $\theta_6 = -0.015625$. Show that a much simpler model would give essentially the same ψ-weights.

4.5. Consider an MA(7) model with $\theta_1 = 1$, $\theta_2 = -0.5$, $\theta_3 = 0.25$, $\theta_4 = -0.125$, $\theta_5 = 0.0625$, $\theta_6 = -0.03125$, and $\theta_7 = 0.015625$. Show that a much simpler model would give essentially the same ψ-weights.

4.6. Describe the important characteristics of the autocorrelation functions for the following models: (a) MA(1), (b) MA(2), (c) AR(1), (d) AR(2), and (e) ARMA(1, 1).

4.7. For the ARMA(1, 2) model

$$Z_t = 0.8Z_{t-1} + a_t + 0.7a_{t-1} + 0.6a_{t-2}$$

show that
a. $\rho_k = 0.8\rho_{k-1}$ for $k \geq 3$ and
b. $\rho_2 = 0.8\rho_1 + 0.6\sigma_a^2/\gamma_0$

4.8. Verify that for an MA(1) process

$$\max_{-\infty < \theta < \infty} \rho_1 = 0.5 \quad \text{and} \quad \min_{-\infty < \theta < \infty} \rho_1 = -0.5$$

4.9. Consider two MA(2) processes, one with $\theta_1 = \theta_2 = \frac{1}{6}$ and another with $\theta_1 = -1$ and $\theta_2 = 6$. Show that these processes have exactly the same autocorrelation function. How do the roots of the corresponding characteristic polynomials compare?

4.10. Suppose $\{Z_t\}$ is a zero-mean, stationary, normal process with $|\rho_1| < 0.5$ and $\rho_k = 0$ for $k > 1$. Show that $\{Z_t\}$ must be representable as an MA(1) process; that is, there is a white noise process $\{a_t\}$ such that $Z_t = a_t - \theta a_{t-1}$ where a_t is uncorrelated with Z_{t-k} for $k > 0$. (*Hint*: Choose θ such that $-1 < \theta < 1$ and $\rho_1 = -\theta/(1 + \theta^2)$; then let $a_t = \sum_{j=0}^{\infty} \theta^j Z_{t-j}$.)

4.11. Consider the "nonstationary" AR(1) model

$$Z_t = 3Z_{t-1} + a_t$$

a. Show that $Z_t = -\sum_{j=1}^{\infty} (\frac{1}{3})^j a_{t+j}$ satisfies the AR(1) equation and is actually stationary.

b. In what way is this solution unsatisfactory?

4.12. Consider the model $Z_t = a_{t-1} - a_{t-2} + \frac{1}{2}a_{t-3}$.

a. Find the autocovariance function for $\{Z_t\}$.

b. Show that $\{Z_t\}$ is a stationary ARMA(p, q) model. Identify p, q, and the θ's and ϕ's.

4.13. Show that the statement "The roots of

$$1 - \phi_1 x - \phi_2 x^2 - \cdots - \phi_p x^p = 0$$

are greater than 1 in absolute value" is equivalent to the statement "The roots of

$$x^p - \phi_1 x^{p-1} - \cdots - \phi_p = 0$$

are less than 1 in absolute value." (*Hint*: If G is a root of one equation, is $1/G$ a root of the other?)

APPENDIX B SIMULATION OF ARMA SERIES WITH MINITAB

The use of stored programs, or **macros,** greatly facilitates the simulation of series from various ARMA models within Minitab. Moving average series are quite easily formed directly from their defining equations. Autoregressive series are more difficult due to the recursive nature of their defining equations.

The reader interested in doing the simulations without necessarily understanding the programs need merely store the macros as listed in Exhibits B.2 through B.6 and then execute them as shown. An example is shown in Exhibit B.1, where we illustrate the simulation of an ARMA(1, 1) series of length 100 with $\phi = 0.6$ and $\theta = -0.9$. In all cases the white noise series is generated according to a normal distribution with a zero mean and unit variance.

The macros may be entered into files within Minitab using the STORE command. However, you must type carefully, since no editing, correcting, or

EXHIBIT B.1 Using the ARMA(1, 1) Macro to Simulate a Series

```
MTB > exec 'arma11'
MTB > Note: This program uses k50-k52 and c50-c51.
MTB > Note
MTB > Note: Enter sample size, phi and theta in that order.
MTB > Note: Type END to end input.
DATA> 100 .6 -.9
DATA> end
MTB > Note: Simulated series is in C50.
```

EXHIBIT B.2 MA(1) Simulation Macro

```
  Macro MA1

NoEcho
Note: This program uses K50,K51,C50 and C51.
Note
Note: Enter sample size and MA1 coefficient(theta) in that order.
Note: Type END to end input.
Set 'terminal' c50
Let k50=c50(1)+1     # Sample size + 1
Let k51=c50(2)       # Theta
Random k50 c50       # Generate N(0,1) White Noise
Lag c50 c51
Let c50=c50-k51*c51
Copy c50 c50;
    Omit '*'.
Erase k50 k51 c51
Note: Simulated MA1 series is in C50.
Echo
```

EXHIBIT B.3 MA(2) Simulation Macro

```
      Macro MA2

NoEcho
Note: This program uses K50-52 and C50-52.
Note
Note: Enter sample size and MA2 coefficients(theta1 and theta2)
Note: in that order. Type END to end input.
Set 'terminal' c50
Let k50=c50(1)+2          # Sample size + 2
Let k51=c50(2)            # Theta1
Let k52=c50(3)            # Theta2
Random k50 c50            # Generate N(0,1) White Noise
Lag c50 c51
Lag c51 c52
Let c50=c50-k51*c51-k52*c52
Copy c50 c50;
   Omit '*'.
Erase k50-k52 c51-c52
Note: Simulated MA2 series is in C50.
Echo
```

reviewing is possible once a line is entered. A better method for creating the macro files is to use a program/text editor outside of Minitab. Check with your computer system personnel to·find out what editor is available on your system.

The macros assume that Minitab is being used interactively. In particular, when the SET 'TERMINAL' command is executed, the computer expects the user to type in data. In batch mode the data input must be handled differently, for example, by setting up the values for K50, K51, . . . before executing the stored commands.

The MA macros given in Exhibits B.2 and B.3 are quite straightforward

EXHIBIT B.4 AR(1) Simulation Macro

```
      Macro AR1

NoEcho
Note: This program uses K50-51 and C50-51.
Note:
Note: Enter sample size and AR1 coefficient(phi) in that order.
Note: Type END to end input.
Set 'terminal' c50
Let k50=c50(1)+51         # Sample size + 51  (extra start-up values)
Let k51=c50(2)           # Phi
NRandom k50 0 1 c50      # Generate N(0,1) White Noise
Lag c50 c51
Exec 'AR1.A' 50
Copy c51 c50;
   Omit '*'.
erase k50 k51 c51
Note: Simulated series is in C50
Echo


      Submacro AR1.A

   Let c51=k51*c51+c50
   Lag c51 c51
```

EXHIBIT B.5 AR(2) Simulation Macro

```
Macro AR2

NoEcho
Note: This program uses K50-52 and C50-C53.
Note:
Note: Enter sample size and AR2 coefficients, phi1 and phi2 in that order
Note: Type END to end input.
Set  'terminal' c50
Let k50=c50(1)+51      # Sample size + 51  (extra start-up values)
Let k51=c50(2)         # Phi1
Let k52=c50(3)         # Phi2
Random k50 c50         # Generate N(0,1) White Noise
Lag c50 c51
Lag c51 c52
Exec 'AR2.A' 50
Copy c53 c50;
   Omit '*'.
erase k50-k52 c51-c53
Note: Simulated series is in C50.
Echo


   Submacro AR2.A

  Let c53=k51*c52+k52*c51+c50
  Lag c52 c51
```

and can easily be generalized to handle higher-order models or seasonal models (see Chapter 10).

The AR macros, Exhibits B.4 and B.5, are somewhat harder to follow. Notice that each executes a (sub) macro 50 times. This is necessary because of the recursiveness of the AR defining equations. The initial 50 transient observations are omitted in the final series. This should work well when values

EXHIBIT B.6 ARMA(1, 1) Simulation Macro

```
Macro ARMA11

NoEcho
Note: This program uses k50-k52 and c50-c51.
Note:
Note: Enter sample size, phi and theta in that order.
Note: Type END to end input.
Set 'terminal' c50
Let k50=c50(1)+52    # Sample size +52 (extra start-up values)
Let k51=c50(2)       # Phi
Let k52=c50(3)       # Theta
Random k50 c50       # Generate N(0,1) White Noise
Lag c50 c51
Let c50=c50-k52*c51
Lag c50 c51
Exec 'AR1.A' 50
Copy c51 c50;
   Omit '*'.
Erase k50-k52 c51
Note: Simulated series is in C50.
Echo
```

are chosen for the parameters that imply stationary models. By closely reading the AR2 macro, the user should be able to construct macros for simulating higher-order AR models.

Alternatively, we can generate the initial observation of an AR(1) series according to its correct distribution, namely, normal with mean μ and variance $\sigma_a^2/(1 - \phi^2)$. However, this method would be difficult to generalize to AR(p) models for $p > 2$. Also, the method as given allows the user to generate nonstationary AR series and, with a minor change, to experiment with nonnormal white noise innovations.

The ARMA11 macro, Exhibit B.6, generates an MA(1) series and then uses it as input to the AR(1) macro to obtain the final ARMA(1, 1) series.

APPENDIX C STATIONARITY CHECKING VIA SCHUR'S THEOREM

Mathematically, conditions for stationarity and invertibility of ARMA models are most easily stated in terms of the roots of the AR or MA characteristic equations. However, for orders higher than 2 the restrictions on those roots are difficult to translate into restrictions on the parameters of the models. Given a particular set of parameter values, it would be useful if we could easily determine whether the model is stationary and/or invertible.

Fortunately, by using a little-known theorem by Schur, we can make such a determination numerically without actually finding any of the roots. A detailed statement of the theorem will not be given here but may be found in Henrici (1974, pp. 491–494). Rather, we give in Exhibit C.1 a program (in BASIC) to perform the required computations. The algorithm is so simple that it can be implemented easily on almost any machine from a programmable calculator on up. It can also be easily translated into FORTRAN, PASCAL, C, and so on. (Unfortunately, it apparently is not possible to do these calculations within Minitab.)

EXHIBIT C.1 A Program in BASIC to Check for Stationarity and Invertibility of Models (Based on Schur's Theorem)

```
10 DIM A(25),B(25)
20 PRINT 'Input order of polynomial (0 to quit)'
30 INPUT P
35 IF P=0 THEN GOTO 220
40 PRINT 'Input coefficients - low to high order'
50 FOR I=2 TO P+1
60 INPUT A(I)
70 NEXT I
80 A(1)=-1
90 FOR K=P TO 1 STEP -1
100 FOR J= 1 TO K
110 B(J)=A(1)*A(J)-A(K+1)*A(K+2-J)
120 NEXT J
130 IF B(1)<=0 THEN 200
140 FOR J=1 TO K
150 A(J)=B(J)
160 NEXT J
170 NEXT K
180 PRINT 'ALL roots lie OUTSIDE the unit circle.'
190 GOTO 20
200 PRINT 'At least one root is NOT outside the unit circle.'
210 GOTO 20
220 END
```

Exhibit C.2 shows an example of using the program to check the roots of several polynomials.

EXHIBIT C.2 Illustration of the Checking of Stationarity Conditions

```
OK, basic roots.bas
BASIC REV19.0
Input order of polynomial (0 to quit)
!1
Input coefficients - low to high order
!.8
ALL roots lie OUTSIDE the unit circle.
Input order of polynomial (0 to quit)
!1
Input coefficients - low to high order
!-.6
ALL roots lie OUTSIDE the unit circle.
Input order of polynomial (0 to quit)
!1
Input coefficients - low to high order
!1.2
At least one root is NOT outside the unit circle.
Input order of polynomial (0 to quit)
!2
Input coefficients - low to high order
!1.4
!-.6
ALL roots lie OUTSIDE the unit circle.
Input order of polynomial (0 to quit)
!2
Input coefficients - low to high order
!1.6
!.7
At least one root is NOT outside the unit circle.
Input order of polynomial (0 to quit)
!4
Input coefficients - low to high order
!0
!0
!0
!.99
ALL roots lie OUTSIDE the unit circle.
Input order of polynomial (0 to quit)
!0

END AT LINE 220
```

APPENDIX D THE AUTOCORRELATION FUNCTION FOR ARMA(p, q)

Let $\{Z_t\}$ be a stationary ARMA(p, q) process. Recall that we can always write such a process in general linear process form

$$Z_t = \sum_{j=0}^{\infty} \psi_j a_{t-j} \tag{D-1}$$

where the ψ-weights are obtained recursively from Equations (4-41).
    We then have

$$E(Z_{t+k}a_t) = E\left[\sum_{j=0}^{\infty} \psi_j a_{t+k-j} a_t\right] = \psi_k \sigma_a^2 \quad \text{for } k \geq 0 \tag{D-2}$$

Thus the autocovariance function must satisfy

$$\gamma_k = E(Z_{t+k}Z_t) = E\left[\left(\sum_{j=1}^{p} \phi_j Z_{t+k-j} - \sum_{j=0}^{q} \theta_j a_{t+k-j}\right)Z_t\right]$$

$$= \sum_{j=1}^{p} \phi_j \gamma_{k-j} - \sigma_a^2 \sum_{j=k}^{q} \theta_j \psi_{j-k} \tag{D-3}$$

where $\theta_0 = -1$ and the last sum is absent if $k > q$. Setting $k = 0, 1, \ldots, p$ and using $\gamma_{-k} = \gamma_k$ leads to $p + 1$ linear equations in $\gamma_0, \gamma_1, \ldots, \gamma_p$:

$$\gamma_0 = \phi_1\gamma_1 \ + \phi_2\gamma_2 \ + \cdots + \phi_p\gamma_p \ - \sigma_a^2(-1 \ + \theta_1\psi_1 \ + \theta_2\psi_2 \ + \cdots + \theta_q\psi_q)$$
$$\gamma_1 = \phi_1\gamma_0 \ + \phi_2\gamma_1 \ + \cdots + \phi_p\gamma_{p-1} - \sigma_a^2(\theta_1 \ + \theta_2\psi_1 \ + \theta_3\psi_2 \ + \cdots + \theta_q\psi_{q-1})$$
$$\vdots$$
$$\gamma_p = \phi_1\gamma_{p-1} + \phi_2\gamma_{p-2} + \cdots + \phi_p\gamma_0 \ - \sigma_a^2(\theta_p \ + \theta_{p+1}\psi_1 + \theta_{p+2}\psi_2 + \cdots + \theta_q\psi_{q-p})$$

$$\tag{D-4}$$

where $\theta_j = 0$ if $j > q$.
    For a given set of parameter values σ_a^2, ϕ's, and θ's (and hence ψ's), we can solve the linear equations to obtain $\gamma_0, \gamma_1, \ldots, \gamma_p$ and hence $\rho_k = \gamma_k/\gamma_0$ for $0 \leq k \leq p$. The values for $k > p$ can then be evaluated from Equation (D-3).

CHAPTER 5 MODELS FOR NONSTATIONARY SERIES

Any time series without a constant mean over time is nonstationary. Models of the form

$$Z_t = \mu_t + X_t \qquad\qquad \textbf{(5-1)}$$

where μ_t is a nonconstant mean function and X_t is a zero-mean stationary series, were considered in Chapter 3. As stated there, such models are reasonable only if there are good reasons for believing that the deterministic trend is appropriate "forever." That is, just because a segment of the series looks like it is increasing approximately linearly, do we believe that the linearity is intrinsic to the process and will persist in the future? Frequently in applications, particularly in business and economics, we cannot legitimately assume a deterministic trend. Exhibit 5.1 shows a time plot of a simulated series that appears to have an upward linear trend; however, the model from which it was generated contains no deterministic trend at all.

Exhibit 5.2 is a time plot of monthly AA railroad bond yields from January 1968 through June 1976. Here we see various short-term "trends," but we would not want to assume an overall deterministic trend for the series. In this chapter

EXHIBIT 5.1 Simulation of a Nonstationary Series: IMA(1, 1) with $\theta = 0.5$

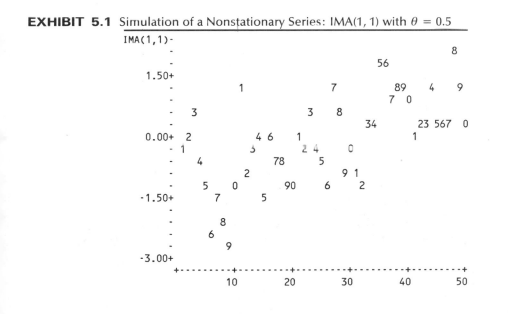

EXHIBIT 5.2 AA Railroad Bond Yields, January 1968 to June 1976

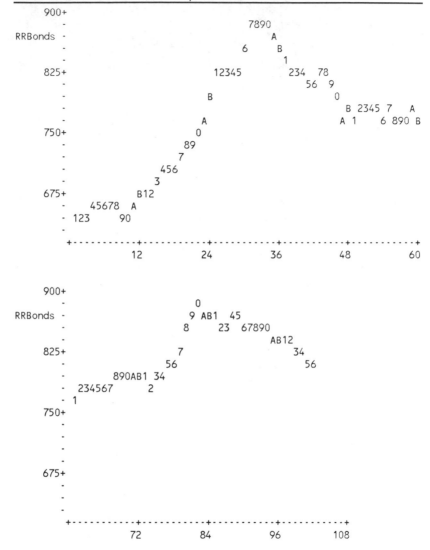

we shall discuss important nonstationary models that have been described as containing stochastic trends but are modeled with relatively few parameters.

5.1 STATIONARITY THROUGH DIFFERENCING

Consider again the AR(1) model

$$Z_t = \phi Z_{t-1} + a_t \qquad \text{(5-2)}$$

We have seen that, assuming a_t is a true "innovation" (that is, a_t is uncorrelated with $Z_{t-1}, Z_{t-2}, \ldots)$, we must have $|\phi| < 1$ in order to have a stationary solution to Equation (5-2). What can we say about the solutions if $|\phi| \geq 1$? Consider in particular the equation

$$Z_t = 3Z_{t-1} + a_t \tag{5-3}$$

Iterating into the past as we have done before yields

$$Z_t = a_t + 3a_{t-1} + 3^2 a_{t-2} + \cdots + 3^{t-1} a_1 + 3^t Z_0 \tag{5-4}$$

We see that the influence of distant past values of Z and a do not die out—indeed, the weights applied to Z_0 and a_1 grow exponentially fast. In Table 5.1 we show a very short simulation of such a series where the white noise sequence was generated as standard normal variables and with $Z_0 = 0$.

TABLE 5.1   Simulation of $Z_t = 3Z_{t-1} + a_t$

t	1	2	3	4	5	6	7	8
a	0.630	−1.25	1.80	1.51	1.56	0.616	0.644	−0.982
Z	0.630	0.640	3.72	12.67	39.57	119.32	358.604	1074.83

Exhibit 5.3 plots the simulated series and clearly shows the explosive nature of the process. This behavior is also reflected in the variance and covariance function, which can easily be found to be

$$\text{Var}\,(Z_t) = \frac{\sigma_a^2}{8}(9^t - 1) \tag{5-5}$$

and

$$\text{Cov}\,(Z_t, Z_{t-k}) = \frac{3^k}{8}(9^{t-k} - 1)\sigma_a^2$$

EXHIBIT 5.3 Explosive AR(1) Series

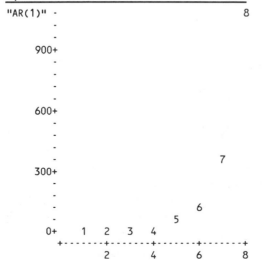

so that

$$\text{Corr}(Z_t, Z_{t-k}) = 3^k \left(\frac{9^{t-k} - 1}{9^t - 1}\right)^{1/2}$$

$$\approx 1 \quad \text{for large } t \text{ and moderate } k$$

(We have taken $Z_0 = 0$ in these calculations.)

The same general exponential growth or explosive behavior will occur for any ϕ such that $|\phi| > 1$. A more reasonable type of nonstationarity obtains when $\phi = 1$. If $\phi = 1$, Equation (5-2) is just

$$Z_t = Z_{t-1} + a_t \tag{5-6}$$

This relationship is satisfied by the random walk process of Chapter Two. We can also write Equation (5-6) as

$$\nabla Z_t = a_t \tag{5-7}$$

where $\nabla Z_t = Z_t - Z_{t-1}$ is the **first difference** of Z_t. This idea can easily be generalized to models whose first difference is any stationary process rather than white noise.

Several somewhat different sets of assumptions lead to models whose differences are stationary processes. Suppose

$$Z_t = M_t + X_t \tag{5-8}$$

where M_t is a series that is changing only slowly over time. Here M_t could be either deterministic or stochastic. If we assume that M_t is approximately constant for every two consecutive time points, we might estimate (predict) M_t at t by choosing $b_{0,t}$ so that

$$\sum_{j=0}^{1} (Z_{t-j} - b_{0,t})^2$$

is minimized. This clearly leads to

$$\hat{M}_t = \frac{Z_t + Z_{t-1}}{2}$$

and the "detrended" series at time t is then

$$Z_t - \hat{M}_t = Z_t - \frac{Z_t + Z_{t-1}}{2}$$

$$= \frac{Z_t - Z_{t-1}}{2}$$

This is a constant multiple of the first difference of Z_t.

An alternative assumption might be that M_t is stochastic and changes slowly over time as governed by a random walk model. Suppose, for example, that

$$Z_t = M_t + a_t \quad \text{with} \quad M_t = M_{t-1} + b_t \tag{5-9}$$

where $\{a_t\}$ and $\{b_t\}$ are independent white noise series. Then

$$\nabla Z_t = \nabla M_t + \nabla a_t$$
$$= b_t + a_t - a_{t-1}$$

which would have the autocorrelation function of an MA(1) series with $\rho_1 = -\{1/[2 + (\sigma_b^2/\sigma_a^2)]\}$.

In either of these situations, we are led to the study of ∇Z_t as a stationary series. In Minitab, differences are obtained with the DIFFERENCE command with format

DIFFERENCE [of lag **K**] for series in **C**, put result in **C**

With the series denoted Z_t, $Z_t - Z_{t-K}$ is the resulting differenced series. If K is omitted, the assumption is that $K = 1$. With a series of length n, only $n - K$ differences of lag K can be calculated; consequently, the code for a missing value, $*$, is placed in the first K rows of the differenced series.

For the railroad bond yield series, the first difference series ∇Z_t is plotted in Exhibit 5.4. After differencing, the series might well be considered as arising from a stationary process.

We can also make assumptions that lead to stationary second-difference models. Again we assume that Equation (5-8) holds, but we now assume that M_t is linear in time over three consecutive time points. We can now estimate (predict) M_t at the middle point t by choosing $b_{0,t}$ and $b_{1,t}$ to minimize

$$\sum_{j=-1}^{1} (Z_{t-j} - b_{0,t} - jb_{1,t})^2$$

The solution yields

$$\hat{M}_t = \frac{Z_{t+1} + Z_t + Z_{t-1}}{3}$$

and thus the detrended series is

$$Z_t - \hat{M}_t = Z_t - \frac{Z_{t+1} + Z_t + Z_{t-1}}{3}$$
$$= -\frac{(Z_{t+1} - 2Z_t + Z_{t-1})}{3}$$

—a constant multiple of the centered *second difference* of Z_t.

Alternately, we might assume that

$$Z_t = M_t + a_t \quad \text{where} \quad M_t = M_{t-1} + b_t \quad \text{and} \quad b_t = b_{t-1} + \varepsilon_t \qquad \text{(5-10)}$$

with $\{a_t\}$ and $\{\varepsilon_t\}$ independent white noise series. Here the stochastic trend M_t is such that its rate of change $\nabla M_t = b_t$ is changing slowly over time. Then

$$\nabla Z_t = \nabla M_t + \nabla a_t \qquad \text{and} \qquad \nabla^2 Z_t = \nabla b_t + \nabla^2 a_t$$
$$= b_t + \nabla a_t \qquad\qquad\qquad = \varepsilon_t + a_t - 2a_{t-1} + a_{t-2}$$

which has the autocorrelation function of an MA(2) process. The important point is that the second difference of the nonstationary process $\{Z_t\}$ is

EXHIBIT 5.4 First Difference of Railroad Bond Yields

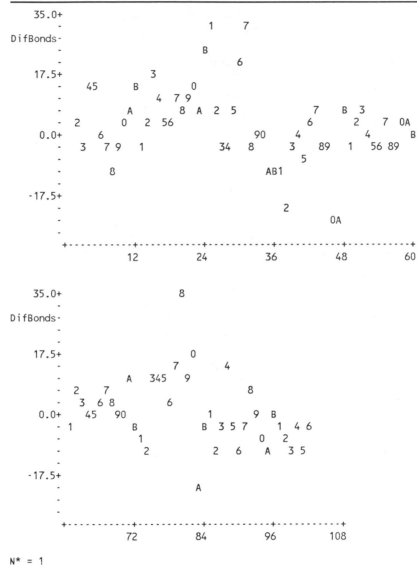

N* = 1

stationary. This leads us to the general definition of integrated autoregressive–moving average models.

5.2 ARIMA MODELS

A series $\{Z_t\}$ is said to follow an **integrated autoregressive–moving average** model if the dth difference $W_t = \nabla^d Z_t$ is a stationary ARMA process. If W_t is ARMA(p, q), we say that Z_t is ARIMA(p, d, q). Fortunately, for practical purposes we can usually take $d = 1$ or at most 2.

Consider then an ARIMA$(p, 1, q)$ process. With $W_t = Z_t - Z_{t-1}$, we have

$$W_t = \phi_1 W_{t-1} + \phi_2 W_{t-2} + \cdots + \phi_p W_{t-p} + a_t - \theta_1 a_{t-1} - \theta_2 a_{t-2} - \cdots - \theta_q a_{t-q}$$

(5-11)

or, in terms of the observed series,

$$Z_t - Z_{t-1} = \phi_1(Z_{t-1} - Z_{t-2}) + \phi_2(Z_{t-2} - Z_{t-3}) + \cdots$$
$$+ \phi_p(Z_{t-p} - Z_{t-p-1}) + a_t - \theta_1 a_{t-1} - \theta_2 a_{t-2} - \cdots - \theta_q a_{t-q}$$

which we may rewrite as

$$Z_t = (1 + \phi_1)Z_{t-1} + (\phi_2 - \phi_1)Z_{t-2} + (\phi_3 - \phi_2)Z_{t-3} + \cdots$$
$$+ (\phi_p - \phi_{p-1})Z_{t-p} - \phi_p Z_{t-p-1} + a_t - \theta_1 a_{t-1} - \theta_2 a_{t-2} - \cdots - \theta_q a_{t-q}$$

(5-12)

We call this the **difference-equation form** of the model. Notice that it appears to be an ARMA$(p + 1, q)$ process. However, the characteristic polynomial satisfies

$$1 - (1 + \phi_1)x - (\phi_2 - \phi_1)x^2 - (\phi_3 - \phi_2)x^3 - \cdots - (\phi_p - \phi_{p-1})x^p + \phi_p x^{p+1}$$
$$= (1 - \phi_1 x - \phi_2 x^2 - \cdots - \phi_p x^p)(1 - x)$$

which can be checked easily. This factorization clearly shows the root at $x = 1$, which implies nonstationarity. The remaining roots, however, are the roots of the characteristic polynomial of the *stationary* ∇Z_t process.

Explicit representations of the observed series Z_t in terms of either W_t or the white noise series underlying W_t are more difficult than in the stationary case. Since nonstationary processes are *not* in statistical equilibrium, we *cannot* assume that they go infinitely into the past or that they start at $t = -\infty$. However, we can and shall assume that they start at some time point $t = -m$, say, where $-m$ is earlier than time $t = 1$, at which point we first observed the series. For convenience we take $Z_t = 0$ for $t < -m$. The equation $Z_t - Z_{t-1} = W_t$ can then be solved by summing both sides from $t = -m$ to $t = t$ to get the explicit representation

$$Z_t = \sum_{j=0}^{t+m} W_{t-j}$$

(5-13)

for the ARIMA$(p, 1, q)$ process.

The ARIMA$(p, 2, q)$ process can be dealt with similarly by summing twice to get the representations

$$Z_t = \sum_{j=-m}^{t} \sum_{l=-m}^{j} W_i$$
$$= \sum_{j=0}^{t+m} (j+1)W_{t-j}$$

(5-14)

These representations have limited use but can be used to investigate the covariance properties of ARIMA models and also to express Z_t in terms of the white noise series $\{a_t\}$. We defer the calculations until we evaluate specific cases.

If the process contains no autoregressive terms, we call it an *integrated moving average* model and abbreviate the name to IMA(d, q). If no moving average terms are present, we denote the model as ARI(p, d). We first consider in detail the important IMA(1, 1) model.

THE IMA(1, 1) MODEL

The simple IMA(1, 1) model satisfactorily represents numerous time series, especially those arising in economics and business. In difference-equation form, the model is

$$Z_t = Z_{t-1} + a_t - \theta a_{t-1} \tag{5-15}$$

The series in Exhibit 5.1 shows a simulation from such a model with $\theta = \frac{1}{2}$ using normally distributed white noise with $\sigma_a^2 = 1$.

To write Z_t explicitly as a function of present and past a-values, we use Equation (5-13) and the fact that $W_t = a_t - \theta a_{t-1}$ in this case. After a little rearrangement, we can write

$$Z_t = a_t + (1 - \theta)a_{t-1} + (1 - \theta)a_{t-2} + \cdots + (1 - \theta)a_{-m} - \theta a_{-m-1} \tag{5-16}$$

Notice that in contrast to our stationary ARMA models, the weights on the noise terms *do not die out* as we go into the past. Since we are assuming that $-m < 1$ and $0 < t$, we may usefully think of Z_t as being an *equally weighted accumulation* of a large number of white noise values.

From Equation (5-16) we can easily derive variances and correlations. We find

$$\text{Var}(Z_t) = [1 + \theta^2 + (1 - \theta)^2(t + m)]\sigma_a^2 \tag{5-17}$$

and

$$\begin{aligned}\text{Corr}(Z_t, Z_{t-k}) &= \frac{1 + \theta^2 + (1 - \theta)^2(t + m - k)}{[\text{Var}(Z_t) \cdot \text{Var}(Z_{t-k})]^{1/2}} \\ &\approx \sqrt{\frac{t + m - k}{t + m}} \\ &\approx 1 \quad \text{for large } m \text{ and moderate } k\end{aligned} \tag{5-18}$$

We see that for $0 < t$, Var(Z_t) will be quite large and Corr(Z_t, Z_{t-k}) will be strongly positive for many lags $k = 1, 2, \ldots$.

THE IMA(2, 2) MODEL

The assumptions of Equation (5-10) led to an IMA(2, 2) model. In difference-equation form, we have

$$\nabla^2 Z_t = a_t - \theta_1 a_{t-1} - \theta_2 a_{t-2}$$

or

$$Z_t = 2Z_{t-1} - Z_{t-2} + a_t - \theta_1 a_{t-1} - \theta_2 a_{t-2} \qquad \text{(5-19)}$$

The representation of Equation (5-14) may be used to express Z_t in terms of $a_t, a_{t-1}, \ldots$. After some tedious algebra we find that

$$Z_t = a_t + \sum_{j=1}^{m} \psi_j a_{t-j} - [(t + m + 1)\theta_1 + (t + m)\theta_2]a_{-m-1} - (t + m + 1)\theta_2 a_{-m-2}$$

(5-20)

where

$$\psi_j = 1 + \theta_2 + (1 - \theta_1 - \theta_2)j \quad \text{for } j = 1, 2, \ldots, t + m$$

Once more we see that the ψ-weights do not die out but form a linear function of j.

Again, variances and correlations for Z_t can be obtained from the representation given in Equation (5-20), but the calculations are tedious. We shall simply note that the variance of Z_t increases rapidly with t and again Corr (Z_t, Z_{t-k}) is nearly 1 for all moderate k.

A computer simulation of an IMA(2, 2) model is shown in Exhibit 5.5. Notice the smooth change in the process and the unimportance of the zero mean function, since the variance is increasing so quickly and the correlations of neighboring observations are so strong. Exhibit 5.6 shows the first difference of the data displayed in Exhibit 5.5. This series is also nonstationary, namely, IMA(1, 2). Exhibit 5.7 is a time plot of the second difference of the original series and would be considered as arising from a stationary model. The true model here is MA(2).

EXHIBIT 5.5 Simulation of an IMA(2, 2) Series with $\theta_1 = 1$ and $\theta_2 = -0.6$

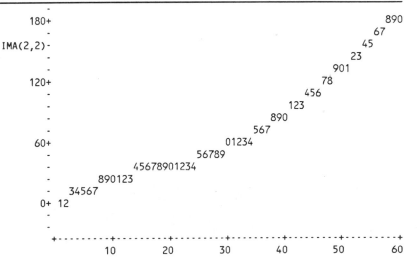

EXHIBIT 5.6 First Difference of an IMA(2, 2) Series

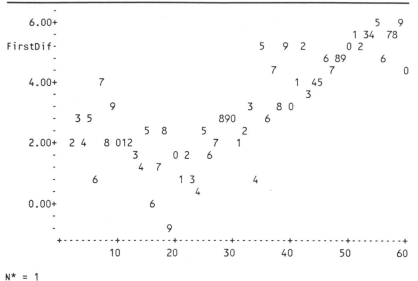

N* = 1

EXHIBIT 5.7 Second Difference of an IMA(2, 2) Series

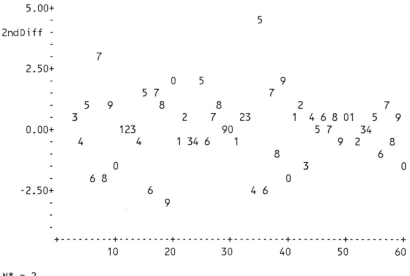

N* = 2

THE ARI(1, 1) MODEL

The ARI(1, 1) process will satisfy

$$Z_t - Z_{t-1} = \phi(Z_{t-1} - Z_{t-2}) + a_t$$

or

$$Z_t = (1 + \phi)Z_{t-1} - \phi Z_{t-2} + a_t$$

(5-21)

where $|\phi| < 1$.

To find the ψ-weights in this case, we shall use a technique that will generalize to arbitrary ARIMA models. It can be shown that the ψ-weights can be obtained by equating like powers of x in the identity:

$$(1 - \phi_1 x - \dot\psi_2 x^2 - \cdots - \phi_p x^p)(1 - x)^d(1 + \psi_1 x + \psi_2 x^2 + \cdots)$$
$$= 1 - \theta_1 x - \theta_2 x^2 - \cdots - \theta_q x^q \quad \text{(5-22)}$$

In our case this relationship reduces to

$$(1 - \phi x)(1 - x)(1 + \psi_1 x + \psi_2 x^2 + \cdots) = 1$$

or

$$[1 - (1 + \phi)x + \phi x^2](1 + \psi_1 x + \psi_2 x^2 + \cdots) = 1$$

By equating like powers of x on both sides we get

$$-(1 + \phi) + \psi_1 = 0$$
$$\phi - (1 + \phi)\psi_1 + \psi_2 = 0$$

and, in general,

$$\psi_k = (1 + \phi)\psi_{k-1} - \phi\psi_{k-2} \quad \text{for } k \geq 2 \qquad \text{(5-23)}$$

with $\psi_0 = 1$ and $\psi_1 = 1 + \phi$. This recursion with starting values allows us to compute as many ψ-weights as necessary.

It can also be shown that in this case the following explicit expression holds:

$$\psi_k = \frac{1 - \phi^{k+1}}{1 - \phi} \quad \text{for } k \geq 1 \qquad \text{(5-24)}$$

(It is easy to show that this expression satisfies Equation (5-23).)

5.3 CONSTANT TERMS IN ARIMA MODELS

For an ARIMA(p, d, q) model, $\nabla^d Z_t = W_t$ is a stationary ARMA(p, q) process. Our standard assumption is that stationary models have a zero mean; that is, we are actually considering deviations from the constant mean. A nonzero fixed mean μ in a stationary ARMA model $\{W_t\}$ can be accommodated in either of two ways. We can assume that

$$W_t - \mu = \phi_1(W_{t-1} - \mu) + \phi_2(W_{t-2} - \mu) + \cdots + \phi_p(W_{t-p} - \mu)$$
$$+ a_t - \theta_1 a_{t-1} - \theta_2 a_{t-2} - \cdots - \theta_q a_{t-q}$$

Alternatively, we can introduce a constant term θ_0 into the model as follows:

$$W_t = \phi_1 W_{t-1} + \phi_2 W_{t-2} + \cdots + \phi_p W_{t-p}$$
$$+ \theta_0 + a_t - \theta_1 a_{t-1} - \theta_2 a_{t-2} - \cdots - \theta_q a_{t-q}$$

Taking expected values in the latter representation, we find that

$$\mu = \mu(\phi_1 + \phi_2 + \cdots + \phi_p) + \theta_0$$

so that

$$\mu = \frac{\theta_0}{1 - \phi_1 - \phi_2 - \cdots - \phi_p} \tag{5-25}$$

or, conversely, that

$$\theta_0 = \mu(1 - \phi_1 - \phi_2 - \cdots - \phi_p) \tag{5-26}$$

Since the alternative representations for W_t are equivalent, we shall use whichever parameterization is convenient.

What will be the effect of a nonzero mean for W_t on the undifferenced series Z_t? Consider the IMA(1, 1) case with a constant term. We have

$$Z_t = Z_{t-1} + \theta_0 + a_t - \theta_1 a_{t-1}$$

or

$$W_t = \theta_0 + a_t - \theta_1 a_{t-1}$$

Either substituting into Equation (5-13) or by iterating into the past, we find that

$$Z_t = a_t + (1 - \theta_1)a_{t-1} + (1 - \theta_1)a_{t-2} + \cdots$$
$$+ (1 - \theta_1)a_{-m} - \theta_1 a_{-m-1} + (t + m + 1)\theta_0 \tag{5-27}$$

Comparing this with Equation (5-16), we see that we have an added *linear deterministic trend* $(t + m + 1)\theta_0$ with *slope* θ_0.

An equivalent representation of such a process would then be

$$Z_t = Z_t' + \beta_0 + \beta_1 t$$

where Z_t' is an IMA(1, 1) series with $E(\nabla Z_t') = 0$ and $E(\nabla Z_t) = \beta_1$.

For a general ARIMA(p, d, q) model where $E(\nabla^d Z_t) \neq 0$, it can be argued that $Z_t = Z_t' + \mu_t$ where μ_t is a deterministic polynomial trend of degree d and Z_t' is ARIMA(p, d, q) with $E(\nabla^d Z_t') = 0$. With $d = 2$ and $\theta_0 \neq 0$, a quadratic trend would be implied.

5.4 OTHER TRANSFORMATIONS

We have seen how differencing can be a useful transformation for achieving stationarity. However, the logarithm transformation, perhaps followed by differencing, is also a useful method in certain circumstances.

We frequently encounter series where increased dispersion seems to be associated with increased levels of the series—the larger the level of the series, the more variation there is around that level and conversely.

Specifically, suppose that $Z_t > 0$ for all t with

$$E(Z_t) = \mu_t \quad \text{and} \quad \sqrt{\text{Var}(Z_t)} = \mu_t \sigma$$

Then

$$E[\log(Z_t)] \approx \log(\mu_t) \quad \text{and} \quad \text{Var}[\log(Z_t)] \approx \sigma^2 \tag{5-28}$$

These results follow from taking expected values and variances of both sides of the (Taylor) approximation

$$\log(Z_t) \approx \log(\mu_t) + \frac{Z_t - \mu_t}{\mu_t}$$

In words, if the standard deviation of the series is proportional to the level of the series, then transforming to logarithms will produce a series with approximately constant variance. Also, if the level of the series is changing roughly exponentially, the log-transformed series will exhibit a linear trend. Thus, we then might want to take first differences. An alternative set of assumptions leading to differences of logged data follows.

PERCENTAGE CHANGES

Suppose Z_t tends to have relatively stable percentage changes from one time period to the next. Specifically, assume that

$$Z_t = (1 + X_t)Z_{t-1}$$

where $100X_t$ is the percentage change (possibly negative) from Z_{t-1} to Z_t.
   Then

$$\log(Z_t) - \log(Z_{t-1}) = \log\left(\frac{Z_t}{Z_{t-1}}\right)$$
$$= \log(1 + X_t)$$

If X_t is restricted to, say, $|X_t| < 0.2$, that is, the percentage changes are at most $\pm 20\%$, then, to a good approximation, $\log(1 + X_t) \approx X_t$. Consequently,

$$\boxed{\nabla[\log(Z_t)] \approx X_t} \qquad \text{(5-29)}$$

will be relatively stable and perhaps modeled by a stationary model. Notice that we take logs first and then compute first differences.
   In Minitab the logarithm can be calculated by simply writing

LET C2 = LOG (C1)

The DIFFERENCE command was introduced earlier in this chapter. Notice that second differences are computed in two steps:

DIFF C1 C2

DIFF C2 C3

Colume C3 will then contain the second difference of column C1. Note that DIFF 2 C1 C2 does *not* compute a second difference but rather a first difference at *lag* 2.
   As an example where transforming to logarithms and taking differences appears to be appropriate, consider the series plotted in Exhibit 5.8. This series consists of the monthly commercial air-passenger-miles traveled in the United States (including Alaska and Hawaii) from January 1960 through December 1977. Notice that the dispersion increases as the level of the series increases.
   Exhibit 5.9 shows the log-transformed series. Here, although the level

EXHIBIT 5.8 U.S. Air-Passenger-Miles, January 1960 to December 1977

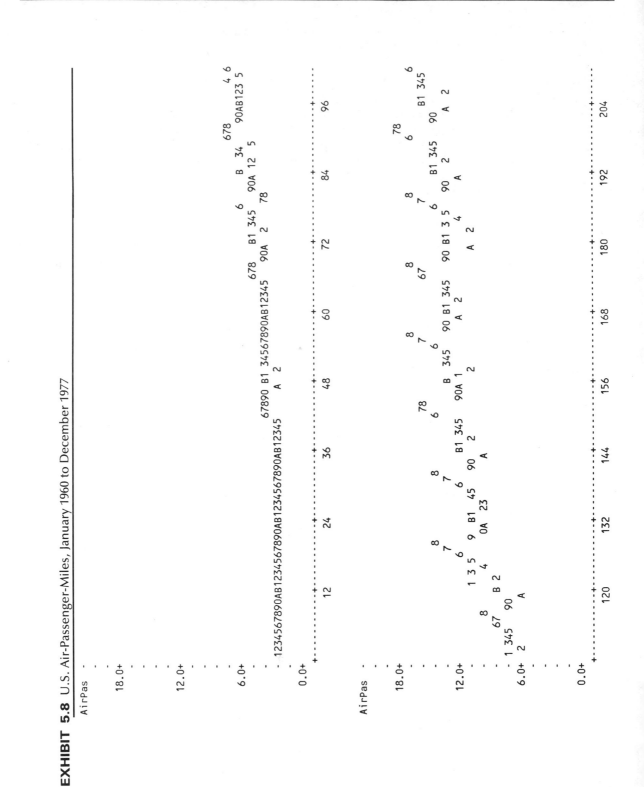

EXHIBIT 5.9 Logarithms of Air-Passenger-Miles

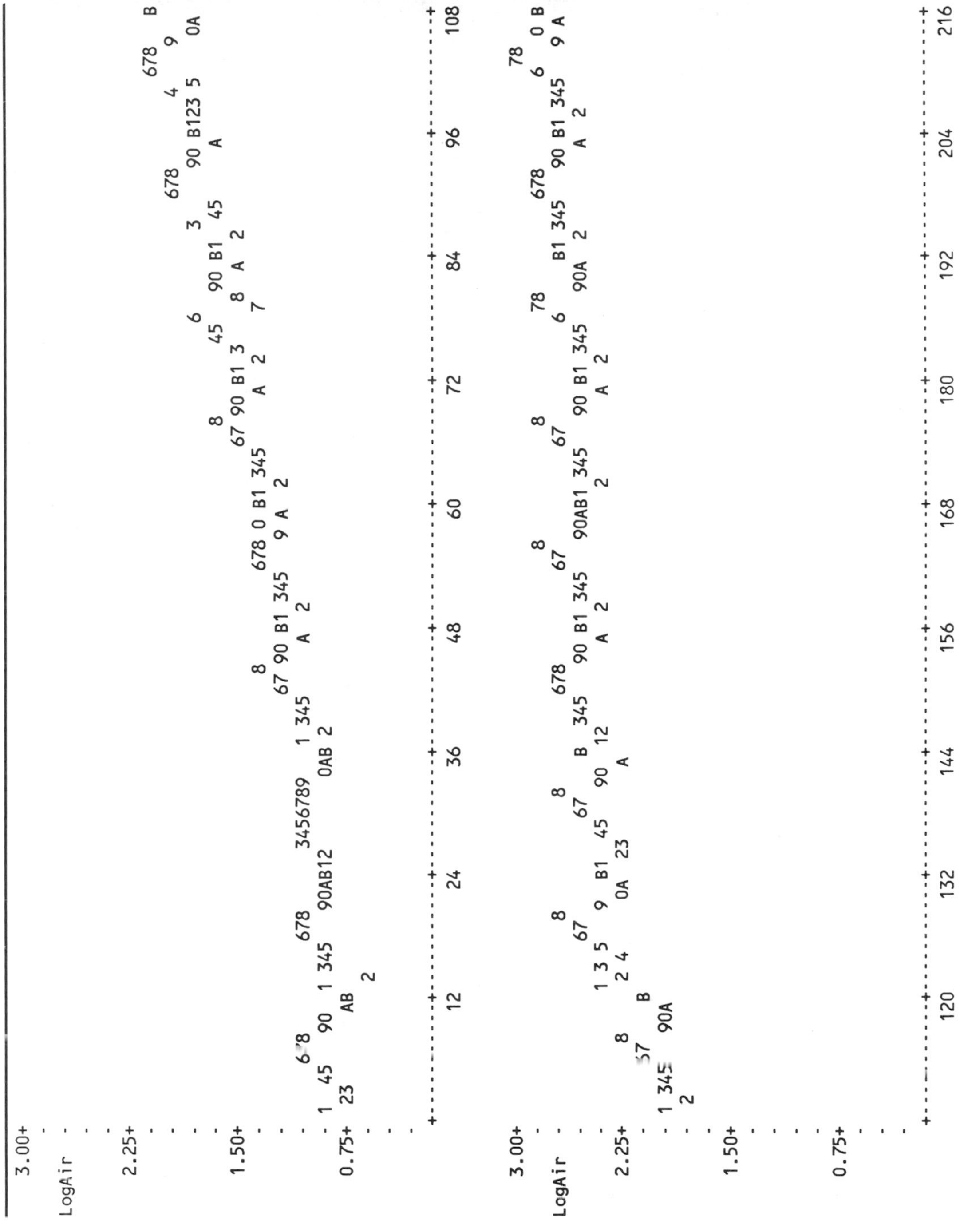

EXHIBIT 5.10 Differences of Logarithms of Air-Passenger-Miles

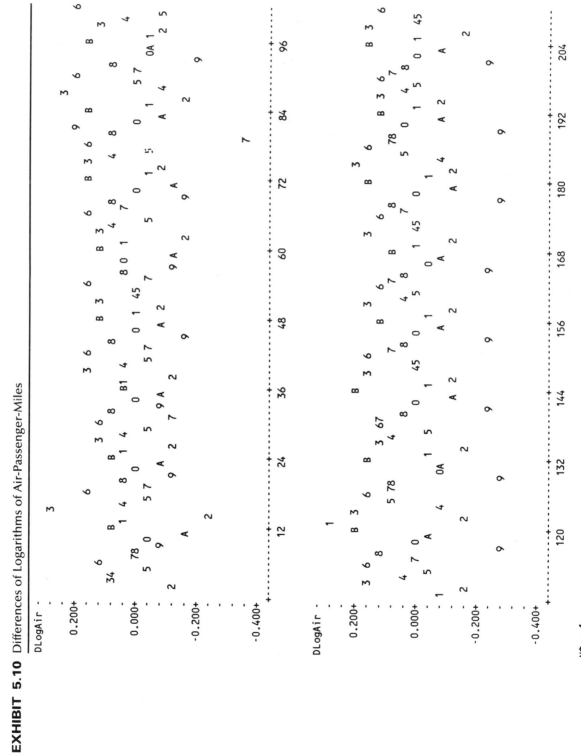

increases, the dispersion around the level remains relatively constant over the length of the series. Finally, Exhibit 5.10 displays the first difference of the logged series, which could be considered stationary. However, there is a seasonal effect present that needs to be modeled. We will return to this series in Chapter 10.

POWER TRANSFORMATIONS

A flexible family of transformations, the **power transformations**, was introduced by Box and Cox (1964). For a given value of the parameter λ, the transformation is defined by

$$g(x) = \begin{cases} \dfrac{x^\lambda - 1}{\lambda}, & \lambda \neq 0 \\ \log(x), & \lambda = 0 \end{cases} \qquad (5\text{-}30)$$

Subtracting 1 and dividing by λ makes $g(x)$ change smoothly as λ approaches 0. A calculus argument shows that as $\lambda \to 0$, $(x^\lambda - 1)/\lambda \to \log(x)$. Notice that $\lambda = \frac{1}{2}$ produces a square root transformation useful with Poisson-like data, and $\lambda = -1$ corresponds to a reciprocal transformation. Of course, $\lambda = 0$ yields the log transformation.

We can consider λ as an additional parameter in the model to be estimated from the observed data. However, precise estimation of λ is usually not warranted. Evaluation of a range of transformations based on a grid of λ values, say, ± 1, $\pm\frac{1}{2}$, $\pm\frac{1}{3}$, $\pm\frac{1}{4}$, and 0 will usually suffice and may have some intuitive meaning.

CHAPTER 5 EXERCISES

5.1. Consider two models:

$$\text{A: } Z_t = 0.9Z_{t-1} + 0.09Z_{t-2} + a_t$$
$$\text{B: } Z_t = Z_{t-1} + a_t - 0.1a_{t-1}$$

a. Identify each as a specific ARIMA model.
b. In what ways are the two models different?
c. In what ways are they similar? (Compare ψ-weights and π-weights.)

5.2. Identify as specific ARIMA models:
a. $Z_t = Z_{t-1} - 0.25Z_{t-2} + a_t + 0.5a_{t-1}$
b. $Z_t = 2Z_{t-1} - Z_{t-2} + a_t$
c. $Z_t = 0.5Z_{t-1} + 0.5Z_{t-2} + a_t - 0.5a_{t-1} + 0.25a_{t-2}$

5.3. Suppose that $\{Z_t\}$ is generated according to

$$Z_t = a_t + ca_{t-1} + ca_{t-2} + \cdots + ca_0 \qquad \text{for } t > 0$$

a. Find the mean and covariance function for Z_t. Is $\{Z_t\}$ stationary?
b. Find the mean and covariance function for ∇Z_t. Is ∇Z_t stationary?
c. Identify Z_t as a specific ARIMA model.

5.4. Suppose that $Z_t = A + Bt + X_t$ where A and B are random variables that are independent of the random walk $\{X_t\}$.

    a. Is $\{Z_t\}$ stationary?

    b. Is $\{\nabla Z_t\}$ stationary?

5.5. Nonstationary ARIMA series can be simulated in Minitab by simulating the corresponding stationary ARMA series, as described in Appendix B, p. 76, and then "integrating it" using the PARSUM (partial sum) command d times. The format is

PARSUM of the column **C**, put results in **C**

    Use Minitab to simulate and plot some IMA(1, 1) and IMA(2, 2) series with a variety of parameter values. Note the stochastic trends in the time plots.

APPENDIX E THE BACKSHIFT OPERATOR

Many other books and much of the time series literature use what is called the *backshift operator* to express and manipulate ARIMA models. The **backshift operator**, denoted B, operates on the time index of a series and shifts time back 1 time unit to form a new series. In particular,

$$B(Y_t) = Y_{t-1}$$

The backshift operator is linear since for any constants a, b, and c and series Y_t and Z_t, it is easy to see that

$$B(aY_t + bZ_t + c) = a[B(Y_t)] + b[B(Z_t)] + c$$

Consider now the MA(1) model. In terms of B, we can write

$$
\begin{aligned}
Z_t &= a_t - \theta a_{t-1} \\
&= a_t - \theta B(a_t) \\
&= (1 - \theta B)(a_t) \\
&= \theta(B)a_t
\end{aligned}
$$

where $\theta(B)$ is the MA characteristic polynomial "evaluated at B."

Since $B(Y_t)$ is itself a time series, it is meaningful to consider $B[B(Y_t)]$. But clearly $B[B(Y_t)] = B(Y_{t-1}) = Y_{t-2}$, and we can write

$$B^2(Y_t) = Y_{t-2}$$

More generally, we have

$$B^m(Y_t) = Y_{t-m}$$

for any positive integer m.

For a general MA(q) model, we can then write

$$
\begin{aligned}
Z_t &= a_t - \theta_1 a_{t-1} - \theta_2 a_{t-2} - \cdots - \theta_q a_{t-q} \\
&= a_t - \theta_1 B(a_t) - \theta_2 B^2(a_t) - \cdots - \theta_q B^q(a_t) \\
&= (1 - \theta_1 B - \theta_2 B^2 - \cdots - \theta_q B^q)(a_t)
\end{aligned}
$$

or

$$Z_t = \theta(B)a_t$$

where again, $\theta(B)$ is the MA characteristic polynomial evaluated at B.

For autoregressive models, say AR(p), we first move all the terms involving Z to the left-hand side:

$$Z_t - \phi_1 Z_{t-1} - \phi_2 Z_{t-2} - \cdots - \phi_p Z_{t-p} = a_t$$

Then we write

$$Z_t - \phi_1 B(Z_t) - \phi_2 B^2(Z_t) - \cdots - \phi_p B^p(Z_t) = a_t$$

or

$$(1 - \phi_1 B - \phi_2 B^2 - \cdots - \phi_p B^p)Z_t = a_t$$

which can be expressed as

$$\phi(B)Z_t = a_t$$

where $\phi(B)$ is the AR characteristic polynomial evaluated at B.

The general ARMA (p, q) model can then be compactly written as

$$\phi(B)Z_t = \theta(B)a_t$$

Differencing can also be conveniently written in terms of B. We have

$$\nabla Z_t = Z_t - Z_{t-1} = Z_t - B(Z_t)$$
$$= (1 - B)Z_t$$

with second differences given by

$$\nabla^2(Z_t) = (1 - B)^2 Z_t$$

Effectively, $\nabla = 1 - B$ and $\nabla^d = (1 - B)^d$.

The ARIMA(p, d, q) model is then expressed as

$$\phi(B)(1 - B)^d Z_t = \theta(B)a_t$$

In the literature one must carefully distinguish from the context the use of B as a backshift operator and its use as an ordinary real (or complex) variable. For example, the stationarity condition is frequently given by stating that the roots of $\phi(B) = 0$ must be greater than 1 in absolute value. Here B is to be treated as a dummy variable in an equation rather than as the backshift operator.

CHAPTER 6 MODEL SPECIFICATION

We have developed a large class of parametric models for both stationary and certain nonstationary time series (the ARIMA models). We can now begin our study and implementation of *statistical inference* for such models. The subjects of the next three chapters, respectively, are:

1. How do we choose appropriate values of p, d, and q for a given series?
2. How do we estimate the parameters of a specific ARIMA(p, d, q) model?
3. How do we check on the appropriateness of the fitted model?

Our overall strategy will first be to decide on reasonable—but tentative—values for p, d, and q. Having done so, we shall estimate the ϕ's, θ's, and σ_a for that model in the most efficient way. Finally, we shall look critically at the model thus obtained to check its adequacy, much in the same way that we did in Section 3.6. If the model appears inadequate in some way, we consider the nature of the inadequacy to help us select another model. We then proceed to estimate that new model and check it for adequacy.

With a few iterations of this model-building strategy, we hope to arrive at the best possible model for a given series. The book by George E. P. Box and G. M. Jenkins (1976) so popularized this technique that many authors call the procedure the "Box–Jenkins method." We begin by continuing our investigation of the properties of the sample autocorrelation function begun in Section 3.6.

6.1 PROPERTIES OF THE SAMPLE AUTOCORRELATION FUNCTION

Recall from Section 3.6 the definition of the sample or estimated autocorrelation function. For the observed series $Z_1, Z_2, \ldots, Z_n$ we have

$$r_k = \frac{\sum\limits_{t=1}^{n-k} (Z_t - \bar{Z})(Z_{t+k} - \bar{Z})}{\sum\limits_{t=1}^{n} (Z_t - \bar{Z})^2}, \qquad k = 0, 1, 2, \ldots \qquad \text{(6-1)}$$

Our goal is to recognize, to the extent possible, patterns in r_k that are characteristic of the known patterns in ρ_k for common ARMA models. For example, we know that $\rho_k = 0$ for $k > q$ in an MA(q) model. However, as the r_k

are only *estimates* of ρ_k, we need to investigate their sampling properties to facilitate the comparison of estimated correlations with theoretical correlations.

From the definition of r_k, a ratio of quadratic functions of the Z's, it should be apparent that the sampling properties of r_k will *not* be easily obtained. Even the expected value of r_k is difficult to find—the expected value of a ratio is *not* the ratio of the respective expected values. We shall have to accept a general large-sample result and consider its implications in special cases. Bartlett (1946) carried out the original work in this area. We shall take a more general result from Anderson (1971).

We suppose that

$$Z_t = \mu + \sum_{j=0}^{\infty} \psi_j a_{t-j}$$

where the a_t are independent and identically distributed with zero means and finite, nonzero variances. We assume that $\sum_{j=0}^{\infty} |\psi_j| < \infty$ and $\sum_{j=0}^{\infty} j\psi_j^2 < \infty$. (This will be satisfied by any stationary ARMA model.)

Then for any fixed m, the joint distribution of

$$\sqrt{n}\,(r_1 - \rho_1), \sqrt{n}\,(r_2 - \rho_2), \ldots, \sqrt{n}\,(r_m - \rho_m)$$

approaches, as $n \to \infty$, a joint normal distribution with zero means, variances c_{ii}, and covariances c_{ij} where

$$c_{ij} = \sum_{k=-\infty}^{\infty} (\rho_{k+i}\rho_{k+j} + \rho_{k-i}\rho_{k+j} - 2\rho_i\rho_k\rho_{k+j} - 2\rho_j\rho_k\rho_{k+i} + 2\rho_i\rho_j\rho_k^2) \quad \text{(6-2)}$$

For large n, we would say that r_k is approximately normally distributed with mean ρ_k and variance c_{kk}/n. Furthermore, $\text{Corr}\,(r_k, r_j) \approx c_{kj}/\sqrt{(c_{kk}c_{jj})}$. Notice that the approximate variance of r_k is inversely proportional to the sample size but that the approximate $\text{Corr}\,(r_k, r_j)$ is *constant* for large n.

Since Equation (6-2) is clearly(!) difficult to interpret in its present generality, we shall consider some important special cases and simplifications. Suppose first that $\{Z_t\}$ is white noise. Then Equation (6-2) reduces considerably, and we obtain

$$\text{Var}\,(r_k) \approx \frac{1}{n} \quad \text{and} \quad \text{Corr}\,(r_k, r_j) \approx 0 \quad \text{for } k \neq j \quad \text{(6-3)}$$

Next, suppose that $\{Z_t\}$ is an AR(1) series with $\rho_k = \phi^k$ for $k > 0$. Then, after considerable algebra and summing of several geometric series, Equation (6-2) yields

$$\text{Var}\,(r_k) \approx \frac{1}{n}\left[\frac{(1 + \phi^2)(1 - \phi^{2k})}{1 - \phi^2} - 2k\phi^{2k}\right] \quad \text{(6-4)}$$

In particular,

$$\text{Var}\,(r_1) \approx \frac{1 - \phi^2}{n} \quad \text{(6-5)}$$

Notice that the closer ϕ is to ± 1, the more precise our estimate of ρ_1 becomes.

For large lags, the terms in Equation (6-4) involving ϕ^{2k} may be ignored, and we have

$$\text{Var}(r_k) \approx \frac{1}{n}\left(\frac{1 + \phi^2}{1 - \phi^2}\right) \quad \text{for large } k \qquad (6\text{-}6)$$

Notice that here, in contrast to Equation (6-5), values of ϕ close to ± 1 imply *large* variances for r_k. Thus we should not expect nearly as precise estimates of $\rho_k = \phi^k \approx 0$ for large k as we do of $\rho_k = \phi^k$ for small k.

For the AR(1) model, Equation (6-2) can also be simplified (after much algebra) for general $0 < i \leq j$ as

$$c_{ij} = \frac{(\phi^{j-i} - \phi^{j+i})(1 + \phi^2)}{1 - \phi^2} + (j - i)\phi^{j-i} - (j + i)\phi^{j+i} \qquad (6\text{-}7)$$

In particular, we find

$$\text{Corr}(r_1, r_2) \approx 2\phi\left(\frac{1 - \phi^2}{1 + 2\phi^2 - 3\phi^4}\right)^{1/2} \qquad (6\text{-}8)$$

Based on Equations (6-4) through (6-8), Table 6.1 gives approximate standard deviations and correlations for the sample autocorrelations for a variety of ϕ-values.

For the MA(1) case, Equation (6-2) simplifies as follows:

$$c_{11} = 1 - 3\rho_1^2 + 4\rho_1^4 \quad \text{and} \quad c_{kk} = 1 + 2\rho_1^2 \quad \text{for } k > 1 \qquad (6\text{-}9)$$

Furthermore,

$$c_{12} = 2\rho_1(1 - \rho_1^2) \qquad (6\text{-}10)$$

Based on these c's, Table 6.2 lists the large-sample standard deviations and correlations for the sample autocorrelations for several θ-values. Notice again

TABLE 6.1   Large-Sample Results for r_k from an AR(1) Model

ϕ	$\sqrt{\text{Var}(r_1)}$	$\sqrt{\text{Var}(r_2)}$	Corr (r_1, r_2)	$\sqrt{\text{Var}(r_{10})}$
0.9	$0.44/\sqrt{n}$	$0.807/\sqrt{n}$	0.97	$2.44/\sqrt{n}$
0.7	$0.71/\sqrt{n}$	$1.12/\sqrt{n}$	0.89	$1.70/\sqrt{n}$
0.4	$0.92/\sqrt{n}$	$1.11/\sqrt{n}$	0.66	$1.18/\sqrt{n}$
0.2	$0.98/\sqrt{n}$	$1.04/\sqrt{n}$	0.38	$1.04/\sqrt{n}$

TABLE 6.2   Large-Sample Results for r_k from an MA(1) Model

θ	$\sqrt{\text{Var}(r_1)}$	$\sqrt{\text{Var}(r_k)}$ $(k > 1)$	Corr (r_1, r_2)
0.9	$0.71/\sqrt{n}$	$1.22/\sqrt{n}$	-0.86
0.7	$0.73/\sqrt{n}$	$1.20/\sqrt{n}$	-0.84
0.5	$0.79/\sqrt{n}$	$1.15/\sqrt{n}$	-0.74
0.4	$0.89/\sqrt{n}$	$1.11/\sqrt{n}$	-0.53

that the sample autocorrelations can be highly correlated and that the standard deviation of r_k is larger for $k > 1$ than for $k = 1$.

For a general MA(q) process, Equation (6-2) reduces to

$$c_{kk} = 1 + 2\sum_{j=1}^{q} \rho_j^2 \quad \text{for } k > q$$

so that

$$\text{Var}(r_k) \approx \frac{1 + 2\sum\limits_{j=1}^{q} \rho_j^2}{n} \quad \text{for } k > q \tag{6-11}$$

For an observed time series, we could replace ρ's by r's to obtain the estimated standard deviation of r_k, that is, the standard error of r_k, for large lags. A test of the hypothesis that the series is MA(q) could then be carried out by comparing

$$\frac{r_{q+1}}{\left(1 + 2\sum\limits_{j=1}^{q} r_j^2\right)^{1/2} \Big/ n^{1/2}} \tag{6-12}$$

with the critical values of a standard normal distribution. Since these calculations are based on large-sample approximations, we could simply observe whether the magnitude of the statistic given in (6-12) exceeds 2 (rather than 1.96) to test at the 5% significance level. In general, we should not expect the sample autocorrelation to mimic the true autocorrelation in great detail. Thus, we should not be surprised to see ripples or trends in the r_k that have no counterparts in the ρ_k.

6.2 THE PARTIAL AUTOCORRELATION FUNCTION

Since for MA(q) series the correlation function is zero for lags beyond q, the sample autocorrelation is a good indicator of the order of the process. However, the autocorrelations of AR(p) series do not remain zero after a certain number of lags, and a different function turns out to be useful for determining the order p. We would like to define the correlation between Z_t and Z_{t-k} *after removing the effect of the intervening variables* $Z_{t-1}, Z_{t-2}, \ldots, Z_{t-k+1}$. For stationary time series, this coefficient is called the **partial autocorrelation** at lag k and will be denoted by ϕ_{kk}.

There are several approaches to making such a definition precise. If $\{Z_t\}$ is a normally distributed time series, we can let

$$\phi_{kk} = \text{Corr}(Z_t, Z_{t-k} \mid Z_{t-1}, Z_{t-2}, \ldots, Z_{t-k+1}) \tag{6-13}$$

that is, ϕ_{kk} is the correlation coefficient in the bivariate distribution of Z_t, Z_{t-k} conditional on $Z_{t-1}, Z_{t-2}, \ldots, Z_{t-k+1}$.

A theory not based on normality can be developed in the following way. Consider predicting Z_t based on a linear function of $Z_{t-1}, Z_{t-2}, \ldots, Z_{t-k+1}$, say,

$\beta_1 Z_{t-1} + \beta_2 Z_{t-2} + \cdots + \beta_{k-1} Z_{t-k+1}$ with the β's chosen to minimize the mean square error of prediction. If we assume that the β's have been so chosen and then think backwards in time, the best predictor of Z_{t-k} based on the same $Z_{t-1}, Z_{t-2}, \ldots, Z_{t-k+1}$ will be

$$\beta_1 Z_{t-k+1} + \beta_2 Z_{t-k+2} + \cdots + \beta_{k-1} Z_{t-1}$$

The **partial autocorrelation function** at lag k is then defined to be the correlation between the two prediction errors, that is,

$$\phi_{kk} = \text{Corr}\,(Z_t - \beta_1 Z_{t-1} - \beta_2 Z_{t-2} - \cdots - \beta_{k-1} Z_{t-k+1},$$
$$Z_{t-k} - \beta_1 Z_{t-k+1} - \beta_2 Z_{t-k+2} - \cdots - \beta_{k-1} Z_{t-1}) \quad \text{(6-14)}$$

(For normally distributed series, it can be shown that the two definitions coincide.) By convention we take $\phi_{11} = \rho_1$.

As an example, consider ϕ_{22}. It is shown in Appendix G that the best linear prediction of Z_t based on Z_{t-1} alone is just $\rho_1 Z_{t-1}$. Thus we need to compute

$$\text{Cov}\,(Z_t - \rho_1 Z_{t-1}, Z_{t-2} - \rho_1 Z_{t-1}) = \gamma_0(\rho_2 - \rho_1^2 - \rho_1^2 + \rho_1^2) = \gamma_0(\rho_2 - \rho_1^2)$$

Since

$$\text{Var}\,(Z_t - \rho_1 Z_{t-1}) = \text{Var}\,(Z_{t-2} - \rho_1 Z_{t-1})$$
$$= \gamma_0(1 + \rho_1^2 - 2\rho_1^2)$$
$$= \gamma_0(1 - \rho_1^2)$$

we have

$$\phi_{22} = \frac{\rho_2 - \rho_1^2}{1 - \rho_1^2} \quad \text{(6-15)}$$

Consider now an AR(1) model. Recall that $\rho_k = \phi^k$ so that

$$\phi_{22} = \frac{\phi^2 - \phi^2}{1 - \phi^2} = 0$$

We shall soon see that for the AR(1) case, $\phi_{kk} = 0$ for all $k > 1$. Thus the partial autocorrelation is nonzero for lag 1, the order of the process, but is zero for lags beyond the order 1. We shall show this to be generally the case for AR(p) models.

Consider a general AR(p) series. It will be shown in Chapter 9 that the best linear predictor of Z_t in terms of $Z_{t-1}, Z_{t-2}, \ldots, Z_{t-p}, \ldots, Z_{t-k+1}$ for $k > p$ is simply

$$\phi_1 Z_{t-1} + \phi_2 Z_{t-2} + \cdots + \phi_p Z_{t-p}$$

Also, the best predictor of Z_{t-k} will be some function $h(Z_{t-k+1}, Z_{t-k+2}, \ldots, Z_{t-1})$, say. So

$$\text{Cov}\,[Z_t - \phi_1 Z_{t-1} - \phi_2 Z_{t-2} - \cdots - \phi_p Z_{t-p}, Z_{t-k} - h(Z_{t-k+1}, Z_{t-k+2}, \ldots, Z_{t-1})]$$
$$= \text{Cov}\,[a_t, Z_{t-k} - h(Z_{t-k+1}, Z_{t-k+2}, \ldots, Z_{t-1})]$$
$$= 0 \quad \text{(since } a_t \text{ is independent of } Z_{t-1}, Z_{t-2}, \ldots)$$

Thus for an AR(p) model,

$$\phi_{kk} = 0 \quad \text{for } k > p \tag{6-16}$$

For an MA(1) series, Equation (6-15) quickly yields

$$\phi_{22} = \frac{-\theta^2}{1 + \theta^2 + \theta^4} \tag{6-17}$$

Furthermore, it can be shown that in this case

$$\phi_{kk} = \frac{-(\theta^k)(1 - \theta^2)}{1 - \theta^{2(k+1)}} \quad \text{for } k \geq 1 \tag{6-18}$$

Notice that the partial autocorrelation of an MA(1) model is never zero but essentially decays exponentially to zero, rather like the autocorrelation for an AR(1) series.

More generally, it can be shown that the partial autocorrelation functions for MA(q) models behave very much like the autocorrelation functions of AR(q) models. A general method for finding the partial autocorrelation function for any stationary process with autocorrelation function ρ_k is as follows (see Anderson, 1971, pp. 187–188): For a given lag k, it can be shown that ϕ_{kk} satisfies the Yule–Walker equations (which first appeared in Chapter Four):

$$\rho_j = \phi_{k1}\rho_{j-1} + \phi_{k2}\rho_{j-2} + \cdots + \phi_{kk}\rho_{j-k}, \qquad j = 1, 2, \ldots, k \tag{6-19}$$

More explicitly,

$$
\begin{aligned}
\rho_1 &= \phi_{k1} &&+ \phi_{k2}\rho_1 &&+ \cdots + \phi_{kk}\rho_{k-1} \\
\rho_2 &= \phi_{k1}\rho_1 &&+ \phi_{k2} &&+ \cdots + \phi_{kk}\rho_{k-2} \\
&\ \ \vdots \\
\rho_k &= \phi_{k1}\rho_{k-1} &&+ \phi_{k2}\rho_{k-2} &&+ \cdots + \phi_{kk}
\end{aligned}
\tag{6-20}
$$

Here we are treating $\rho_1, \rho_2, \ldots, \rho_k$ as given and wish to solve for the unknowns $\phi_{k1}, \phi_{k2}, \ldots, \phi_{kk}$ (discarding all but ϕ_{kk}).

These equations yield ϕ_{kk} for *any* stationary time series. However, if the model is in fact AR(p), then since for $k = p$ Equations (6-20) are just the Yule–Walker equations of Chapter Four, we must have $\phi_{pp} = \phi_p$. In addition, as we have already seen by an alternative derivation, $\phi_{kk} = 0$ for $k > p$. Thus the partial autocorrelation effectively displays the correct order p of an autoregressive process as the last lag k before ϕ_{kk} becomes zero.

THE SAMPLE PARTIAL AUTOCORRELATION FUNCTION

For an observed time series, we need to be able to estimate the partial autocorrelation function. Given the relationships in Equation (6-20), an obvious method is to estimate the ρ's by r's and then to solve the resulting linear equations for $k = 1, 2, \ldots$ to get estimates of ϕ_{kk}. We call this estimated function the **sample partial autocorrelation function** and denote it $\hat{\phi}_{kk}$.

Levinson (1947) and Durbin (1960) gave an efficient method for obtaining the solutions to Equation (6-20), either for theoretical or sample partial autocorrelations. They showed independently that Equations (6-20) can be solved recursively as follows:

$$\phi_{kk} = \frac{\rho_k - \sum\limits_{j=1}^{k-1} \phi_{k-1,j}\rho_{k-j}}{1 - \sum\limits_{j=1}^{k-1} \phi_{k-1,j}\rho_j} \qquad (6\text{-}21)$$

where

$$\phi_{kj} = \phi_{k-1,j} - \phi_{kk}\phi_{k-1,k-j} \quad \text{for } j = 1, 2, \ldots, k-1$$

For example, using $\phi_{11} = \rho_1$ to get started, we have

$$\phi_{22} = \frac{\rho_2 - \phi_{11}\rho_1}{1 - \phi_{11}\rho_1} = \frac{\rho_2 - \rho_1^2}{1 - \rho_1^2}$$

(as before) with

$$\phi_{21} = \phi_{11} - \phi_{22}\phi_{11} \quad \text{(needed for } k = 3\text{)}$$

Then

$$\phi_{33} = \frac{\rho_3 - \phi_{21}\rho_2 - \phi_{22}\rho_1}{1 - \phi_{21}\rho_1 - \phi_{22}\rho_2}$$

We may thus numerically calculate as many values for ϕ_{kk} as desired. As stated, these equations give us the theoretical partial autocorrelations, but by replacing the ρ's with r's, we obtain the sample partial autocorrelations $\hat{\phi}_{kk}$. The Minitab command PACF performs these calculations. The format is:

PACF [out to lag **K**] for the series in **C**

To assess the possible magnitude of the partial autocorrelations, Quenouille (1949) has shown that, under the hypothesis that an AR(p) model is correct, the estimated partial autocorrelations at lags greater than p are approximately independently normally distributed with zero means and variance $1/n$. Thus $\pm 2/\sqrt{n}$ can be used as critical limits on $\hat{\phi}_{kk}$ for $k > p$ to test the hypothesis of an AR(p) model.

6.3 SIMULATED SERIES

To illustrate the theory of Sections 6.1 and 6.2, we shall consider the sample autocorrelation and sample partial autocorrelation functions of some simulated time series.

Exhibit 6.1 displays the sample autocorrelation function (ACF) for a simulated normal white noise series of length $n = 121$. From Equations (6-3) we calculate the standard deviation of r_k to be about $1/\sqrt{121} = 0.09$. We would therefore expect about 95% of the estimates to lie within ± 0.18, that is, within two standard deviations of the true value of zero. In our particular sample, all of the 21 sample correlations are within ± 0.18.

Exhibit 6.2 gives the sample partial autocorrelation function (PACF) for the same white noise series. Since white noise can be considered as AR(p) with

EXHIBIT 6.1 Sample Autocorrelation Function for White Noise

```
ACF of WhiteN

               1.0 -0.8 -0.6 -0.4 -0.2  0.0  0.2  0.4  0.6  0.8  1.0
               +----+----+----+----+----+----+----+----+----+----+
    1   0.151                                XXXXX
    2  -0.077                              XXX
    3   0.039                               XX
    4   0.077                               XXX
    5   0.083                               XXX
    6   0.023                               XX
    7   0.019                               X
    8   0.047                               XX
    9   0.039                               XX
   10  -0.105                             XXXX
   11  -0.019                               X
   12  -0.060                              XX
   13   0.049                               XX
   14   0.062                               XXX
   15  -0.092                             XXX
   16  -0.060                             XXX
   17   0.035                               XX
   18   0.106                               XXXX
   19   0.032                               XX
   20   0.066                               XXX
   21  -0.058                              XX
```

$p = 0$, Quenouille's (1949) result applies, and $\pm2/\sqrt{n} = \pm0.18$ can be used to judge the significance of the estimates. Here again none of the 21 PACF values exceed these limits.

Exhibit 6.3 shows the sample ACF for a series of length $n = 59$ simulated according to an AR(1) model with $\phi = 0.9$. From Table 6.1 the standard

EXHIBIT 6.2 Sample Partial Autocorrelation Function for White Noise

```
PACF of WhiteN

              -1.0 -0.8 -0.6 -0.4 -0.2  0.0  0.2  0.4  0.6  0.8  1.0
               +----+----+----+----+----+----+----+----+----+----+
    1   0.151                                XXXXX
    2  -0.102                             XXXX
    3   0.070                               XXX
    4   0.053                               XX
    5   0.073                               XXX
    6   0.008                               X
    7   0.023                               XX
    8   0.034                               XX
    9   0.019                               X
   10  -0.122                            XXXX
   11   0.016                               X
   12  -0.097                             XXX
   13   0.081                               XXX
   14   0.034                               XX
   15  -0.078                             XXX
   16  -0.020                               X
   17   0.037                               XX
   18   0.097                               XXX
   19   0.024                               XX
   20   0.083                               XXX
   21  -0.091                             XXX
```

EXHIBIT 6.3 Sample Autocorrelation Function for an AR(1) Series with $\phi = 0.9$

```
ACF of AR(1)

             -1.0 -0.8 -0.6 -0.4 -0.2  0.0  0.2  0.4  0.6  0.8  1.0
             +----+----+----+----+----+----+----+----+----+----+
    1   0.855                          XXXXXXXXXXXXXXXXXXXXXX
    2   0.726                          XXXXXXXXXXXXXXXXXXX
    3   0.580                          XXXXXXXXXXXXXXX
    4   0.463                          XXXXXXXXXXXX
    5   0.317                          XXXXXXXXX
    6   0.185                          XXXXX
    7   0.097                          XXX
    8   0.068                          XXX
    9   0.062                          XXX
   10   0.048                          XX
   11   0.023                          XX
   12   0.005                          X
   13   0.020                          X
   14   0.004                          X
   15  -0.027                          XX
   16  -0.039                          XX
   17  -0.032                          XX
```

deviation of r_1 is about $0.44/\sqrt{59} = 0.057$, and of r_2 about $0.81/\sqrt{59} = 0.105$. The values being estimated are $\rho_1 = 0.9$ and $\rho_2 = 0.81$, respectively. The estimates $r_1 = 0.855$ and $r_2 = 0.726$ are thus well within two standard deviations of the correct values. For lag 10, $\rho_{10} = 0.349$ and $r_{10} = 0.048$, but $\sqrt{\text{Var}(r_{10})} \approx 2.44/\sqrt{59} = 0.318$, so again the estimate is easily within two standard deviations of the true value. In general, the plot shows a tendency toward exponential decay with increasing lag.

The sample PACF given in Exhibit 6.4 dramatically conveys the fact that we are dealing with an AR(1) model. Note that $2/\sqrt{59} = 0.26$, thus only partials outside of ± 0.26 would be considered significant at the 5% level, and none are that large for any lag greater than 1.

EXHIBIT 6.4 Sample Partial Autocorrelation Function for an AR(1) Series with $\phi = 0.9$

```
PACF of AR(1)

             -1.0 -0.8 -0.6 -0.4 -0.2  0.0  0.2  0.4  0.6  0.8  1.0
             +----+----+----+----+----+----+----+----+----+----+
    1   0.855                          XXXXXXXXXXXXXXXXXXXXXX
    2  -0.019                         X
    3  -0.133                     XXXX
    4   0.008                         X
    5  -0.176                    XXXXX
    6  -0.075                      XXX
    7   0.082                          XXX
    8   0.138                          XXXX
    9   0.058                          XX
   10  -0.058                       XX
   11  -0.104                     XXXX
   12  -0.048                       XX
   13   0.103                          XXXX
   14  -0.056                       XX
   15  -0.047                       XX
   16   0.080                          XXX
   17   0.012                          X
```

EXHIBIT 6.5 Sample Autocorrelation Function for an AR(1) Series with $\phi = 0.4$

```
ACF of AR(1)

        -1.0 -0.8 -0.6 -0.4 -0.2  0.0  0.2  0.4  0.6  0.8  1.0
        +----+----+----+----+----+----+----+----+----+----+
   1   0.409                         XXXXXXXXXX
   2   0.033                         XX
   3  -0.038                         XX
   4  -0.081                         XXX
   5  -0.146                         XXXXX
   6  -0.099                         XXX
   7  -0.053                         XX
   8  -0.035                         XX
   9  -0.003                         X
  10  -0.028                         XX
  11   0.020                         XX
  12  -0.094                         XXX
  13  -0.060                         XX
  14  -0.018                         X
  15  -0.114                         XXXX
  16  -0.138                         XXXX
  17  -0.021                         XX
  18   0.055                         XX
  19   0.191                         XXXXXX
  20   0.126                         XXXX
```

Exhibits 6.5 and 6.6 give the ACF and PACF from a simulated AR(1) model with $\phi = 0.4$ and $n = 119$. Here $\sqrt{\text{Var}(r_1)} \approx 0.084$ and $\sqrt{\text{Var}(r_2)} \approx 0.102$. The true autocorrelations are 0.4 and 0.16, which are estimated as $r_1 = 0.409$ and $r_2 = 0.033$. The PACF's in Exhibit 6.6 clearly indicate that an AR(1) model is appropriate; however, here $2/\sqrt{119} = 0.183$ and the lag 19 partial is just significant. One significant result among the 19 lags $2, 3, \ldots, 20$ is not

EXHIBIT 6.6 Sample Partial Autocorrelation Function for an AR(1) Series with $\phi = 0.4$

```
PACF of AR(1)

        -1.0 -0.8 -0.6 -0.4 -0.2  0.0  0.2  0.4  0.6  0.8  1.0
        +----+----+----+----+----+----+----+----+----+----+
   1   0.409                         XXXXXXXXXX
   2  -0.162                         XXXXX
   3   0.015                         X
   4  -0.078                         XXX
   5  -0.105                         XXXX
   6   0.000                         X
   7  -0.035                         XX
   8  -0.020                         X
   9   0.005                         X
  10  -0.064                         XXX
  11   0.056                         XX
  12  -0.176                         XXXXX
  13   0.058                         XX
  14  -0.038                         XX
  15  -0.151                         XXXXX
  16  -0.037                         XX
  17   0.014                         X
  18   0.021                         XX
  19   0.188                         XXXXXX
  20  -0.105                         XXXX
```

EXHIBIT 6.7 Sample Autocorrelation Function for an AR(1) Series with $\phi = -0.7$

```
ACF of AR(1)

                 -1.0 -0.8 -0.6 -0.4 -0.2  0.0  0.2  0.4  0.6  0.8  1.0
                 +----+----+----+----+----+----+----+----+----+----+
    1  -0.731        XXXXXXXXXXXXXXXXXX
    2   0.514                         XXXXXXXXXXXXXX
    3  -0.394             XXXXXXXXXXX
    4   0.313                         XXXXXXXXX
    5  -0.233                XXXXXX
    6   0.185                         XXXXX
    7  -0.194                XXXXX
    8   0.249                         XXXXXXX
    9  -0.223                XXXXXX
   10   0.188                         XXXXX
   11  -0.237                XXXXXXX
   12   0.237                         XXXXXXX
   13  -0.199                XXXXX
   14   0.121                         XXXX
   15  -0.092                 XXX
   16   0.046                         XX
   17   0.015                         X
   18  -0.037                 XX
   19   0.045                         XX
   20  -0.032                 XX
```

surprising, and we would certainly want to consider an AR(1) model for these data.

Exhibits 6.7 and 6.8 give similar output for an AR(1) simulation with $\phi = -0.7$ and $n = 119$. Here we see a definite oscillation in the ACF, but the decay in magnitude is not as fast as predicted by theory. However, the PACF again indicates an AR(1) model but with a negative ϕ this time.

EXHIBIT 6.8 Sample Partial Autocorrelation Function for AR(1) Series with $\phi = -0.7$

```
PACF of AR(1)

                 -1.0 -0.8 -0.6 -0.4 -0.2  0.0  0.2  0.4  0.6  0.8  1.0
                 +----+----+----+----+----+----+----+----+----+----+
    1  -0.731        XXXXXXXXXXXXXXXXXX
    2  -0.044                      XX
    3  -0.073                     XXX
    4   0.018                         X
    5   0.024                         XX
    6   0.028                         XX
    7  -0.096                    XXX
    8   0.138                         XXXX
    9   0.075                         XXX
   10   0.010                         X
   11  -0.163                   XXXXX
   12  -0.039                      XX
   13   0.010                         X
   14  -0.094                    XXX
   15  -0.025                      XX
   16  -0.093                    XXX
   17   0.053                         XX
   18   0.035                         XX
   19   0.081                         XXX
   20   0.022                         XX
```

EXHIBIT 6.9 Sample Autocorrelation Function for an AR(2) Series with $\phi_1 = 1.5$ and $\phi_2 = -0.75$

```
ACF of AR(2)

            -1.0 -0.8 -0.6 -0.4 -0.2  0.0  0.2  0.4  0.6  0.8  1.0
            +----+----+----+----+----+----+----+----+----+----+
    1   0.851                              XXXXXXXXXXXXXXXXXXXXXX
    2   0.516                              XXXXXXXXXXXXX
    3   0.104                              XXXX
    4  -0.268                       XXXXXXX
    5  -0.506                 XXXXXXXXXXXXX
    6  -0.574                 XXXXXXXXXXXXXX
    7  -0.468                 XXXXXXXXXXXX
    8  -0.243                       XXXXXX
    9   0.024                            XX
   10   0.231                              XXXXXX
   11   0.341                              XXXXXXXXX
   12   0.315                              XXXXXXXXX
   13   0.168                              XXXXX
   14  -0.043                           XX
   15  -0.257                       XXXXXX
   16  -0.402                 XXXXXXXXXX
   17  -0.449                 XXXXXXXXXXXX
   18  -0.378                 XXXXXXXXXX
   19  -0.214                       XXXXX
   20   0.003                            X
```

Similar results can be shown for AR(2) models. Exhibits 6.9 and 6.10 give the ACF and PACF for an AR(2) simulation with $\phi_1 = 1.5$, $\phi_2 = -0.75$, and $n = 119$. The theoretical ACF was given in Figure 4.2(c) and exhibited a damped sine wave decay with period 12 and damping factor 0.866. The sample ACF also oscillates with a period of about 11 or 12 but does not damp out nearly so

EXHIBIT 6.10 Sample Partial Autocorrelation Function for an AR(2) Series with $\phi_1 = 1.5$ and $\phi_2 = -0.75$

```
PACF of AR(2)

            -1.0 -0.8 -0.6 -0.4 -0.2  0.0  0.2  0.4  0.6  0.8  1.0
            +----+----+----+----+----+----+----+----+----+----+
    1   0.851                              XXXXXXXXXXXXXXXXXXXXXX
    2  -0.751           XXXXXXXXXXXXXXXXXXXX
    3  -0.213                       XXXXXX
    4  -0.021                           XX
    5   0.079                            XXX
    6  -0.040                           XX
    7   0.131                            XXXX
    8   0.005                            X
    9   0.017                            X
   10  -0.215                       XXXXX
   11   0.157                            XXXXX
   12  -0.230                       XXXXXX
   13  -0.071                          XXX
   14  -0.077                          XXX
   15  -0.041                           XX
   16  -0.069                          XXX
   17  -0.103                         XXXX
   18   0.050                            XX
   19  -0.006                            X
   20   0.020                            X
```

EXHIBIT 6.11 Sample Autocorrelation Function for an MA(1) Series with $\theta = 0.9$

```
ACF of MA(1)

              -1.0 -0.8 -0.6 -0.4 -0.2  0.0  0.2  0.4  0.6  0.8  1.0
              +----+----+----+----+----+----+----+----+----+----+
     1 -0.519                XXXXXXXXXXXXX
     2 -0.043                          XX
     3  0.130                          XXXX
     4 -0.093                         XXX
     5 -0.014                          X
     6  0.098                          XXX
     7 -0.103                        XXXX
     8  0.028                          XX
     9  0.098                          XXX
    10 -0.199                     XXXXX
    11  0.249                          XXXXXXX
    12 -0.097                        XXX
    13 -0.046                         XX
    14 -0.077                        XXX
    15  0.155                          XXXXX
    16 -0.118                       XXXX
    17  0.038                          XX
    18  0.101                          XXXX
    19 -0.142                       XXXXX
    20  0.025                          XX
```

quickly. However, the PACF still gives a fairly clear indication of an AR(2) model. The values of the PACF at lags 3, 10, and 12 exceed the limits $\pm 2/\sqrt{119} = \pm 0.183$, and we would also consider an AR(3) model. Unless there are substantial reasons to justify a very high order model, we would likely discount the somewhat high values at lags 10 and 12. However, for monthly data we

EXHIBIT 6.12 Sample Partial Autocorrelation Function for an MA(1) Series with $\theta = 0.9$

```
PACF of MA(1)

              -1.0 -0.8 -0.6 -0.4 -0.2  0.0  0.2  0.4  0.6  0.8  1.0
              +----+----+----+----+----+----+----+----+----+----+
     1 -0.519                XXXXXXXXXXXXX
     2 -0.427                XXXXXXXXXXXX
     3 -0.206                    XXXXX
     4 -0.204                    XXXXX
     5 -0.228                  XXXXXXX
     6 -0.101                      XXXX
     7 -0.154                     XXXXX
     8 -0.156                     XXXXX
     9 -0.001                          X
    10 -0.200                     XXXXX
    11  0.088                          XXX
    12  0.127                          XXXX
    13  0.170                          XXXXX
    14 -0.106                       XXXX
    15  0.002                          X
    16 -0.082                        XXX
    17 -0.137                       XXXX
    18  0.024                          XX
    19 -0.049                         XX
    20 -0.164                       XXXXX
```

would carefully consider correlations at lags $12, 24, 36, \ldots$ to check for seasonal effects.

Consider now some moving average simulations. Exhibit 6.11 is the sample ACF of an MA(1) series with $\theta = 0.9$ and $n = 120$. Here $\rho_1 = -0.497$, and from Table 6.2 the standard deviation of r_1 is about $0.71/\sqrt{120} = 0.06$. Thus we would be 95% confident that r_1 estimates ρ_1 to within ± 0.12. In fact, $r_1 = -0.519$, which is well within expectations. For lags greater than 1, Table 6.2 gives the standard deviation of r_k as $1.22/\sqrt{120} = 0.11$; thus the estimates are expected to lie within ± 0.22. Because only $r_{11} = 0.249$ exceeds these limits of the 19 values given, the data certainly suggest the MA(1) model.

Exhibit 6.12 shows the sample PACF for this series. Note the approximate exponential decay as suggested by Equation (6-18), which reinforces our MA(1) specification.

Lastly, we consider a mixed model—an ARMA(1, 1) with $\phi = 0.8$, $\theta = 0.4$, and $n = 99$. From Section 4.4 we have that $\rho_1 = 0.523$ and then $\rho_k = 0.8\rho_{k-1}$ for $k > 1$. Furthermore, because of the moving average term in the model, it can be shown that the theoretical PACF also decays (essentially) exponentially but never becomes zero. Hence looking for a cutoff in either the ACF or PACF will be fruitless for mixed models. Exhibits 6.13 and 6.14 display the sample ACF and PACF of the simulated ARMA(1, 1) series. Neither function is especially dramatic, and considerable experience and skill may be needed to specify mixed models. However, we reemphasize that we are only *tentatively* specifying models and may need several iterations of specification, estimation, and diagnostic checking to reach our final mixed model. (See also Section 6.6.)

EXHIBIT 6.13 Sample Autocorrelation Function for an ARMA(1, 1) Series with $\phi = 0.8$ and $\theta = 0.4$

```
ACF of ARMA11

          -1.0 -0.8 -0.6 -0.4 -0.2  0.0  0.2  0.4  0.6  0.8  1.0
          +----+----+----+----+----+----+----+----+----+----+
    1    0.542                         XXXXXXXXXXXXXX
    2    0.370                         XXXXXXXXX
    3    0.371                         XXXXXXXXX
    4    0.310                         XXXXXXXXX
    5    0.267                         XXXXXXXX
    6    0.103                         XXXX
    7    0.014                         X
    8    0.009                         X
    9   -0.011                         X
   10   -0.140                      XXXX
   11   -0.124                      XXXX
   12   -0.161                     XXXXX
   13   -0.107                      XXXX
   14   -0.099                       XXX
   15   -0.174                     XXXXX
   16   -0.138                      XXXX
   17   -0.095                       XXX
   18   -0.081                       XXX
   19   -0.126                      XXXX
```

EXHIBIT 6.14 Sample Partial Autocorrelation Function for an ARMA(1, 1) Series with $\phi = 0.8$ and $\theta = 0.4$

```
PACF of ARMA11

              -1.0 -0.8 -0.6 -0.4 -0.2  0.0  0.2  0.4  0.6  0.8  1.0
              +----+----+----+----+----+----+----+----+----+----+
     1   0.542                         XXXXXXXXXXXXXX
     2   0.108                         XXXX
     3   0.192                         XXXXXX
     4   0.043                         XX
     5   0.050                         XX
     6  -0.175                     XXXXX
     7  -0.091                       XXX
     8  -0.024                        XX
     9   0.004                         X
    10  -0.154                     XXXXX
    11   0.055                         XX
    12  -0.089                       XXX
    13   0.094                         XXX
    14  -0.023                        XX
    15  -0.060                        XX
    16  -0.022                        XX
    17   0.017                         X
    18  -0.020                        XX
    19  -0.071                       XXX
```

6.4 NONSTATIONARITY

As indicated in Chapter 5, many series exhibit nonstationarity that can be explained by integrated ARMA models. The nonstationarity will frequently be apparent in the time plot of the series. A review of Exhibits 5.1, 5.2, and 5.5 is in order here.

The sample ACF computed for a nonstationary series will also usually indicate nonstationarity. The definition of the sample ACF *assumes* a stationary series; for example, we use lagged products of deviations from the grand mean $\bar{Z}$ in the numerator, and the denominator assumes a constant variance. Thus it is not at all clear what the sample ACF is estimating for a nonstationary series. Nevertheless, the sample ACF typically fails to die out rapidly for nonstationary series. The values of r_k need not be high even for low lags but frequently are.

Exhibit 6.15 gives the sample ACF for the IMA(1, 1) series first shown in Exhibit 5.1. After taking first differences of the series, the ACF is again computed and displayed in Exhibit 6.16. In addition, we plot ∇Z_t versus time to have a visual check of stationarity. At this point, we clearly would specify an IMA(1, 1) model for this series.

If the first difference of the series and its sample ACF do not appear to support a stationary ARMA model, then we would take another difference and again compute the ACF to look for the characteristics of a stationary ARMA process. Usually one or two differences, perhaps combined with a logarithm transformation, will accomplish this reduction to stationarity. Additional properties of the sample ACF computed on nonstationary data are given in

EXHIBIT 6.15 Sample Autocorrelation Function for an IMA(1, 1) Series with $\theta = 0.4$

```
ACF of IMA(1,1)

          -1.0 -0.8 -0.6 -0.4 -0.2  0.0  0.2  0.4  0.6  0.8  1.0
           +----+----+----+----+----+----+----+----+----+----+
  1  0.575                          XXXXXXXXXXXXXX
  2  0.391                          XXXXXXXXXX
  3  0.309                          XXXXXXXX
  4  0.193                          XXXXX
  5  0.167                          XXXX
  6  0.147                          XXXX
  7  0.142                          XXXX
  8  0.165                          XXXX
  9  0.199                          XXXXX
 10  0.223                          XXXXXX
 11  0.244                          XXXXXX
 12  0.238                          XXXXXX
 13  0.190                          XXXXX
 14  0.092                          XXX
 15 -0.040                        XX
 16 -0.043                        XX
 17 -0.034                        XX
```

EXHIBIT 6.16 Sample Autocorrelation Function for the First Difference of an IMA(1, 1) Series with $\theta = 0.4$

```
ACF of FirstDif

          -1.0 -0.8 -0.6 -0.4 -0.2  0.0  0.2  0.4  0.6  0.8  1.0
           +----+----+----+----+----+----+----+----+----+----+
  1 -0.291                 XXXXXXX
  2 -0.128                    XXXX
  3  0.051                       XX
  4 -0.105                    XXXX
  5 -0.011                     X
  6 -0.024                    XX
  7 -0.026                    XX
  8 -0.017                     X
  9  0.017                      X
 10 -0.002                     X
 11  0.029                       XX
 12  0.048                       XX
 13  0.063                       XXX
 14  0.028                       XX
 15 -0.150                  XXXXX
 16 -0.001                     X
 17  0.177                       XXXXX
```

Wichern (1973), Roy (1977), and Hasza (1980). See also Box and Jenkins (1976, pp. 200–201).

OVERDIFFERENCING

From Exercise 2.6 we know that the difference of a stationary process is also stationary. However, overdifferencing tends to introduce unnecessary correlations into the model and may complicate a relatively simple model.

For example, suppose our observed series is, in fact, a random walk so that

$$W_t = Z_t - Z_{t-1} = a_t$$

If we unnecessarily difference one more time, we get

$$\nabla W_t = a_t - a_{t-1}$$

which is an MA(1) model with $\theta = 1$. With an observed series we would unnecessarily have to estimate the unknown value of θ. (Overdifferencing also creates a noninvertible model; see Chapter 4.)

6.5 SPECIFICATION OF SOME ACTUAL TIME SERIES

Consider now the specification of models for some actual time series. Return first to the U.S. quarterly rate of unemployment of Chapter 1. The series is plotted in Exhibit 1.1. The plot exhibits slow changes over time, and we expect high positive correlations at low lags. This is borne out in the sample ACF given in Exhibit 6.17, which suggests approximate exponential decay. The PACF of Exhibit 6.18 clearly suggests an AR(2) model. Here $n = 121$ and $\pm 2/\sqrt{n} = \pm 0.18$; thus none of the PACF values are significantly different from zero for lags beyond 2.

With such a strong lag 1 correlation, we would also want to consider a nonstationary model with $d = 1$, but AR(2) appears to be our first choice.

The time plot of railroad bond yields in Exhibit 5.2 strongly indicates a

EXHIBIT 6.17 Sample Autocorrelation Function for the U.S. Unemployment Rate Series

```
ACF of Unemp

       -1.0 -0.8 -0.6 -0.4 -0.2  0.0  0.2  0.4  0.6  0.8  1.0
       +----+----+----+----+----+----+----+----+----+----+
  1  0.936                          XXXXXXXXXXXXXXXXXXXXXXXX
  2  0.802                          XXXXXXXXXXXXXXXXXXXXX
  3  0.646                          XXXXXXXXXXXXXXXX
  4  0.502                          XXXXXXXXXXXXX
  5  0.392                          XXXXXXXXXX
  6  0.317                          XXXXXXXX
  7  0.263                          XXXXXXX
  8  0.223                          XXXXXXX
  9  0.197                          XXXXX
 10  0.172                          XXXXX
 11  0.144                          XXXX
 12  0.115                          XXXX
 13  0.098                          XXX
 14  0.095                          XXX
 15  0.109                          XXXX
 16  0.127                          XXXX
 17  0.126                          XXXX
 18  0.099                          XXX
 19  0.046                          XX
 20 -0.025                         XX
 21 -0.102                        XXXX
```

EXHIBIT 6.18 Sample Partial Autocorrelation Function for the U.S. Unemployment Rate Series

```
PACF of Unemp

            -1.0 -0.8 -0.6 -0.4 -0.2  0.0  0.2  0.4  0.6  0.8  1.0
            +----+----+----+----+----+----+----+----+----+----+
   1   0.936                          XXXXXXXXXXXXXXXXXXXXXXXX
   2  -0.607            XXXXXXXXXXXXXXX
   3   0.101                          XXXX
   4   0.052                          XX
   5   0.117                          XXXX
   6  -0.034                          XX
   7  -0.046                          XX
   8   0.021                          XX
   9   0.097                          XXX
  10  -0.140                        XXXX
  11   0.021                          XX
  12   0.003                          X
  13   0.159                          XXXXX
  14  -0.020                          XX
  15   0.051                          XX
  16  -0.088                        XXX
  17  -0.136                        XXXX
  18  -0.044                          XX
  19  -0.070                        XXX
  20  -0.070                        XXX
  21  -0.051                          XX
```

nonstationary model. The sample ACF for this series is given in Exhibit 6.19, and its appearance strengthens the suggestion of nonstationarity. The ACF dies out quite slowly and more linearly than exponentially.

The time plot of the first difference of the yields, given in Exhibit 5.4, is much more nearly stationary. Exhibits 6.20 and 6.21 show the ACF and PACF of

EXHIBIT 6.19 Sample Autocorrelation Function for the Railroad Bond Yield Series

```
ACF of RRBonds

            -1.0 -0.8 -0.6 -0.4 -0.2  0.0  0.2  0.4  0.6  0.8  1.0
            +----+----+----+----+----+----+----+----+----+----+
   1   0.963                          XXXXXXXXXXXXXXXXXXXXXXXXX
   2   0.916                          XXXXXXXXXXXXXXXXXXXXXXXX
   3   0.863                          XXXXXXXXXXXXXXXXXXXXXXX
   4   0.811                          XXXXXXXXXXXXXXXXXXXXXX
   5   0.759                          XXXXXXXXXXXXXXXXXXXX
   6   0.704                          XXXXXXXXXXXXXXXXXX
   7   0.645                          XXXXXXXXXXXXXXXXX
   8   0.580                          XXXXXXXXXXXXXXX
   9   0.511                          XXXXXXXXXXXXX
  10   0.437                          XXXXXXXXXXX
  11   0.361                          XXXXXXXXX
  12   0.288                          XXXXXXX
  13   0.212                          XXXXX
  14   0.137                          XXXX
  15   0.069                          XXX
  16   0.006                          X
  17  -0.051                          XX
  18  -0.105                        XXXX
  19  -0.155                        XXXXX
  20  -0.203                        XXXXX
```

EXHIBIT 6.20 Sample Autocorrelation Function for the First Difference of the Railroad Bond Yield Series

```
ACF of DifBonds

              -1.0 -0.8 -0.6 -0.4 -0.2  0.0  0.2  0.4  0.6  0.8  1.0
               +----+----+----+----+----+----+----+----+----+----+
    1   0.468                           XXXXXXXXXXXXX
    2   0.168                           XXXXX
    3   0.114                           XXXX
    4   0.093                           XXX
    5   0.118                           XXXX
    6   0.094                           XXX
    7   0.043                           XX
    8   0.089                           XXX
    9   0.202                           XXXXXX
   10   0.126                           XXXX
   11   0.083                           XXX
   12   0.041                           XX
   13  -0.055                        XX
   14  -0.067                       XXX
   15  -0.130                      XXXX
   16  -0.199                     XXXXX
   17  -0.098                       XXX
   18  -0.048                        XX
   19  -0.045                        XX
   20  -0.035                        XX
```

the first difference of yields and suggest an AR(1) model. Our sample size is $n = 102$ and so $\pm 2/\sqrt{n} = \pm 0.20$, and no significant partials occur after lag 1. The model specified for the original series is thus ARI(1, 1).

EXHIBIT 6.21 Sample Partial Autocorrelation Function for the First Difference of the Railroad Bond Yield Series

```
PACF of DifBonds

              -1.0 -0.8 -0.6 -0.4 -0.2  0.0  0.2  0.4  0.6  0.8  1.0
               +----+----+----+----+----+----+----+----+----+----+
    1   0.468                           XXXXXXXXXXXXX
    2  -0.064                        XXX
    3   0.078                           XXX
    4   0.022                           XX
    5   0.080                           XXX
    6   0.002                           X
    7  -0.014                          X
    8   0.087                           XXX
    9   0.154                           XXXXX
   10  -0.056                        XX
   11   0.035                           XX
   12  -0.030                        XX
   13  -0.105                       XXXX
   14  -0.040                        XX
   15  -0.129                      XXXX
   16  -0.118                      XXXX
   17   0.048                          XX
   18  -0.039                        XX
   19   0.005                          X
   20  -0.003                         X
```

6.6 OTHER SPECIFICATION METHODS

A number of other approaches to model specification have been investigated since Box and Jenkin's seminal work. One of the most studied is Akaike's (1973, 1974) information criteria, **AIC**. Here we select the model that minimizes

$$\text{AIC} = -2 \log (\text{maximum likelihood}) + 2k \qquad \text{(6-22)}$$

where k is the total number of AR and MA parameters in the model. Maximum likelihood estimation is discussed in Section 7.3. The addition of the term $2k$ serves as a "penalty function" to avoid consideration of models with too many parameters and to help ensure the selection of parsimonious models. Hannan and Rissanen (1982) investigate a related idea.

Parzen (1974) introduced his so-called criterion autoregressive transfer function, **CAT,** to be optimized. See also Parzen (1981). Cleveland (1972) defined the concept of **inverse autocorrelations** to aid in specifying models, but the recent work of Abraham and Ledolter (1984) seems to reveal substantial weaknesses in this approach.

Tiao and Tsay (1981) have recently defined and investigated **extended autocorrelations** (see also Tsay and Tiao, 1983a, 1983b). Another approach using **S** and **R arrays** has been advocated by Gray, Kelley, and McIntire (1978). See also Woodward and Gray (1981).

Generally speaking, it is our opinion that model specification is always *tentative*. Having chosen a model for consideration, estimate the parameters of that model (Chapter 7) as efficiently as possible and then consider the implications of that model (Chapter 8). If the model is inappropriate for some reason, we should detect that fact at this stage in the modeling and respecify the model in the direction suggested.

CHAPTER 6 EXERCISES

6.1. From a series of 100 observations, we calculate $r_1 = -0.49$, $r_2 = 0.31$, $r_3 = -0.21$, $r_4 = 0.11$, and $|r_k| \leq 0.09$ for $k > 4$. On the basis of this information alone, what ARIMA model would we tentatively specify for the series?

6.2. A stationary time series of length 121 produced sample partial autocorrelations of $\hat{\phi}_{11} = 0.8$, $\hat{\phi}_{22} = -0.6$, $\hat{\phi}_{33} = 0.08$, and $\hat{\phi}_{44} = 0.00$. Based on this information alone, what model would we tentatively specify for the series?

6.3. For a series of length 169, we find that $r_1 = 0.41$, $r_2 = 0.32$, $r_3 = 0.26$, $r_4 = 0.21$, and $r_5 = 0.16$. What ARIMA model fits this pattern of autocorrelations?

6.4. The sample ACFs for a series and its first difference are given in the following table ($n = 100$):

ACF	Lag 1	2	3	4	5	6
Z_t	0.97	0.97	0.93	0.85	0.80	0.71
∇Z_t	−0.42	0.18	−0.02	0.07	−0.10	−0.09

Based on this information, which ARIMA model(s) would you consider for the series?

6.5. For a series of length 64, the sample partial autocorrelation function is given by:

Lag 1	2	3	4	5	6
0.47	−0.34	0.20	0.02	0.15	−0.06

Which model(s) should be considered in this case?

6.6. Consider an AR(1) series of length 100 with $\phi = 0.7$.
a. Would you be surprised if $r_1 = 0.6$?
b. Would $r_{10} = -0.15$ be unusual?

6.7. Suppose that $\{X_t\}$ is a stationary AR(1) process with parameter ϕ but that we can only observe

$$Z_t = X_t + N_t$$

where $\{N_t\}$ is a white noise measurement error assumed to be independent of $\{X_t\}$.

a. Find the autocorrelation function for the observed process in terms of ϕ, σ_x^2, and σ_N^2.
b. Which ARMA model might we specify for $\{Z_t\}$?

6.8. The time plots of two time series are shown below.
a. For each of the series, decide whether r_1 could best be described as: strongly positive, moderately positive, near zero, moderately negative, or strongly negative. Or do you need to know the scale of measurement?
b. Repeat (a) for r_2.

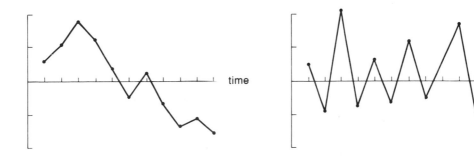

6.9. Use Minitab to specify a tentative model for the Portland, Oregon, gasoline price series given in Appendix K, "Data Sets." Is a log transformation needed?

6.10. Consider the Iowa nonfarm income time series given in Appendix K. Specify a tentative ARIMA model for this series. (Do not fail to look at second differences.)

CHAPTER 7 PARAMETER ESTIMATION

This chapter deals with the problem of estimating the parameters of an ARIMA model based on the observed time series $Z_1, Z_2, \ldots, Z_n$. We assume that the model has already been (tentatively) specified; that is, we have specified values for p, d, and q using the methods of Chapter 6. With regard to non-stationarity, since the dth difference of the observed series is assumed to be a stationary ARMA(p, q) process, we need only concern ourselves with the problem of estimating the parameters in such stationary models. In practice, then, we (or the computer) treat the dth difference of the original time series as the time series from which we estimate the parameters of the complete model. For simplicity, we shall let $Z_1, Z_2, \ldots, Z_n$ denote our observed *stationary* process even though it may be an appropriate difference of the original series. We first discuss the method-of-moments estimators, then least-squares estimators, and finally full maximum likelihood estimators.

7.1 THE METHOD OF MOMENTS

The method of moments is frequently one of the easiest, if not the most efficient, methods for obtaining parameter estimates. The method consists of equating sample moments to theoretical moments and solving the resultant equations to obtain estimates of unknown parameters. The simplest example of the method is to estimate the series mean μ by the sample mean $\bar{Z}$. The properties of this estimator were considered extensively in Chapter 3.

AUTOREGRESSIVE MODELS

Consider first the AR(1) case. For this model we have the simple relationship $\rho_1 = \phi$. In the method of moments, ρ_1 is equated to r_1, the lag 1 sample autocorrelation. Thus we can estimate ϕ by simply

$$\hat{\phi} = r_1 \tag{7-1}$$

Now consider the AR(2) case. The relationship between the parameters ϕ_1 and ϕ_2 and various moments is given by the Yule–Walker equations (4-33):

$$\rho_1 = \phi_1 + \rho_1\phi_2 \quad \text{and} \quad \rho_2 = \rho_1\phi_1 + \phi_2$$

The method of moments replaces ρ_1 by r_1 and ρ_2 by r_2 to obtain

$$r_1 = \phi_1 + r_1\phi_2 \quad \text{and} \quad r_2 = r_1\phi_1 + \phi_2$$

which are then solved to obtain $\hat{\phi}_1$ and $\hat{\phi}_2$:

$$\hat{\phi}_1 = \frac{r_1(1 - r_2)}{1 - r_1^2} \tag{7-2}$$

$$\hat{\phi}_2 = \frac{r_2 - r_1^2}{1 - r_1^2} \tag{7-3}$$

The general AR(p) case proceeds similarly: Replace ρ_k by r_k in the Yule–Walker equations to obtain

$$
\begin{aligned}
r_1 &= \phi_1 + r_1\phi_2 + \cdots + r_{p-1}\phi_p \\
r_2 &= r_1\phi_1 + \phi_2 + \cdots + r_{p-2}\phi_p \\
&\ \ \vdots \\
r_p &= r_{p-1}\phi_1 + r_{p-2}\phi_2 + \cdots + \phi_p
\end{aligned}
\tag{7-4}
$$

These linear equations are then solved for $\hat{\phi}_1, \hat{\phi}_2, \ldots, \hat{\phi}_p$ in terms of $r_1, r_2, \ldots, r_p$. The Durbin–Levinson recursion of Equation (6-21) provides a convenient method of solution but is subject to substantial round-off errors if the solution is close to the stationarity boundary.

MOVING AVERAGE MODELS

Surprisingly, the method of moments is not nearly as convenient when applied to moving average models. Consider the MA(1) process. From Equations (4-6), we know that

$$\rho_1 = -\frac{\theta}{1 + \theta^2}$$

Equating ρ_1 to r_1, we are led to solve a quadratic equation in θ. If $|r_1| < .5$; then the two real roots are given by

$$-\frac{1}{2r_1} \pm \left[\frac{1}{2r_1^2} - 1\right]^{1/2}$$

As can easily be checked, the product of the two solutions is always equal to 1; therefore, only one of the solutions satisfies the invertibility condition $|\theta| < 1$.

After further algebraic manipulation, we can see that the invertible solution can be written as

$$\hat{\theta} = \frac{-1 + [1 - 4r_1^2]^{1/2}}{2r_1} \tag{7-5}$$

If $r_1 = \pm .5$, unique, real solutions exist, namely ± 1, but neither is invertible. If $|r_1| > .5$ (which is possible even though $|\rho_1| < .5$), no real solutions exist, and so the method of moments fails to yield an estimator of θ. Of course, if $|r_1| > .5$, the specification of an MA(1) model would be in some doubt.

For higher-order MA(q) models, the method of moments quickly gets complicated. We can use Equation (4-8) for lags $1, 2, \ldots, q$, replacing ρ_k by r_k to obtain q equations in q unknowns $\theta_1, \theta_2, \ldots, \theta_q$. The resulting equations are highly nonlinear in the θ's, however, and their solution would of necessity be numerical. In addition, there would be multiple solutions, of which only one would be invertible. We shall not pursue this further, since we shall see in Section 7.4 that for MA models, method-of-moments estimators are not particularly good.

MIXED MODELS

We consider only the ARMA(1, 1) case. Recall Equation (4-39):

$$\rho_k = \frac{(1 - \theta\phi)(\phi - \theta)}{1 - 2\phi\theta + \theta^2} \phi^{k-1}$$

Noting that $\rho_2/\rho_1 = \phi$, we can first estimate ϕ as

$$\hat{\phi} = \frac{r_2}{r_1} \tag{7-6}$$

Having done so, we can then use

$$r_1 = \frac{(1 - \theta\hat{\phi})(\hat{\phi} - \theta)}{1 - 2\hat{\phi}\theta + \theta^2} \tag{7-7}$$

to solve for $\hat{\theta}$. Note again that a quadratic equation must be solved and only the invertible solution retained.

ESTIMATES OF THE NOISE VARIANCE

The final parameter to be estimated is the noise variance σ_a^2. In all cases, we can first estimate $\gamma_0 = \text{Var}(Z_t)$ by the sample variance

$$S^2 = \frac{\sum_{i=1}^{n} (Z_t - \bar{Z})^2}{n - 1} \tag{7-8}$$

and use known relationships from Chapter 4 among γ_0, σ_a^2, and the θ's and ϕ's to estimate σ_a^2.

For the AR(p) models, Equation (4-34) yields

$$\hat{\sigma}_a^2 = (1 - \hat{\phi}_1 r_1 - \hat{\phi}_2 r_2 - \cdots - \hat{\phi}_p r_p) S^2 \tag{7-9}$$

In particular, for an AR(1) process,

$$\hat{\sigma}_a^2 = [1 - r_1^2] S^2$$

since $\hat{\phi}_1 = r_1$.

For the MA(q) case, we have, using Equation (4-9),

$$\hat{\sigma}_a^2 = \frac{S^2}{1 + \hat{\theta}_1^2 + \hat{\theta}_2^2 + \cdots + \hat{\theta}_q^2}$$

(7-10)

For the ARMA(1, 1) process, Equation (4-38) yields

$$\hat{\sigma}_a^2 = \frac{(1 - \hat{\phi}^2)}{1 - 2\hat{\theta}\hat{\phi} + \hat{\theta}^2} S^2$$

(7-11)

NUMERICAL EXAMPLES

Table 7.1 presents the results of using the method of moments to estimate the parameters of the simulated series first discussed in Sections 6.3 and 6.4. Notice that the estimates are quite good in all of the autoregressive models but not in the moving average and mixed models. It can be shown that theory confirms this observation—method-of-moments estimators are not very efficient for models containing moving average terms.

Consider now some actual time series. In Section 6.5 we investigated the quarterly U.S. unemployment rate series and were led to specifying an AR(2) model. From Exhibit 6.17, we read $r_1 = 0.936$ and $r_2 = 0.802$. Thus from Equations (7-2) and (7-3) we have the estimates

$$\hat{\phi}_1 = \frac{0.936(1 - 0.802)}{1 - (0.936)^2} = 1.50$$

(7-12)

and

$$\hat{\phi}_2 = \frac{0.802 - (0.936)^2}{1 - (0.936)^2} = -0.598$$

(7-13)

The sample mean and variance of this series are found to be 5.1069 and

TABLE 7.1   Parameter Estimates for Some Simulated Series

Model	True Parameters			Method-of-Moments Estimates			Sample Size
	ϕ_1	ϕ_2	θ	ϕ_1	ϕ_2	θ	
AR(1)	0.9			0.855			59
AR(1)	0.4			0.409			119
AR(1)	−0.7			−0.731			119
AR(2)	1.5	−0.75		1.49	−0.75		119
MA(1)			0.9			*	120
ARMA(1, 1)	0.8		0.4	0.683		0.202	99
IMA(1, 1)			0.4			0.321	50

Note: The simulated series used were described in Sections 6.3 and 6.4.
* The method-of-moments estimate for the MA(1) simulation *does not exist* since $r_1 = -0.519$.

1.99487, respectively. Using Equations (7-9), (7-12), and (7-13), we then obtain

$$\hat{\sigma}_a = 0.388 \qquad \text{(7-14)}$$

so that the estimated model for the unemployment rate series is given as

$$Z_t - 5.1069 = 1.50(Z_{t-1} - 5.1069) - 0.598(Z_{t-2} - 5.1069) + a_t \qquad \text{(7-15)}$$

where $\{a_t\}$ is zero mean white noise with standard deviation 0.388. Alternatively, in constant-term form, we can write

$$Z_t = 1.50Z_{t-1} - 0.598Z_{t-2} + 0.5005 + a_t \qquad \text{(7-16)}$$

For the railroad bond series, we specified an ARI(1, 1) model based on Exhibits 6.19, 6.20, and 6.21. Since the lag 1 sample ACF calculated from the differenced series is 0.468, we would estimate

$$\hat{\phi} = 0.468 \qquad \text{(7-17)}$$

for this series.

For the differenced series, the mean and variance are 1.7129 and 108.16, respectively. Using Equations (7-9) and (7-17), we have

$$\hat{\sigma}_a = 9.19 \qquad \text{(7-18)}$$

Is the mean of the differenced series, 1.7129, small enough that we should not include a constant term in our model? To assess the magnitude of this sample mean, we refer back to Equation (3-6), which gives the approximate variance of the sample mean. In our case we use the estimates $\hat{\gamma}_0 = 108.16$, $\hat{\phi} = 0.468$, and $n = 101$ (for the differenced series) to estimate the standard deviation of the sample mean of ∇Z_t as

$$\sqrt{\left(\frac{1 + 0.468}{1 - 0.468}\right)\left(\frac{108.16}{101}\right)} = 1.719 \qquad \text{(7-19)}$$

Thus our sample mean of 1.7129 is only about one standard deviation from zero; consequently, there is no indication that the population mean of the differenced series is not zero.

Thus our estimated model for the railroad bond series is

$$Z_t = 1.400Z_{t-1} - 0.468Z_{t-2} + a_t \qquad \text{(7-20)}$$

As we have noted before, this looks like an AR(2) model, but it is not stationary since one of the roots of the characteristic polynomial is equal to 1.

7.2 LEAST-SQUARES ESTIMATION

Because the method of moments is unsatisfactory for models with moving average terms, we must consider other methods of estimation. We begin with least squares. For autoregressive models, the ideas are quite straightforward.

AUTOREGRESSIVE MODELS

Consider the first-order case where

$$Z_t - \mu = \phi(Z_{t-1} - \mu) + a_t \tag{7-21}$$

We can view this as a regression model with predictor variable Z_{t-1} and response variable Z_t. Least-squares estimation then proceeds by minimizing the sum of squares of the differences

$$(Z_t - \mu) - \phi(Z_{t-1} - \mu)$$

Since only $Z_1, Z_2, \ldots, Z_n$ are observed, we can only sum from $t = 2$ to $t = n$. Let

$$S_*(\phi, \mu) = \sum_{t=2}^{n} [(Z_t - \mu) - \phi(Z_{t-1} - \mu)]^2 \tag{7-22}$$

This is usually called the **conditional sum-of-squares function.** (The reason for the use of the term *conditional* will become apparent later on.) According to the principle of least squares, we estimate ϕ and μ by the respective values that minimize $S_*(\phi, \mu)$ given the observed series $Z_1, Z_2, \ldots, Z_n$.

Consider the equation $\partial S_* / \partial \mu = 0$. We have

$$\frac{\partial S_*}{\partial \mu} = \sum_{t=2}^{n} 2[(Z_t - \mu) - \phi(Z_{t-1} - \mu)](-1 + \phi) = 0$$

or, simplifying and solving for μ,

$$\mu = \frac{\sum\limits_{t=2}^{n} Z_t - \phi \sum\limits_{t=2}^{n} Z_{t-1}}{(n - 1)(1 - \phi)} \tag{7-23}$$

Now for large n,

$$\sum_{t=2}^{n} \frac{Z_t}{n - 1} \approx \sum_{t=2}^{n} \frac{Z_{t-1}}{n - 1} \approx \bar{Z}$$

Thus, regardless of the value of ϕ, Equation (7-23) reduces to

$$\hat{\mu} \approx \frac{\bar{Z} - \phi\bar{Z}}{1 - \phi} = \bar{Z} \tag{7-24}$$

We sometimes say, except for end effects, $\hat{\mu} = \bar{Z}$.

Consider now the minimization of $S_*(\phi, \bar{Z})$ with respect to ϕ. We have

$$\frac{\partial S_*(\phi, \bar{Z})}{\partial \phi} = -\sum_{t=2}^{n} 2[(Z_t - \bar{Z}) - \phi(Z_{t-1} - \bar{Z})](Z_{t-1} - \bar{Z})$$

Setting this to zero and solving for ϕ yields

$$\hat{\phi} = \frac{\sum\limits_{t=2}^{n} (Z_t - \bar{Z})(Z_{t-1} - \bar{Z})}{\sum\limits_{t=2}^{n} (Z_{t-1} - \bar{Z})^2}$$

Except for one term missing in the denominator, namely $(Z_n - \bar{Z})^2$, this is the same as r_1; thus the least-squares and method-of-moments estimators are nearly identical, especially for large samples.

For the general AR(p) process, the methods used to obtain Equations (7-23) and (7-24) can easily be extended to yield the same result, namely

$$\hat{\mu} \approx \bar{Z} \qquad \text{(7-25)}$$

To generalize the estimation of the ϕ's, we consider the second-order model. In accordance with Equation (7-25), we replace μ by $\bar{Z}$ in the conditional sum-of-squares function. So

$$S_*(\phi_1, \phi_2) = \sum_{t=3}^{n} [(Z_t - \bar{Z}) - \phi_1(Z_{t-1} - \bar{Z}) - \phi_2(Z_{t-2} - \bar{Z})]^2 \qquad \text{(7-26)}$$

Setting $\partial S_*/\partial \phi_1 = 0$, we have

$$-2 \sum_{t=3}^{n} [(Z_t - \bar{Z}) - \phi_1(Z_{t-1} - \bar{Z}) - \phi_2(Z_{t-2} - \bar{Z})](Z_{t-1} - \bar{Z}) = 0 \qquad \text{(7-27)}$$

which we can rewrite as

$$\sum_{t=3}^{n} (Z_t - \bar{Z})(Z_{t-1} - \bar{Z}) = \sum_{t=3}^{n} (Z_{t-1} - \bar{Z})^2 \phi_1 + \sum_{t=3}^{n} (Z_{t-1} - \bar{Z})(Z_{t-2} - \bar{Z}) \phi_2 \qquad \text{(7-28)}$$

The sum of lagged products $\sum_{t=3}^{n} (Z_t - Z)(Z_{t-1} - Z)$ is very nearly the numerator of r_1—we are missing one product $(Z_2 - \bar{Z})(Z_1 - \bar{Z})$. A similar situation exists for $\sum_{t=3}^{n} (Z_{t-1} - \bar{Z})(Z_{t-2} - \bar{Z})$, but here we are missing $(Z_n - \bar{Z})(Z_{n-1} - \bar{Z})$. If we divide both sides of Equation (7-28) by $\sum_{t=3}^{n} (Z_t - Z)^2$, then, except for end effects, we obtain

$$r_1 = \phi_1 + r_1 \phi_2 \qquad \text{(7-29)}$$

Approximating in a similar way with the equation $\partial S_*/\partial \phi_2 = 0$ leads to

$$r_2 = r_1 \phi_1 + \phi_2 \qquad \text{(7-30)}$$

But Equations (7-29) and (7-30) are just the sample Yule–Walker equations for an AR(2) model.

Entirely analogous results follow for the general AR(p) case. To an excellent approximation, the least-squares estimates of the ϕ's are obtained by solving the sample Yule–Walker equations (7-4).

Within Minitab the exact conditional least-squares estimates can easily be obtained using the LAG and REGRESS commands. For example, Exhibit 7.1 displays the appropriate commands and output for fitting an AR(2) model to the unemployment rate series. The estimated model is

$$Z_t = 1.55Z_{t-1} - 0.651Z_{t-2} + 0.505 + a_t \qquad \text{(7-31)}$$

with $\hat{\sigma}_a = 0.3632$. These results can be compared with those obtained using the method of moments, shown in Equations (7-14) and (7-16).

EXHIBIT 7.1 Fitting an AR(2) Model to the Unemployment Rate Series with Regression

```
MTB > lag c1 c20
MTB > lag c20 c21
MTB > name c20 'Lag1' c21 'Lag2'
MTB > brief
MTB > regress 'Unemp' on 2 variables: 'Lag1' and 'Lag2'

The regression equation is
Unemp = 0.505 + 1.55 Lag1 - 0.651 Lag2

119 cases used 2 cases contain missing values
```

Predictor	Coef	Stdev	t-ratio
Constant	0.5051	0.1267	3.99
Lag1	1.55368	0.07074	21.96
Lag2	-0.65148	0.07081	-9.20

```
s = 0.3632    R-sq = 93.499    R-sq(adj) = 93.387
```

Analysis of Variance

SOURCE	DF	SS	MS
Regression	2	220.05	110.02
Error	116	15.30	0.13
Total	118	235.35	

MOVING AVERAGE MODELS

Consider now the least-squares estimation of θ in the MA(1) model:

$$Z_t = a_t - \theta a_{t-1} \tag{7-32}$$

At first glance it is not apparent how a least-squares or regression method can be applied to such models. However, recall from Section 4.5 that invertible MA(1) models can be expressed as

$$Z_t = -\theta Z_{t-1} - \theta^2 Z_{t-2} - \cdots + a_t \tag{7-33}$$

that is, as an autoregressive model but of infinite order. Thus least squares can be meaningfully carried out by choosing a value for the parameter that minimizes

$$S_*(\theta) = \sum a_t^2 \tag{7-34}$$

where, implicitly, $a_t = a_t(\theta)$ is a function of the observed series and the parameter θ.

It is clear from Equation (7-33) that the least-squares problem is *nonlinear* in the parameters. Thus for even the simple MA(1) case, S_* cannot be minimized analytically, and we must resort to techniques of numerical optimization.

For a given observed series $Z_1, Z_2, \ldots, Z_n$ and a *particular value* of θ, consider the problem of evaluating $S_*(\theta)$. Equation (7-33) is not useful for this purpose, so we rewrite Equation (7-32) as

$$a_t = Z_t + \theta a_{t-1} \tag{7-35}$$

Using Equation (7-35), $a_1, a_2, \ldots, a_n$ can be calculated recursively if we have the initial value a_0. A common approximation is to set a_0 equal to its expectation of zero. Then, *conditional on* $a_0 = 0$, we can obtain

$$
\begin{aligned}
a_1 &= Z_1 \\
a_2 &= Z_2 + \theta a_1 \\
a_3 &= Z_3 + \theta a_2 \\
&\ \vdots \\
a_n &= Z_n + \theta a_{n-1}
\end{aligned} \tag{7-36}
$$

and thus calculate $S_*(\theta) = \sum_{t=1}^{n} a_t^2$, conditional on $a_0 = 0$ for that particular value of θ.

For the simple case of one parameter, we can carry out a grid search over the invertible range $(-1, 1)$ for θ to find the minimum sum of squares. For more general MA models, a numerical optimization algorithm, such as Gauss–Newton, is preferable to a grid search. The Gauss–Newton approach consists of approximating $a_t = a_t(\theta)$ by a *linear* function of θ around an initial estimate of θ, say θ^*. That is

$$a_t(\theta) \approx a_t(\theta^*) + (\theta - \theta^*)\frac{da_t(\theta^*)}{d\theta} \tag{7-37}$$

We note that $da_t(\theta)/d\theta$ can be computed recursively by differentiating both sides of Equation (7-35) to obtain

$$\frac{da_t(\theta)}{d\theta} = \frac{\theta da_{t-1}(\theta)}{d\theta} + a_{t-1}(\theta) \tag{7-38}$$

with initial value $da_0(\theta)/d\theta = 0$.

Since the approximation in (7-37) is linear in θ, the sum of squares computed from it can be minimized analytically to get a new (and, one hopes, improved) estimate of θ. This process can then be repeated with θ^* replaced by the new estimate, and we end the computation when either the change in the estimate of θ or the change in the sum of squares is sufficiently small. The method of moments can be used to obtain the first guess for θ, but in most cases the procedure quickly converges to the minimum sum of squares from an arbitrary initial estimate such as $\theta = 0.1$.

For higher-order moving average models, the ideas are analogous and no new difficulties arise. We compute $a_t = a_t(\theta_1, \theta_2, \ldots, \theta_q)$ recursively from

$$a_t = Z_t + \theta_1 a_{t-1} + \theta_2 a_{t-2} + \cdots + \theta_q a_{t-q} \tag{7-39}$$

with $a_0 = a_{-1} = \cdots = a_{1-q} = 0$. The sum of squares is minimized jointly in $\theta_1, \theta_2, \ldots, \theta_q$ using a multivariate Gauss–Newton algorithm. (See Box and Jenkins, 1976, pp. 231–238, for more details.)

MIXED MODELS

To begin our study of mixed models, consider the ARMA(1, 1) case:

$$Z_t = \phi Z_{t-1} + a_t - \theta a_{t-1} \tag{7-40}$$

As in the pure MA case, we consider $a_t = a_t(\phi, \theta)$ and wish to minimize $S_*(\phi, \theta) = \sum_{t=1}^{n} a_t^2$. We can rewrite Equation (7-40) as

$$a_t = Z_t - \phi Z_{t-1} + \theta a_{t-1} \tag{7-41}$$

To obtain a_1, we now have an additional "start-up" problem, namely Z_0. One approach is to set $Z_0 = 0$ (or to $\bar{Z}$ if our model includes a nonzero mean). However, a better approach is to begin the recursion at $t = 2$, thus avoiding Z_0, and simply minimize $\sum_{t=2}^{n} a_t^2$. The derivatives needed for the Gauss–Newton algorithm can again be obtained from recursions that follow from Equation (7-41).

For the general ARMA(p, q) model, we compute

$$a_t = a_t(\phi_1, \phi_2, \ldots, \phi_p, \theta_1, \theta_2, \ldots, \theta_q) \quad \text{for } t = p + 1, p + 2, \ldots, n$$

from

$$a_t = Z_t - \phi_1 Z_{t-1} - \phi_2 Z_{t-2} - \cdots - \phi_p Z_{t-p}$$
$$+ \theta_1 a_{t-1} + \theta_2 a_{t-2} + \cdots + \theta_q a_{t-q} \tag{7-42}$$

with $a_p = a_{p-1} = \cdots = a_{1-q} = 0$ and then numerically minimize $\sum a_t^2$ to obtain the conditional least-squares estimates of $\phi_1, \phi_2, \ldots, \phi_p, \theta_1, \ldots, \theta_q$. A nonzero constant term can be included in the model without difficulty.

For parameter sets $\theta_1, \theta_2, \ldots, \theta_q$ corresponding to invertible models, the start-up values $a_p = a_{p-1} = \cdots = a_{1-q} = 0$ will have very little influence on the final estimates of the parameters for large sample sizes.

7.3 MAXIMUM LIKELIHOOD AND UNCONDITIONAL LEAST SQUARES

For series of moderate length and also for the stochastic seasonal models to be discussed in Chapter 10, the start-up values $a_p = a_{p-1} = \cdots = a_{1-q} = 0$ will have a more pronounced effect on the final estimates for the parameters. Thus we are led to consider the more difficult problem of minimizing the unconditional sum-of-squares function and the related problem of maximum likelihood estimation.

The advantage of the method of maximum likelihood is that all of the information available in the data is used rather than just the first and second moments, as is the case with least squares. Another advantage is that many large-sample results are known under very general conditions. However, one disadvantage is that we must for the first time work specifically with the joint probability density functions.

MAXIMUM LIKELIHOOD ESTIMATION

For any set of observations $Z_1, Z_2, \ldots, Z_n$ (time series or otherwise), the **likelihood function** L is defined to be the probability (density) of obtaining the data actually observed; however, it is considered a function of the parameters in the model with the observations held fixed. For ARIMA models, L will be a function of the ϕ's, θ's, μ, and σ_a^2 given the observations $Z_1, Z_2, \ldots, Z_n$. The maximum likelihood estimators are then defined as those values of the parameters for which the data actually observed are *most likely*, that is, the values that maximize the likelihood function.

We begin by looking in detail at the AR(1) model. The most common assumption is that the white noise terms have normal distributions. In particular, we assume that

$$\ldots, a_{-1}, a_0, a_1, a_2, \ldots$$

are independent, normally distributed, random variables, each with a zero mean and variance σ_a^2. The probability density function (p.d.f.) for each a is then given by

$$(2\pi\sigma_a^2)^{-1/2} \exp\left(-\frac{a^2}{2\sigma_a^2}\right) \quad \text{for } -\infty < a < \infty$$

and, by independence, the joint p.d.f. for $a_2, a_3, \ldots, a_n$ is

$$(2\pi\sigma_a^2)^{-(n-1)/2} \exp\left(-\frac{1}{2\sigma_a^2}\sum_{t=2}^{n} a_t^2\right) \tag{7-43}$$

Now consider

$$
\begin{aligned}
Z_2 - \mu &= \phi(Z_1 - \mu) & + a_2 \\
Z_3 - \mu &= \phi(Z_2 - \mu) & + a_3 \\
&\quad\vdots \\
Z_n - \mu &= \phi(Z_{n-1} - \mu) & + a_n
\end{aligned}
\tag{7-44}
$$

If we condition on $Z_1 = z_1$, (7-44) defines a linear transformation between $a_2, a_3, \ldots, a_n$ and $Z_2, Z_3, \ldots, Z_n$ (with Jacobian equal to 1). Thus the joint p.d.f. of $Z_2, Z_3, \ldots, Z_n$ given $Z_1 = z_1$ can be obtained by using (7-44) to substitute for the a's in terms of the z's in (7-43). Thus we get

$$
f(z_2, z_3, \ldots, z_n \mid z_1) = (2\pi\sigma_a^2)^{-(n-1)/2}
$$
$$
\times \exp\left\{\left(\frac{-1}{2\sigma_a^2}\right)\sum_{t=2}^{n}[(z_t - \mu) - \phi(z_{t-1} - \mu)]^2\right\} \tag{7-45}
$$

Now consider the (marginal) distribution of Z_1. It follows from the linear process representation of the AR(1) process, Equation (4-16), that Z_1 will also have a normal distribution, namely normal with mean μ and variance $\gamma_0 = \sigma_a^2/(1 - \phi^2)$ from Equation (4-12). Multiplying the conditional p.d.f. given in Equation (7-45) by the (marginal) p.d.f. of Z_1 gives us the joint p.d.f. of $Z_1, Z_2, \ldots, Z_n$ that we require. Interpreted as a function of the parameters ϕ, μ, and σ_a^2, the likelihood function for an AR(1) model is given by

$$L(\phi, \mu\, \sigma_a^2) = (2\pi\sigma_a^2)^{-n/2}(1 - \phi^2)^{1/2} \exp\left[\frac{-1}{2\sigma_a^2} S(\phi, \mu)\right] \tag{7-46}$$

where

$$S(\phi, \mu) = \sum_{t=2}^{n} [(Z_t - \mu) - \phi(Z_{t-1} - \mu)]^2 + (1 - \phi^2)(Z_1 - \mu)^2 \tag{7-47}$$

The function $S(\phi, \mu)$ is called the **unconditional sum-of-squares function.**

As a general rule, the logarithm of the likelihood function is more convenient mathematically than the likelihood function itself. For the AR(1) case, the log-likelihood function, denoted $\lambda(\phi, \mu, \sigma_a^2)$, is given by

$$\lambda(\phi, \mu, \sigma_a^2) = -\frac{n}{2}\log 2\pi - \frac{n}{2}\log \sigma_a^2 + \frac{1}{2}\log (1 - \phi^2) - \frac{1}{2\sigma_a^2} S(\phi, \mu) \tag{7-48}$$

For given values of ϕ and μ, $\lambda(\phi, \mu, \sigma_a^2)$ can be maximized analytically with respect to σ_a^2. Setting $\partial\lambda/\partial\sigma_a^2 = 0$, we readily find the maximum likelihood estimator of σ_a^2 in terms of the yet-to-be-determined estimators of ϕ and μ:

$$\hat{\sigma}_a^2 = \frac{S(\hat{\phi}, \hat{\mu})}{n} \tag{7-49}$$

As in many other similar contexts, we usually divide by $n - 2$ (since we are estimating *two* parameters, ϕ and μ) rather than n to obtain an estimator with less bias. For typical sample sizes of time series, there will be very little difference.

Consider now the estimation of ϕ and μ. A comparison of the unconditional sum-of-squares function $S(\phi, \mu)$ with the earlier conditional sum-of-squares function $S_*(\phi, \mu)$ of Equation (7-22) reveals one simple difference:

$$S(\phi, \mu) = S_*(\phi, \mu) + (1 - \phi^2)(Z_1 - \mu)^2 \tag{7-50}$$

Since $S_*(\phi, \mu)$ involves a sum of $n - 1$ similar terms, whereas $(1 - \phi^2)(Z_1 - \mu)^2$ does not involve n, we shall have $S(\phi, \mu) \approx S_*(\phi, \mu)$. Thus the values of ϕ and μ that minimize S or S_* should be very similar. The effect of the rightmost term in (7-50) will be more substantial when the minimum for ϕ occurs near the stationarity boundary of ± 1.

Similar comments apply to the maximization of the full log-likelihood of Equation (7-48) with respect to the term $\frac{1}{2}\log (1 - \phi^2)$. For large n this term will be dominated by the sum-of-squares function $S(\phi, \mu)$ except in the case where ϕ is close to ± 1.

UNCONDITIONAL LEAST SQUARES

As a compromise between conditional least-squares estimates and full maximum likelihood estimates, we often compute unconditional least-squares estimates, that is, estimates minimizing $S(\phi, \mu)$. Unfortunately, the term

$(1 - \phi^2)(Z_1 - \mu)^2$ causes the equations $\partial S/\partial \phi = 0$ and $\partial S/\partial \mu = 0$ to be non-linear in ϕ and μ, and reparameterizing to a constant term $\theta_0 = \mu(1 - \phi)$ does not improve the situation substantially. Thus the minimization must be carried out by numerical iteration, as discussed in the previous section. The resulting estimates are called the **unconditional least-squares estimates.**

The derivation of the likelihood function and the associated unconditional sum-of-squares function for more general ARMA models is considerably more involved. We shall merely quote one useful form of the unconditional sum-of-squares function and consider its implications. From Box and Jenkins (1976, p. 213), we have for a general ARMA(p, q) model

$$S(\phi, \theta, \theta_0) = \sum_{t=-\infty}^{n} \hat{a}_t^2 \tag{7-51}$$

where

$$\hat{a}_t = E(a_t \,|\, Z_1, Z_2, \ldots, Z_n) \tag{7-52}$$

Notice that for $t \leq 0$, the $\hat{a}_t$ could be viewed as "forecasts" *backward in time* of the a_t terms, given the data $Z_1, Z_2, \ldots, Z_n$. For this reason, they frequently have been called **back forecasts,** but a more appropriate term seems to be **backcasts.** Since the major topic of Chapter 9 is forecasting, we shall defer the discussion of the calculation of the $\hat{a}_t$ to Appendix I of Chapter 9. Further references on the calculation of the likelihood function include Ljung and Box (1979), Ansley (1979), Newbold (1974), and Harvey (1981a, 1981b).

7.4 PROPERTIES OF THE ESTIMATES

The large-sample properties of the maximum likelihood and least-squares (conditional or unconditional) estimators are identical and can be obtained by modifying standard maximum likelihood theory for large samples. Details can be found in Box and Jenkins (1976) or Fuller (1976). We shall look at the results and their implications for simple ARMA models.

For large n, the estimators are approximately unbiased and follow normal distributions. The variances and correlations are as follows:

$$\text{AR(1):} \quad \text{Var}(\hat{\phi}) \approx \frac{1 - \phi^2}{n} \tag{7-53}$$

$$\text{AR(2):} \quad \text{Var}(\hat{\phi}_1) \approx \text{Var}(\hat{\phi}_2) \approx \frac{1 - \phi_2^2}{n} \tag{7-54}$$

$$\text{Corr}(\hat{\phi}_1, \hat{\phi}_2) \approx -\frac{\phi_1}{1 - \phi_2} = -\rho_1 \tag{7-55}$$

$$\text{MA(1):} \quad \text{Var}(\hat{\theta}) \approx \frac{1 - \theta^2}{n} \tag{7-56}$$

$$\text{MA(2):} \quad \text{Var}\,(\hat{\theta}_1) \approx \text{Var}\,(\hat{\theta}_2) \approx \frac{1 - \theta_2^2}{n} \tag{7-57}$$

$$\text{Corr}\,(\hat{\theta}_1, \hat{\theta}_2) \approx \frac{\theta_1}{1 - \theta_2} \tag{7-58}$$

$$\text{ARMA(1, 1):} \quad \text{Var}\,(\hat{\phi}) \approx \frac{1 - \phi^2}{n}\left(\frac{1 - \phi\theta}{\phi - \theta}\right)^2 \tag{7-59}$$

$$\text{Var}\,(\hat{\theta}) \approx \frac{1 - \theta^2}{n}\left(\frac{1 - \phi\theta}{\phi - \theta}\right)^2 \tag{7-60}$$

$$\text{Corr}\,(\hat{\phi}, \hat{\theta}) \approx \frac{[(1 - \phi^2)(1 - \theta^2)]^{1/2}}{1 - \phi\theta} \tag{7-61}$$

(see Box and Jenkins, 1976; Harvey, 1981b, pp. 130–132.)

Notice that in the AR(1) case, the variance of the estimator of ϕ decreases as ϕ approaches ± 1. Also notice that even though an AR(1) model is a special case of an AR(2) model, Equation (7-54) implies that our estimate of ϕ_1 may suffer if we erroneously fit an AR(2) model when in fact $\phi_2 = 0$. Similar comments could be made about fitting an MA(2) model when an MA(1) would suffice or fitting ARMA(1, 1) when either AR(1) or MA(1) applies.

For the ARMA(1, 1) case, note the denominator of $\phi - \theta$ in Equations (7-59) and (7-60). If ϕ and θ are nearly equal, the variability in the estimators of ϕ and θ can be extremely large.

Note that in all of the two-parameter models, the estimates can be highly correlated, even for very large sample sizes.

Table 7.2 gives numerical values for the approximate standard deviation $[(1 - \phi^2)/n]^{1/2}$ for several values of ϕ and n. The numbers apply equally well to standard deviations computed according to Equations (7-53), (7-54), (7-56), and (7-57).

Thus in estimating an AR(1) model with, for example, $n = 100$ and ϕ of about 0.7, we can be 95% confident that our estimate of ϕ is in error by no more than $\pm 2(0.07) = \pm 0.14$.

For pure autoregressive models, the method of moments yields estimates equivalent to least-squares or maximum likelihood estimators, at least for large samples. For models containing moving average terms, such is not the case. For an MA(1) model, it can be shown (see Fuller, 1976, p. 343) that the large-sample variance of the method-of-moments estimator of θ is equal to

$$\frac{1 + \theta^2 + 4\theta^4 + \theta^6 + \theta^8}{n(1 - \theta^2)^2} \tag{7-62}$$

TABLE 7.2   Values of $[(1 - \phi^2)/n]^{1/2}$

		n	
ϕ	50	100	200
0.4	0.13	0.09	0.06
0.7	0.10	0.07	0.05
0.9	0.06	0.04	0.03

TABLE 7.3   Comparison of Standard Deviations for the Method of Moments and the Method of Maximum Likelihood for MA(1) Models

θ	SD_{MM}/SD_{ML}
0.25	1.07
0.50	1.42
0.75	2.66
0.90	5.33

Comparing Expression (7-62) with that of Equation (7-56), we see that the variance for the method-of-moments estimator is always larger than the variance for the maximum likelihood estimator. Table 7.3 displays the ratio of the large-sample standard deviations for the two methods for several values of θ. It is clear from these values that the method-of-moments estimator for the MA(1) model should not be used. However, it may be employed to obtain starting values for an iterative procedure for finding least-squares or maximum likelihood estimates.

7.5 THE MINITAB ARIMA COMMAND

The Minitab ARIMA command in its simplest form has the format

ARIMA p = **K**, d = **K**, q = **K** for the series in **C**

This command uses the criterion of unconditional least squares to obtain estimates of the parameters in an ARIMA(p, d, q) model.

As an example, consider Exhibit 7.2. Here C1 contains data on the quarterly U.S. unemployment rate. We have fit the AR(2) model

$$Z_t = \phi_1 Z_{t-1} + \phi_2 Z_{t-2} + \theta_0 + a_t$$

From the printout we see that the unconditional least-squares estimates are

$$\hat{\phi}_1 = 1.5629, \quad \hat{\phi}_2 = -0.6583, \quad \text{and} \quad \hat{\theta}_0 = 0.48467 \qquad \text{(7-63)}$$

These estimates should be compared with those obtained earlier by regression and by the method of moments, shown in Equations (7-16) and (7-31).

The estimate for the mean is obtained by substituting estimates into the relationship (see Equation (5-25)):

$$\hat{\mu} = \frac{\hat{\theta}_0}{1 - \hat{\phi}_1 - \hat{\phi}_2}$$

The value of the sum of squares (SS) is $\sum_{t-1}^{n} a_t^2$ evaluated at the final estimates of ϕ_1, ϕ_2, and θ_0. The degrees of freedom (DF) are the true sample size after differencing minus the number of estimated parameters, that is, $(n - d) - p - q - 1$ (the last 1 is for the constant term); the mean square (MS) is the estimate of σ_a^2 given by SS/DF.

The comment, backcasts excluded, on the printout refers to the fact that throughout the iterations, we are computing (and are attempting to minimize)

EXHIBIT 7.2 Fitting an AR(2) Model to the Unemployment Rate Series with Unconditional Least Squares

```
MTB > arima (2,0,0) 'Unemp'

Estimates at each iteration
Iteration        SSE      Parameters
        0     166.026    0.100    0.100    4.166
        1     135.995    0.250    0.025    3.774
        2     109.731    0.400   -0.052    3.392
        3      86.851    0.550   -0.129    3.014
        4      67.240    0.700   -0.208    2.639
        5      50.835    0.850   -0.286    2.266
        6      37.579    1.000   -0.365    1.895
        7      26.675    1.150   -0.435    1.476
        8      19.237    1.300   -0.500    1.032
        9      16.164    1.450   -0.591    0.728
       10      15.441    1.558   -0.657    0.505
       11      15.437    1.563   -0.659    0.486
       12      15.437    1.563   -0.658    0.484
Relative change in each estimate less than  0.0010

Final Estimates of Parameters
Type       Estimate    St. Dev.   t-ratio
AR    1      1.5630     0.0700     22.32
AR    2     -0.6583     0.0702     -9.37
Constant    0.48430    0.03280    14.76
Mean        5.0810     0.3442

No. of obs.:  121
Residuals:    SS = 15.3345  (backforecasts excluded)
              MS =  0.1300  DF = 118

Modified Box-Pierce chisquare statistic
Lag              12          24          36          48
Chisquare   18.0(DF=10)  33.0(DF=22)  44.1(DF=34)  50.1(DF=46)

MTB > cdf 18.0;
SUBC> chisquare 10.
     0.945068
```

$\sum_{t=-\infty}^{n} \hat{a}_t^2$, which includes backcasted values $\hat{a}_t$. However, the backcasted values are excluded in the SS used to estimate σ_a^2.

The values given for the standard deviation are from the general theory that produced Equations (7-53) through (7-61) with estimates replacing all unknown parameters. Using $\phi_2 = -0.6583$ and $n = 121$ in Equation (7-54) yields

$$\sqrt{\text{Var}(\hat{\phi}_1)} \approx \sqrt{\text{Var}(\hat{\phi}_2)} \approx \left[\frac{1 - (-0.6583)^2}{121} \right]^{1/2} = 0.068$$

[The small discrepancy between this value and the 0.070 given on the printout is due to the fact that the computer uses a method applicable to general ARMA(p, q) models rather than Equation (7-54), which pertains specifically to an AR(2) case.]

As in the regression command, the t-ratios are the estimates of the ϕ's and θ_0 divided by their respective estimated standard deviations. For large sample sizes, the t-ratios may be used to test the significance of the estimates with respect to the null hypothesis that the corresponding parameter is zero. In

EXHIBIT 7.3 Estimated ARI(1, 1) Model for the Railroad Bond Yield Series with Unconditional Least Squares

```
MTB > arima (1,1,0) 'RRBonds'

Estimates at each iteration
Iteration        SSE      Parameters
    0         10153.3     0.100
    1          9129.8     0.250
    2          8605.5     0.400
    3          8530.1     0.478
    4          8529.9     0.482
    5          8529.9     0.483
Relative change in each estimate less than  0.0010

Final Estimates of Parameters
Type     Estimate    St. Dev.   t-ratio
AR   1     0.4825     0.0876      5.51

Differencing: 1 regular difference
No. of obs.:  Original series 102, after differencing 101
Residuals:    SS = 8526.97  (backforecasts excluded)
              MS =   85.27  DF = 100

Modified Box-Pierce chisquare statistic
Lag                12           24           36           48
Chisquare    6.4(DF=11)   18.9(DF=23)  27.7(DF=35)  37.8(DF=47)

MTB > cdf 6.4;
SUBC> chisq 11.
    0.154625
```

the unemployment rate example, all t-ratios are large, 22.32, -9.37, and 14.76, respectively, so there is no indication that ϕ_1, ϕ_2, or θ_0 should be excluded from the model. The modified Box–Pierce chi-square statistic appearing on the printout is a diagnostic that will be discussed in Chapter 8.

Exhibit 7.3 shows the Minitab printout when using the ARIMA command to fit an ARI(1, 1) model to the railroad bond yield data. The estimate $\hat{\phi} = 0.4825$ should be compared with the estimate of 0.468 in Equation (7-20), obtained by the method of moments.

Notice that there is no constant term estimated in Exhibit 7.3. If $d = 0$, the constant term θ_0 is automatically included in the model unless the subcommand NOCONSTANT is used. If $d > 0$, the constant term is excluded unless the subcommand CONSTANT is given.

Recall that to use a subcommand, type a semicolon (;) at the end of the main command line. Type the subcommand(s) on the following line(s) and end each subcommand line with a semicolon. End the final subcommand line with a period or use END as the last subcommand.

Notice that in Exhibit 7.3 we have specified $d = 1$, used the original data, and let the ARIMA command take the first difference internally. The same results would be obtained by the following Minitab commands:

DIFF 'RRBONDS' C2

ARIMA (1, 0, 0) C2;

NOCONSTANT.

EXHIBIT 7.4 Using the START Subcommand with the ARIMA Command

```
MTB > set 'Start'
DATA> 0.468
DATA> end
MTB > arima (1,1,0) 'RRBonds';
SUBC> start with 'Start'.

Estimates at each iteration
Iteration        SSE      Parameters
    0         8532.22      0.468
    1         8529.88      0.482
    2         8529.87      0.483
    3         8529.87      0.483
Relative change in each estimate less than  0.0010

Final Estimates of Parameters
Type       Estimate    St. Dev.   t-ratio
AR    1     0.4826      0.0876      5.51

Differencing: 1 regular difference
No. of obs.:  Original series 102, after differencing 101
Residuals:    SS = 8526.98  (backforecasts excluded)
              MS =   85.27  DF = 100

Modified Box-Pierce chisquare statistic
Lag            12          24          36          48
Chisquare   6.4(DF=11)  18.9(DF=23)  27.7(DF=35)  37.8(DF=47)
```

The first method is preferable when we progress to forecasting, where the extended ARIMA command needs to access the original series rather than the differenced series.

Another subcommand can be used to specify starting values (initial values) for the parameters to begin the iterative estimation procedure. The desired starting values must first be entered into a column prior to invoking the ARIMA command. They are entered in the same order as they appear on the printout: AR, MA, and constant. The starting value for the constant term is optional. Having stored the starting values in 'START', say, use the subcommand

STARTING VALUES IN 'START'

with the ARIMA command. If you do not use START, Minitab uses 0.1 as initial values for all of the ϕ's and θ's (except θ_0). This usually works well.

As an example of the START subcommand, consider estimating the ARI(1, 1) model for the railroad bond series but starting at the method-of-moments estimate $\hat{\phi} = 0.468$. Exhibit 7.4 shows the Minitab commands and results.

Other options of the ARIMA command pertain to residuals, forecasting, and seasonal models. These will be discussed in Chapters 8, 9, and 10, respectively.

The amount of output produced by the ARIMA command can be controlled by using the BRIEF command. Four different levels are possible and are fully specified in the *Minitab Reference Manual*. They can also be obtained on-line with the HELP BRIEF command.

CHAPTER 7 EXERCISES

7.1. From a series of length 100, we have computed $r_1 = 0.8$, $r_2 = 0.5$, $r_3 = 0.4$, $\bar{Z} = 2$, and a sample variance of 5. If we assume that an AR(2) model with a constant term is appropriate, how can we get (simple) estimates of ϕ_1, ϕ_2, θ_0, and σ_a^2?

7.2. Assuming that the following data arise from a stationary process, calculate estimates of μ, γ_0, and ρ_1:

$$6, 5, 4, 6, 4$$

7.3. If $\{Z_t\}$ satisfies an AR(1) model with ϕ of about 0.7, how long a series do we need to estimate the true $\phi = \rho_1$ with 95% confidence that our error is no more than ± 0.1?

7.4. Consider an MA(1) process for which it is *known* that the mean is zero. Based on a series of length 3, we observe $Z_1 = 0$, $Z_2 = -1$, and $Z_3 = \frac{1}{2}$.

a. Show that the conditional least-squares estimate of θ is $\frac{1}{2}$.

b. Find an estimate of the noise variance σ_a^2. (*Hint*: Iterative methods are not needed in this simple case.)

7.5. Given the data $Z_0 = 10$, $Z_1 = 10$, $Z_2 = 9$, and $Z_3 = 9.5$, we wish to fit an IMA(1, 1) model without a constant term.

a. Find the conditional least-squares estimate of θ.

b. Estimate σ_a^2. (*Hint*: Do Exercise 4 first.)

7.6. Consider two parameterizations of the AR(1) model:

> I. $Z_t - \mu = \phi(Z_{t-1} - \mu) + a_t$
>
> II. $Z_t \quad = \phi Z_{t-1} + \theta_0 + a_t$ where $\theta_0 = \mu(1 - \phi)$

We want to estimate ϕ and μ or ϕ and θ_0 using conditional least squares conditional on Z_1.

Show that with model I we are led to solving nonlinear equations to obtain the estimates, while in model II we need only solve linear equations.

7.7. Use Minitab to simulate a variety of ARIMA series with various values for the parameters and various sample sizes. For each series use the ARIMA command to estimate the parameters. Compare the estimates to the known parameter values.

7.8. Verify Equation (7-5).

7.9. Use Minitab to estimate the parameters of the ARIMA model specified in Exercise 6.9 for the Portland, Oregon, gasoline price series.

7.10. Estimate the parameters of the ARIMA model specified in Exercise 6.10 for the Iowa nonfarm income series.

CHAPTER 8 MODEL DIAGNOSTICS

We have now discussed methods for specifying models and for efficiently estimating the parameters in those models. Model diagnostics, or model criticism, is concerned with testing the goodness-of-fit of a model and, if the fit is poor, suggesting appropriate modifications. We shall present two complementary approaches: analysis of the residuals from the fitted model and analysis of overparameterized models, that is, models that are more general than the specified model but contain the specified model as a special case.

8.1 RESIDUAL ANALYSIS

We have already used the basic ideas of residual analysis in Section 3.6 when we checked the adequacy of fitted deterministic trends. With autoregressive models residuals are defined in direct analogy to that earlier work. Consider in particular an AR(2) model with a constant term:

$$Z_t = \phi_1 Z_{t-1} + \phi_2 Z_{t-2} + \theta_0 + a_t \tag{8-1}$$

Having estimated ϕ_1, ϕ_2, and θ_0, the **residuals** are defined as

$$\hat{a}_t = Z_t - \hat{\phi}_1 Z_{t-1} - \hat{\phi}_2 Z_{t-2} - \hat{\theta}_0, \qquad t = 3, 4, \ldots, n \tag{8-2}$$

The initial residuals $\hat{a}_1$ and $\hat{a}_2$ can be obtained from the estimation procedure using backcasted values for Z_0 and Z_{-1}.

For general ARMA models containing moving average terms, we must recall the autoregressive form of the model to define the residuals. From Equation (4-48) we have

$$Z_t = \sum_{j=1}^{\infty} \pi_j Z_{t-j} + a_t$$

so that the residuals are defined as

$$\hat{a}_t = Z_t - \sum_{j=1}^{\infty} \hat{\pi}_j Z_{t-j} \tag{8-3}$$

Here the π_j are not estimated directly but rather implicitly as functions of the ϕ's and θ's. In fact, the residuals are not calculated using Equation (8-3) at all but are obtained as a by-product of the estimation procedure itself [see Section 7.2, in particular Equation (7-42) evaluated at $\phi_1, \phi_2, \ldots, \phi_p, \theta_1, \theta_2, \ldots, \theta_q$].

In Chapter 9 we shall argue that

$$\hat{Z}_t = \sum_{j=1}^{\infty} \hat{\pi}_j Z_{t-j}$$

is the best forecast of Z_t based on $Z_{t-1}, Z_{t-2}, \ldots$. Thus Equation (8-3) can be written as

$$\text{Residual} = \text{actual} - \text{predicted}$$

in direct analogy with regression models. Compare with Section 3.6.

The Minitab ARIMA command can be extended to store residuals and

EXHIBIT 8.1   Time Plot of Residuals for Unemployment Rate Series with an AR(2) Model

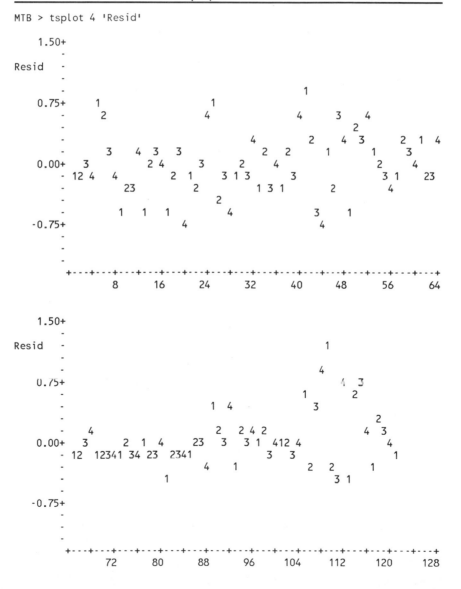

predicted values in specified columns as follows:

ARIMA p = **K**, d = **K**, q = **K**, series in **C** [resid. in **C** [pred. in **C**]]

If the ARMA model is correct and if the parameter estimates are close to the true values, then the residuals should have nearly the properties of independent, identically distributed, normal random variables with zero means and standard deviations σ_a.

Our first diagnostic check is to inspect the plot of residuals over time. If the model is adequate, we expect the plot to suggest a rectangular scatter around a zero horizontal level with no trends whatsoever.

Exhibit 8.1 shows the time plot of residuals from the AR(2) model

EXHIBIT 8.2 Time Plot of Residuals for Railroad Bond Yields Series with ARI(1, 1) Model

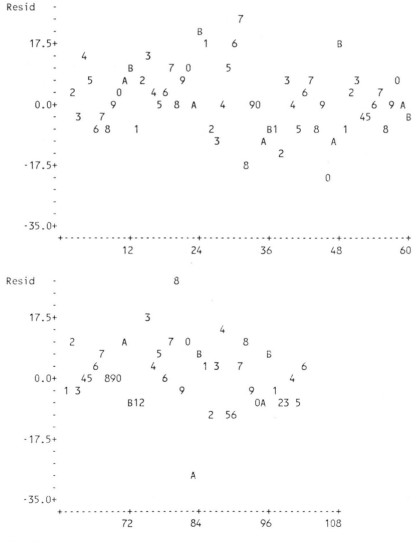

N* = 1

estimated for the U.S. unemployment data. Exhibit 8.2 gives a similar plot for the residuals from the ARI(1, 1) fit of the railroad bond yield series. Neither of these plots shows evidence that their respective model fits poorly.

Normality of the residuals can be checked by computing a histogram of the residuals (or of the standardized residuals), by plotting normal scores versus residuals, and, more formally, by performing the normal-scores correlation test, as described in Section 3.6. Exhibits 8.3 and 8.4 present these checks for the unemployment series residuals and the railroad bond series residuals, respectively. Note that the standardized residuals are formed by dividing the residuals

EXHIBIT 8.3 Distribution of Residuals for Unemployment Rate Series with AR(2) Model

```
MTB > Let 'StdResid'='Resid'/sqrt(0.13)
MTB > histogram of 'StdResid'

Histogram of StdResid   N = 121

Midpoint    Count
   -2.0        4   ****
   -1.5        7   *******
   -1.0       11   ***********
   -0.5       30   ******************************
    0.0       29   *****************************
    0.5       18   ******************
    1.0        7   *******
    1.5        6   ******
    2.0        6   ******
    2.5        2   **
    3.0        0
    3.5        1   *

MTB > nscores of 'Resid' into 'NScores'
MTB > correlation of 'Resid' and 'NScores'

Correlation of Resid and NScores = 0.980

MTB > plot 'NScores' vs 'Resid'


           -                                          *
NScores -
           -                                       2
           -                                 * **
    1.50+                                  22*
           ,                            ?**?*
           -                       *332
           -                     453
           -                   833
   -0.00+                78
           -               +3
           -              *+
           -           233*
           -          6*
   -1.50+        *22
           -      ***
           -     **
           -
           -   *
           - --+--------+---------+---------+---------+---------+----Resid
             -0.800   -0.400    0.000    0.400    0.800    1.200
```

EXHIBIT 8.4 Distribution of Residuals for Railroad Bond Yield Series with ARI(1, 1) Model

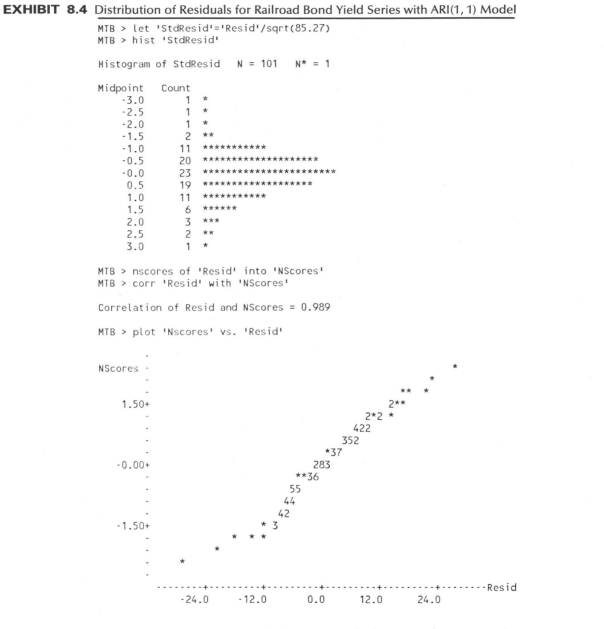

```
MTB > let 'StdResid'='Resid'/sqrt(85.27)
MTB > hist 'StdResid'

Histogram of StdResid   N = 101   N* = 1

Midpoint   Count
    -3.0       1   *
    -2.5       1   *
    -2.0       1   *
    -1.5       2   **
    -1.0      11   ***********
    -0.5      20   ********************
    -0.0      23   ***********************
     0.5      19   *******************
     1.0      11   ***********
     1.5       6   ******
     2.0       3   ***
     2.5       2   **
     3.0       1   *

MTB > nscores of 'Resid' into 'NScores'
MTB > corr 'Resid' with 'NScores'

Correlation of Resid and NScores = 0.989

MTB > plot 'Nscores' vs. 'Resid'
```

by their estimated standard deviation with the Minitab command

LET 'StdResid' = 'Resid'/SQRT(MS)

where MS is replaced by the numerical value of the mean square taken from the ARIMA command output.

In both of the examples above, the normality of the residuals is reasonably well achieved. For the normal-scores correlation test, we refer to Table 3.1 and find 5% critical values of 0.986 for $n = 100$ and 0.991 for $n = 150$. The corresponding 1% critical values are 0.981 and 0.987. Our observed

EXHIBIT 8.5 Residuals versus Fitted Values for Unemployment Rate Series with AR(2) Model

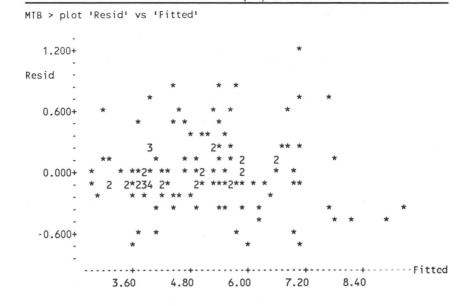

```
MTB > plot 'Resid' vs 'Fitted'

          -
  1.200+                                       *
          -
  Resid   -
          -                  *         *  *
          -            *                      *    *
  0.600+      *             *      *  *          *
          -          *       *  *     *
          -                    *  **   *
          -            3           2*  *        ** *
          -     **         *     * *    * * 2     2        *
  0.000+  *     * **2* **   * *2 * *   2     *  *
          - *  2  2*234 2*     2* ***2** * *     **
          -    *      * *  * ** *           *
          -            *  *   *   **  *  *          *          *
          -                            *        *  *    *
 -0.600+        *  *          *          *
          -        *                 *        *
          -
          ----------+---------+---------+---------+---------+--------Fitted
                   3.60      4.80      6.00      7.20      8.40
```

correlations are 0.980 with $n = 121$ and 0.989 with $n = 101$ for the unemployment rate and railroad bond series, respectively. Thus we would certainly not reject normality in the residuals for the railroad bond data. The unemployment rate example is less clear-cut, mostly due to the one standardized residual at about 3.5, which would warrant further investigation.

The residuals versus the corresponding fitted or predicted values should also be plotted to assess the model. Again we are looking for patterns, the absence of which indicates a good fit for the model. Exhibits 8.5 and 8.6 present

EXHIBIT 8.6 Residuals versus Fitted Values for Railroad Bond Yield Series with ARI(1, 1) Model

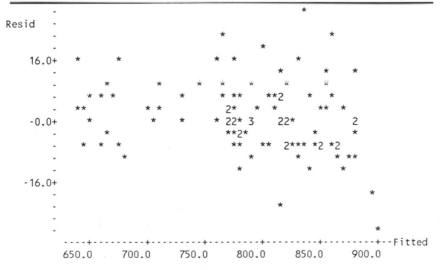

```
         -
  Resid  -                                       *
         -                          *                   *
         -                                *
  16.0+  *      *            *   *           *
         -            *          *    *    *    *    *
         -        * * *         *       * **    **2    *   *
         -      **          * *        2*    *  *      ** *
 -0.0+        *             *    *   * 22* 3   22*         2
         -         *                   **2*             *   *
         -      *  *  *              **    **   2*** *2 *2
         -            *                 *         *       * **
         -                             *              *   *
 -16.0+
         -                                  *                *
         -                              *
         -
         -                                             *
         -
         ----+---------+---------+---------+---------+---------+--Fitted
            650.0     700.0     750.0     800.0     850.0     900.0
```

the results for our two example series, and they indicate no striking difficulties with the residuals.

AUTOCORRELATION OF RESIDUALS

To check on the independence of the noise terms in the model, we consider the sample autocorrelation function of the residuals, denoted $\hat{r}_k$. From Equation (6-3) we know that for true white noise and large n, the sample autocorrelations are approximately uncorrelated and normally distributed with zero means and variance $1/n$. Unfortunately, residuals, even from a correctly specified model with efficiently estimated parameters, have somewhat different properties. This was first explored for multiple-regression models in a series of papers by Durbin and Watson (1950, 1951, 1971) and for autoregressive models in Durbin (1970). The key reference on the distribution of residual autocorrelations in ARIMA models is Box and Pierce (1970), whose results were generalized in McLeod (1978).

Generally speaking the residuals are approximately normally distributed with zero means; however, for small lags k and j, the variance of $\hat{r}_k$ *can be substantially less than* $1/n$ and the estimates $\hat{r}_k$ and $\hat{r}_j$ can be highly correlated. For larger lags, the approximate variance $1/n$ does apply, and $\hat{r}_k$ and $\hat{r}_j$ are approximately uncorrelated.

As an example of these results, consider a correctly specified and efficiently estimated AR(1) model. It can be shown that for large n,

$$\text{Var}(\hat{r}_1) \approx \frac{\phi^2}{n}, \tag{8-4}$$

$$\text{Var}(\hat{r}_k) \approx \frac{1 - (1 - \phi^2)\phi^{2k-2}}{n}, \qquad k > 1, \tag{8-5}$$

and

$$\text{Corr}(\hat{r}_1, \hat{r}_k) \approx -\text{sign}(\phi)\frac{(1 - \phi^2)\phi^{k-2}}{1 - (1 - \phi^2)\phi^{2k-2}}, \qquad k > 1 \tag{8-6}$$

where

$$\text{sign}(\phi) = \begin{cases} 1 & \text{if } \phi > 0 \\ 0 & \text{if } \phi = 0 \\ -1 & \text{if } \phi < 0 \end{cases}$$

Table 8.1 numerically illustrates these formulas for a variety of values of ϕ and k. Notice that $\text{Var}(\hat{r}_k) \approx 1/n$ is a reasonable approximation for $k \geq 2$ over a wide range of ϕ-values.

If we apply these results to the time series for the railroad bond yields using the estimate $\hat{\phi} = 0.483$, we have

$$2\sqrt{\text{Var}(\hat{r}_1)} \approx 0.096, \qquad 2\sqrt{\text{Var}(\hat{r}_2)} \approx 0.18,$$
$$2\sqrt{\text{Var}(\hat{r}_3)} \approx 0.19, \qquad 2\sqrt{\text{Var}(\hat{r}_4)} \approx 0.198,$$

and

$$2\sqrt{\text{Var}(\hat{r}_k)} \approx 0.199, \qquad k > 4 \tag{8-7}$$

Table 8.1   Approximate Standard Deviations and Correlations of Residual Autocorrelations from AR(1) Models

	ϕ				ϕ			
k	0.3	0.5	0.7	0.9	0.3	0.5	0.7	0.9
	Standard Deviation of $\hat{r}_k$ (times $\sqrt{n}$)				Correlation of $\hat{r}_1$ with $\hat{r}_k$			
1	0.30	0.50	0.70	0.90	1.00	1.00	1.00	1.00
2	0.96	0.90	0.87	0.92	−0.95	−0.83	−0.59	−0.21
3	1.00	0.98	0.94	0.94	−0.27	−0.38	−0.38	−0.18
4	1.00	0.99	0.97	0.95	−0.08	−0.19	−0.26	−0.16
5	1.00	1.00	0.99	0.96	−0.02	−0.09	−0.18	−0.14
6	1.00	1.00	0.99	0.97	−0.01	−0.05	−0.12	−0.13
7	1.00	1.00	1.00	0.97	−0.00	−0.02	−0.09	−0.12
8	1.00	1.00	1.00	0.98	−0.00	−0.01	−0.06	−0.10
9	1.00	1.00	1.00	0.99	−0.00	−0.00	−0.03	−0.08

Exhibit 8.7 gives the sample autocorrelations out to lag 20 of the residuals from the ARI(1, 1) fit to the railroad bond yield series. Comparing the observed values with the two standard deviation values given by (8-7) gives us no reason to question the adequacy of the fitted model. Even the somewhat striking value at lag 9 of 0.187 is within the error bounds of ±0.199 applicable at that lag under the null hypothesis of white noise.

For an AR(2) model it can be shown that we have

$$\text{Var}(\hat{r}_1) \approx \frac{\phi_2^2}{n} \tag{8-8}$$

and

$$\text{Var}(\hat{r}_2) \approx \frac{\phi_2^2 + \phi_1^2(1 + \phi_2)^2}{n} \tag{8-9}$$

EXHIBIT 8.7 Sample Autocorrelation Function of Residuals for Railroad Bond Yield Series with ARI(1, 1) Model

```
ACF of Resid

             -1.0 -0.8 -0.6 -0.4 -0.2  0.0  0.2  0.4  0.6  0.8  1.0
             +----+----+----+----+----+----+----+----+----+----+
  1   0.016                              X
  2  -0.092                             XXX
  3   0.019                              X
  4   0.005                              X
  5   0.073                             XXX
  6   0.049                             XX
  7  -0.046                             XX
  8  -0.009                              X
  9   0.187                             XXXXXX
 10   0.025                             XX
 11   0.026                             XX
 12   0.047                             XX
 13  -0.069                            XXX
 14   0.011                              X
 15  -0.043                             XX
 16  -0.174                           XXXXX
 17   0.001                              X
 18   0.011                              X
 19  -0.020                             XX
 20   0.082                             XXX
```

If the parameters ϕ_1 and ϕ_2 are not too close to the nonstationarity boundary, then

$$\text{Var}\,(\hat{r}_k) \approx \frac{1}{n} \quad \text{for } k \geq 3 \tag{8-10}$$

Exhibit 8.8 shows the autocorrelations of the residuals from the AR(2) model estimated for the U.S. unemployment rate series. Recall that $\hat{\phi}_1 = 1.56$ and $\hat{\phi}_2 = 0.66$; thus

$$2\sqrt{\text{Var}\,(\hat{r}_1)} \approx \frac{2(0.66)}{\sqrt{120}} = 0.120$$

and the observed $\hat{r}_1 = 0.107$ is without suspicion. Also,

$$2\sqrt{\text{Var}\,(\hat{r}_2)} \approx 2\left[\frac{(0.66)^2 + (1.56)^2(1 - 0.66)^2}{120}\right]^{1/2} = 0.155$$

and

$$\frac{2}{\sqrt{120}} = 0.183$$

so that the other residual autocorrelations are within plus or minus two standard deviations of zero except at lag 7, where $\hat{r}_7 = 0.191$. Discounting lags 1 and 2 to avoid the dependence problem, we should not be too surprised to find one $\hat{r}_k$ out of the remaining ten outside two standard deviations of zero even if the null hypothesis of independence is true.

With quarterly data, however, we would pay special attention to possible

EXHIBIT 8.8 Sample Autocorrelation Function of Residuals for Unemployment Rate Series with AR(2) Model

```
ACF of Resid

             -1.0 -0.8 -0.6 -0.4 -0.2  0.0  0.2  0.4  0.6  0.8  1.0
             +----+----+----+----+----+----+----+----+----+----+
   1   0.107                               XXXX
   2  -0.058                               XX
   3  -0.041                               XX
   4  -0.112                             XXXX
   5  -0.037                               XX
   6   0.132                               XXXX
   7   0.191                               XXXXXX
   8  -0.137                             XXXX
   9   0.073                               XXX
  10   0.094                               XXX
  11   0.110                               XXXX
  12  -0.094                             XXX
  13   0.057                               XX
  14  -0.112                             XXXX
  15  -0.094                             XXX
  16   0.091                               XXX
  17   0.107                               XXXX
  18   0.100                               XXX
  19  -0.031                               XX
  20   0.047                               XX
  21  -0.175                            XXXXX
```

correlation in the residuals at lags 4, 8, 12, and so on. With monthly series, lags 12, 24, 36, . . . should be checked carefully.

It can be shown that results analogous to those for AR models obtain for MA models also. In particular, replacing ϕ by θ in Equations (8-4), (8-5), and (8-6) give the results for the MA(1) case. Similarly, the MA(2) results can be stated by replacing ϕ_1 and ϕ_2 by θ_1 and θ_2, respectively, in Equations (8-8), (8-9), and (8-10). Results for general ARMA models can be found in Box and Pierce (1970) and McLeod (1978).

THE PORTMANTEAU TEST

In addition to looking at the residual autocorrelations at individual lags, it is useful to have a test that takes into consideration their magnitudes as a group. Box and Pierce (1970) proposed the statistic

$$Q = n \sum_{k=1}^{K} \hat{r}_k^2 \qquad \text{(8-11)}$$

They show that if the correct ARMA(p, q) model is estimated, then, for large n, Q has a chi-square distribution with $K - p - q$ degrees of freedom. Fitting an erroneous model would tend to inflate Q. Thus a general or **portmanteau** test would reject the ARMA(p, q) model if the observed value of Q exceeded an appropriate critical value in a chi-squared distribution with $K - p - q$ degrees of freedom. (Here the maximum lag K is taken that is large enough so that the ψ_j-weights are negligible for $j > K$.)

The chi-square distribution for Q is based on a limit theorem as $n \rightarrow \infty$, but Ljung and Box (1978) subsequently discovered that even for $n = 100$, the approximation of the chi-square distribution is not satisfactory. By modifying the Q statistic slightly, we can define a test statistic whose null distribution is much closer to $\chi^2(K - p - q)$ for typical sample sizes. The **modified Box–Pierce statistic** (or Ljung–Box–Pierce statistic) is given by

$$Q^* = n(n + 2) \sum_{k=1}^{K} \frac{\hat{r}_k^2}{n - k} \qquad \text{(8-12)}$$

Notice that since $(n + 2)/(n - k) > 1$, we have $Q^* > Q$, which partly explains why the original statistic Q tends to overlook inadequate models. More details on the exact distributions of Q and Q^* for finite samples can be found in Ljung and Box (1978); see also Davies, Triggs, and Newbold (1977).

In Minitab the modified Box–Pierce statistic is computed within the ARIMA command out to lags 12, 24, 36, and 48.

Exhibit 7.2 illustrates the use of the Box–Pierce statistic on the residuals of the AR(2) fit to the U.S. unemployment data. Here, for example, with $K = 12$, the approximate null distribution of Q^* is χ^2 with 10 degrees of freedom. Since $\chi^2_{0.05}(10) = 18.307$, the observed value of $Q^* = 18.0$ is not quite significant at the 5% level.

Exhibit 7.3 gives the results for the residuals of the ARI(1, 1) model estimated for the railroad bond yields. With $K = 12$ here, $Q^* = 6.4$; this result can be compared with $\chi^2_{0.05}(11) = 19.7$.

Chi-square tables can be avoided by using the Minitab command CDF (cumulative distribution function) with the CHISQUARE subcommand. The format is

CDF for values in **E** [put results in **E**];

CHISQUARE with degrees of freedom **K.**

This command calculates the cumulative probability to the left of the value specified and prints answers if they are not stored. For the modified Box–Pierce statistic, the observed significance level, or *P*-value, will be 1 minus the value computed by this command. This is illustrated in Exhibits 7.2 and 7.3. Other related commands include PDF, INVCDF, and RANDOM with subcommands NORMAL, T, F, and BINOMIAL and many others for other types of distributions.

8.2 OVERFITTING AND PARAMETER REDUNDANCY

Our second basic diagnostic tool is that of **overfitting**. After specifying and fitting what we believe to be an adequate model, we fit a more general model, that is, a model that contains the original model as a special case. For example, if an AR(2) model seems appropriate, we might overfit with an AR(3) model. The original AR(2) model would be confirmed if:

1. the estimate of the additional parameter, ϕ_3, is not significantly different from zero, and
2. the estimates for the parameters in common, ϕ_1 and ϕ_2, do not change significantly from their original estimates.

Exhibit 8.9 shows the results of overfitting an AR(3) to the U.S. unemployment series. Notice that the *t*-ratio for $\hat{\phi}_3$ is 1.94, which is not large enough to reject $\phi_3 = 0$. Under the AR(3) assumption, we have the estimates

$$\hat{\phi}_1 = 1.6770 \quad \text{and} \quad \hat{\phi}_2 = -0.9319$$

The somewhat striking change in $\hat{\phi}_2$ is not really that large, because the estimated standard deviation of $\hat{\phi}_2$ in the AR(3) model is 0.1577. We also note that, as predicted by Equations (7-53) through (7-61), the standard deviations of the estimated parameters are *larger* with the AR(3) assumption. Thus, the overfit confirms the original AR(2) model.

Of course, there is no unique way to overfit any given model. Another logical model for overfitting an AR(2) is an ARMA(2, 1). Exhibit 8.10 illustrates this model with the unemployment data. Again notice that the extra parameter estimate $\hat{\theta} = -0.2350$ has a *t*-ratio of -1.77, which is insignificant, and the estimates of ϕ_1 and ϕ_2 are not changed significantly. Also, there is very little change in the sum of squares of the residuals among the various models.

Exhibits 8.11 and 8.12 give the results of similar overfits for the ARI(1, 1) model and the data for the railroad bond yields. Both results confirm the original model.

EXHIBIT 8.9 AR(3) Overfit of the Unemployment Rate Series

```
MTB > arima (3,0,0) 'Unemp'

Estimates at each iteration
Iteration      SSE      Parameters
    0        145.247    0.100    0.100    0.100    3.645
    1        120.348    0.250    0.014    0.099    3.318
    2         98.883    0.400   -0.081    0.104    3.003
    3         80.130    0.550   -0.178    0.111    2.694
    4         63.886    0.700   -0.278    0.119    2.387
    5         50.048    0.850   -0.379    0.128    2.082
    6         38.549    1.000   -0.480    0.138    1.779
    7         29.342    1.150   -0.582    0.147    1.477
    8         22.399    1.300   -0.683    0.156    1.175
    9         17.714    1.450   -0.784    0.165    0.873
   10         15.302    1.600   -0.883    0.173    0.567
   11         14.966    1.673   -0.929    0.176    0.410
   12         14.964    1.677   -0.932    0.177    0.396
   13         14.964    1.677   -0.932    0.177    0.394
   14         14.964    1.677   -0.932    0.178    0.393
Relative change in each estimate less than  0.0010

Final Estimates of Parameters
Type      Estimate    St. Dev.   t-ratio
AR   1      1.6770     0.0910     18.43
AR   2     -0.9319     0.1577     -5.91
AR   3      0.1775     0.0915      1.94
Constant   0.39311    0.03245    12.11
Mean       5.0818     0.4195

No. of obs.:  121
Residuals:    SS = 14.8663  (backforecasts excluded)
              MS =  0.1271  DF = 117

Modified Box-Pierce chisquare statistic
Lag              12          24          36          48
Chisquare   15.5(DF= 9)  31.2(DF=21)  39.8(DF=33)  45.9(DF=45)
```

As we have noted, any ARMA(p, q) model can be considered as a special case of a more general ARMA model with the additional parameters equal to zero. However, when generalizing ARMA models, we must be aware of the problem of **parameter redundancy** or **lack of identifiability**.

To make these points clear, consider an ARMA(1, 2) model:

$$Z_t = \phi Z_{t-1} + a_t - \theta_1 a_{t-1} - \theta_2 a_{t-2} \tag{8-13}$$

Now replace t by $t - 1$ to obtain

$$Z_{t-1} = \phi Z_{t-2} + a_{t-1} - \theta_1 a_{t-2} - \theta_2 a_{t-3} \tag{8-14}$$

If we multiply both sides of Equation (8-14) by *any constant c* and then subtract (8-14) from (8-13), we obtain (after rearranging)

$$Z_t - (\phi + c)Z_{t-1} + \phi c Z_{t-2}$$
$$= a_t - (\theta_1 + c)a_{t-1} - (\theta_2 + c\theta_1)a_{t-2} + c\theta_2 a_{t-3} \tag{8-15}$$

This defines, apparently, an ARMA(2, 3) process. But notice that

$$(1 - \phi x)(1 - cx) = 1 - (\phi + c)x + \phi c x^2$$

EXHIBIT 8.10 ARMA(2, 1) Overfit of the Unemployment Rate Series

```
MTB > arima (2,0,1) 'Unemp'

Estimates at each iteration
Iteration       SSE      Parameters
     0       201.402    0.100    0.100    0.100    4.166
     1       160.881    0.250    0.015    0.078    3.824
     2       141.568    0.400   -0.043    0.141    3.344
     3       136.413    0.280   -0.021   -0.009    3.859
     4       129.487    0.173   -0.003   -0.159    4.320
     5        97.899    0.242   -0.057   -0.309    4.239
     6        79.644    0.392   -0.112   -0.289    3.746
     7        62.670    0.542   -0.170   -0.279    3.267
     8        48.028    0.692   -0.230   -0.273    2.796
     9        36.104    0.842   -0.291   -0.267    2.331
    10        26.861    0.992   -0.352   -0.260    1.871
    11        20.280    1.142   -0.415   -0.252    1.417
    12        16.340    1.292   -0.479   -0.242    0.969
    13        15.032    1.429   -0.537   -0.233    0.555
    14        15.024    1.433   -0.535   -0.234    0.523
    15        15.024    1.432   -0.534   -0.235    0.519
Unable to reduce sum of squares any further

Final Estimates of Parameters
Type       Estimate    St. Dev.   t-ratio
AR    1      1.4318     0.1162     12.32
AR    2     -0.5339     0.1139     -4.69
MA    1     -0.2350     0.1329     -1.77
Constant   0.51896     0.04014    12.93
Mean       5.0826      0.3932

No. of obs.:  121
Residuals:    SS = 14.9264  (backforecasts excluded)
              MS =  0.1276  DF = 117

Modified Box-Pierce chisquare statistic
Lag                 12           24          36           48
Chisquare   15.8(DF= 9)  31.5(DF=21)  40.6(DF=33)  46.7(DF=45)
```

and

$$(1 - \theta_1 x - \theta_2 x^2)(1 - cx) = 1 - (\theta_1 + c)x - (\theta_2 + c\theta_1)x^2 + c\theta_2 x^3$$

Thus the AR and MA characteristic polynomials in the ARMA(2, 3) process are obtained from the corresponding characteristic polynomials of the original ARMA(1, 2) model by multiplying both polynomials by the common factor $1 - cx$.

Even though $\{Z_t\}$ does satisfy the ARMA(2, 3) model of Equation (8-15), the parameters in that model are clearly *not unique*, since c was arbitrary. If the true model is ARMA(1, 2) and we attempt to estimate an ARMA(2, 3) model, the parameter estimates vary substantially as a rule. We don't even know what we are estimating!

Exhibits 8.13 and 8.14 illustrate the problem with a simulated white noise series. If we estimate this series as an AR(1) model (Exhibit 8.13), no difficulties arise—the generalization from white noise to AR(1) is unique with $\phi = 0$. Notice that the estimate $\hat{\phi} = 0.1545$ is not significantly different from zero. However, if we attempt to fit an ARMA(1, 1) model (Exhibit 8.14), the estimates change dramatically to $\hat{\phi} = -0.4067$ and $\hat{\theta} = 0.5822$ with large estimated

EXHIBIT 8.11 ARI(2, 1) Overfit of the Railroad Bond Yield Series

```
MTB > arima (2,1,0) 'RRBonds'

Estimates at each iteration
Iteration        SSE      Parameters
    0         9942.10    0.100    0.100
    1         9079.84    0.250    0.045
    2         8606.86    0.400   -0.011
    3         8507.29    0.502   -0.049
    4         8506.98    0.507   -0.052
    5         8506.98    0.508   -0.052
    6         8506.98    0.508   -0.052
Relative change in each estimate less than  0.0010

Final Estimates of Parameters
Type     Estimate    St. Dev.   t-ratio
AR   1     0.5077     0.1004      5.06
AR   2    -0.0521     0.1007     -0.52

Differencing: 1 regular difference
No. of obs.:  Original series 102, after differencing 101
Residuals:    SS = 8503.25  (backforecasts excluded)
              MS =   85.89  DF = 99

Modified Box-Pierce chisquare statistic
Lag              12          24          36          48
Chisquare    6.0(DF=10)  18.3(DF=22)  27.3(DF=34)  36.7(DF=46)
```

EXHIBIT 8.12 ARIMA(1, 1, 1) Overfit of the Railroad Bond Yield Series

```
MTB > arima (1,1,1) 'RRBonds'

Estimates at each iteration
Iteration        SSE      Parameters
    0         11113.0    0.100    0.100
    1          8925.2    0.250   -0.048
    2          8499.0    0.335   -0.165
    3          8495.1    0.357   -0.161
    4          8495.1    0.359   -0.160
    5          8495.1    0.359   -0.160
    6          8495.1    0.359   -0.160
Relative change in each estimate less than  0.0010

Final Estimates of Parameters
Type     Estimate    St. Dev.   t-ratio
AR   1     0.3592     0.1917      1.87
MA   1    -0.1595     0.2030     -0.79

Differencing: 1 regular difference
No. of obs.:  Original series 102, after differencing 101
Residuals:    SS = 8490.32  (backforecasts excluded)
              MS =   85.76  DF = 99

Modified Box-Pierce chisquare statistic
Lag              12          24          36          48
Chisquare    5.8(DF=10)  18.1(DF=22)  27.2(DF=34)  36.3(DF=46)
```

EXHIBIT 8.13 Estimation without Parameter Redundancy

```
MTB > arima (1,0,0) 'WhiteN'

Estimates at each iteration
Iteration        SSE        Parameters
     0        134.968     0.100    -0.043
     1        133.559     0.150    -0.119
     2        133.546     0.154    -0.128
     3        133.546     0.154    -0.128
     4        133.546     0.154    -0.128
Relative change in each estimate less than  0.0010

Final Estimates of Parameters
Type       Estimate    St. Dev.   t-ratio
AR    1      0.1545      0.0913      1.69
Constant   -0.12831     0.09635    -1.33
Mean       -0.1517      0.1140

No. of obs.:  121
Residuals:    SS = 133.526  (backforecasts excluded)
              MS =   1.122  DF = 119

Modified Box-Pierce chisquare statistic
Lag                12          24          36          48
Chisquare   5.8(DF=11)  14.8(DF=23)  19.2(DF=35)  24.9(DF=47)
```

EXHIBIT 8.14 Estimation with Parameter Redundancy

```
MTB > arima (1,0,1) 'WhiteN'

Estimates at each iteration
Iteration        SSE        Parameters
     0        137.951     0.100    0.100    -0.043
     1        133.756     0.164    0.031    -0.109
     2        133.190     0.020   -0.119    -0.132
     3        132.647    -0.123   -0.269    -0.154
     4        132.176    -0.264   -0.419    -0.178
     5        131.877    -0.399   -0.569    -0.205
     6        131.867    -0.407   -0.582    -0.208
     7        131.867    -0.407   -0.582    -0.209
Unable to reduce sum of squares any further

Final Estimates of Parameters
Type       Estimate    St. Dev.   t-ratio
AR    1     -0.4067      0.3666     -1.11
MA    1     -0.5822      0.3264     -1.78
Constant   -0.2092      0.1526     -1.37
Mean       -0.1487      0.1085

No. of obs.:  121
Residuals:    SS = 131.861  (backforecasts excluded)
              MS =   1.117  DF = 118

Modified Box-Pierce chisquare statistic
Lag                12          24          36          48
Chisquare   4.6(DF=10)  13.5(DF=22)  17.2(DF=34)  22.7(DF=46)
```

standard deviations. Notice also that the estimates ϕ and θ are roughly equal to each other, which indicate the problem of a common factor. Geometrically speaking, the sum-of-squares function is very flat along the line $\phi = \theta$ with no clear-cut minimum. Notice the small change in the error sum of squares (SSE) from the first iteration at $\phi = 0.164$, $\theta = 0.031$, to the seventh and final iteration at $\phi = -0.407$, $\theta = -0.582$.

We should also recall Equations (7-59) and (7-60), giving Var $(\hat{\phi})$ and Var $(\hat{\theta})$ for the ARMA(1, 1) case. Each has a denominator of $\phi - \theta$, again pointing to estimation difficulties even if ϕ and θ are only *nearly* equal.

In general, any ARMA model will also satisfy a more general ARMA model where the AR and MA characteristic polynomials are obtained by multiplying the original polynomials by the *same arbitrary polynomial*.* Clearly, the parameters in the more general model are *not unique* and would lead to serious problems in estimation if we attempted to estimate the more general model.

Thus we are led to the following rules of thumb:

1. Specify the original model carefully. If a simpler model seems at all promising, check it out before trying a more complicated model.

2. When overfitting, *do not* increase the orders of the AR and MA parts of the model simultaneously.

3. Extend the model in directions suggested by an analysis of the residuals. If after fitting an MA(1) model, substantial correlation remains at lag 2 in the residuals, try an MA(2), not an ARMA(1, 1).

Box and Jenkins (1976, pp. 298–299) give some results concerning the use of the residuals that suggest modification of the original model. As an example of these results, suppose that we (erroneously) fit an MA(1) model $Z_t = a_t - \theta a_{t-1}$ and find that the residuals a_t look like an MA(1) process: $a_t = b_t - \theta^* b_{t-1}$. Then, in fact,

$$Z_t = b_t - \theta^* b_{t-1} - \theta(b_{t-1} - \theta^* b_{t-2})$$
$$= b_t + (\theta + \theta^*)b_{t-1} + \theta\theta^* b_{t-2}$$

or an MA(2) process. We would then specify an MA(2) model for the next iteration of the model-building process.

CHAPTER 8 EXERCISES

8.1. For an AR(1) model with $n = 100$, $\phi \approx 0.5$, and $\mu \approx 10.8$, the lag 1 sample autocorrelation of the residuals is 0.5. Should we consider this unusual? (Why or why not?)

8.2. Repeat Exercise 8.1 for an MA(1) model with $n = 100$, $\theta \approx 0.5$, and $\mu \approx 10.8$.

---

* Employing the backshift operator of Appendix E, we note that if the correct model is $\phi(B)Z_t = \theta(B)a_t$, then $\{Z_t\}$ also satisfies $[P(B)\phi(B)]Z_t = [P(B)\theta(B)]a_t$, where $P(B)$ is *any* polynomial. In other words, the model $\phi(B)Z_t = \theta(B)a_t$ is uniquely parameterized only when $\phi(B)$ and $\theta(B)$ contain *no common factors*.

8.3. Based on a series of length $n = 200$, we fit an AR(2) model and obtain residual autocorrelations of $\hat{r}_1 = 0.13$, $\hat{r}_2 = 0.13$, and $\hat{r}_3 = 0.12$. If $\hat{\phi}_1 = 1.1$ and $\hat{\phi}_2 = -0.8$, do these residual autocorrelations support the AR(2) specification? Individually? Jointly?

8.4. We have seen that an AR(2) model is adequate for the U.S. unemployment rate series. To illustrate model misspecification, attempt to fit an AR(1) model to this series. Check the residuals for "whiteness" by looking at their autocorrelation function, their partial autocorrelation function, the modified Box–Pierce statistic, the runs test, the normality test, and so on. Comment on the results of each test or calculation.

8.5. Illustrate the problem of parameter redundancy by attempting to fit an ARMA(3, 1) model to the U.S. unemployment rate series [realizing that an AR(2) is adequate]. Comment on your results.

8.6. Use Minitab to simulate various ARMA models using a variety of models, parameter values, and sample sizes. For each such series, practice the diagnostic techniques given in this chapter.

8.7. Perform diagnostic checking on the model estimated in Exercise 7.9 for the Portland, Oregon, gasoline price series.

8.8. Using the Iowa nonfarm income series, continue the model-building process begun in Exercises 6.10 and 7.10.

CHAPTER 9 FORECASTING

One of the primary objectives of building a model for a time series is to be able to forecast the values for that series at future times. We also want to assess the precision of those forecasts. In this chapter we shall consider the calculation of forecasts and their properties for both deterministic trend models and for ARIMA models. Forecasts for models that combine deterministic trends with ARIMA stochastic components are considered briefly in Chapter 11.

For the most part we shall assume that the model is known *exactly*, including the specific values for all parameters. Although this is never true in practice, for large sample sizes the use of estimated parameters does not seriously affect forecasts.

9.1 MINIMUM MEAN SQUARE ERROR FORECASTS

Based on the available history of the series up to time t, $Z_t, Z_{t-1}, \ldots, Z_1$, we would like to forecast the value $Z_{t+\ell}$ ℓ time units into the future. We call time t the **origin** of the forecast and ℓ the **lead time** for the forecast.

The **minimum mean square error forecast,** denoted $\hat{Z}_t(\ell)$, is then given by

$$\hat{Z}_t(\ell) = E(Z_{t+\ell} \mid Z_t, Z_{t-1}, \ldots, Z_1) \tag{9-1}$$

(Appendices F and G review the properties of conditional expectation and minimum mean square error prediction.)

The computation and properties of this conditional expectation as related to forecasting will be our concern for the remainder of this chapter.

9.2 DETERMINISTIC TRENDS

Consider once more the deterministic trend model of Chapter 3,

$$Z_t = \mu_t + X_t \tag{9-2}$$

where the stochastic component, X_t, has a mean of zero. For this section we shall assume that $\{X_t\}$ is, in fact, white noise with variance γ_0.

For the model in Equation (9-2), we have

$$
\begin{aligned}
\hat{Z}_t(\ell) &= E(\mu_{t+\ell} + X_{t+\ell} \mid Z_t, Z_{t-1}, \ldots, Z_1) \\
&= E(\mu_{t+\ell} \mid Z_t, Z_{t-1}, \ldots, Z_1) + E(X_{t+\ell} \mid Z_t, Z_{t-1}, \ldots, Z_1) \\
&= \mu_{t+\ell} + E(X_{t+\ell})
\end{aligned}
$$

or

$$\hat{Z}_t(\ell) = \mu_{t+\ell}, \quad \ell \geq 1 \tag{9-3}$$

since for $\ell \geq 1$, $X_{t+\ell}$ is independent of $Z_t, Z_{t-1}, \ldots, Z_1$.

Thus in this simple case, forecasting amounts to extrapolating the deterministic trend.

For the linear trend case, $\mu_t = \beta_0 + \beta_1 t$, the forecast is

$$\hat{Z}_t(\ell) = \beta_0 + \beta_1(t + \ell) \tag{9-4}$$

As we emphasized in Chapter 3, the model assumes that the *same* linear trend persists into the future, and the forecast reflects that assumption. Note that it is the lack of statistical dependence between $Z_{t+\ell}$ and $Z_t, Z_{t-1}, \ldots, Z_1$ that prevents us from improving on $\mu_{t+\ell}$ as a forecast.

For seasonal trends where, say, $\mu_t = \mu_{t+12}$, our forecast is $\hat{Z}_t(\ell) = \mu_{t+\ell} = \mu_{t+12+\ell} = \hat{Z}_t(\ell + 12)$. Thus the forecast will also be periodic, as desired.

The **forecast error,** $e_t(\ell)$, is given by

$$
\begin{aligned}
e_t(\ell) &= Z_{t+\ell} - \hat{Z}_t(\ell) \\
&= \mu_{t+\ell} + X_{t+\ell} - \mu_{t+\ell} \\
&= X_{t+\ell}
\end{aligned}
$$

so that

$$
\begin{aligned}
E[e_t(\ell)] &= E(X_{t+\ell}) \\
&= 0
\end{aligned}
$$

that is, the forecasts are **unbiased.** Also,

$$
\begin{aligned}
\text{Var}\,[e_t(\ell)] &= \text{Var}\,(X_{t+\ell}) \\
&= \gamma_0
\end{aligned} \tag{9-5}
$$

is the **forecast error variance** for all lead times ℓ.

Applying these results to the cosine trend model for the average monthly temperature series (Exhibit 3.5) gives the estimated trend as

$$\hat{\mu}_t = 46.2659 - 15.2318 \sin\frac{\pi t}{6} - 22.0457 \cos\frac{\pi t}{6} \tag{9-6}$$

with an estimated standard deviation of the forecast error of $s = 3.706$.

In this example, January corresponds to $t = 1, 13, 25$, and so on. Thus our forecast for the average monthly temperature of any future January is given by $\hat{\mu}_1$, which is

$$\hat{\mu}_1 = 46.2659 - 15.2318 \sin\frac{\pi}{6} - 22.0457 \cos\frac{\pi}{6}$$

$$= 19.6^0\,\text{F}$$

with a forecast error standard deviation of $3.7°$ F.

For a future June, the forecast is

$$\hat{\mu}_6 = 46.2659 - 15.2318 \sin \pi - 22.0457 \cos \pi$$

$$= 68.3°\,\text{F}$$

9.3 ARIMA FORECASTING

For ARIMA models the forecasts can be expressed in several different ways. Each expression contributes to our understanding of the overall forecasting procedure with respect to computing, updating, assessing precision, or long-term forecasting behavior.

AR(1)

We shall first illustrate many of the ideas with the simple AR(1) process with a nonzero mean that satisfies

$$Z_t - \mu = \phi(Z_{t-1} - \mu) + a_t \tag{9-7}$$

Consider the problem of forecasting 1 time unit into the future. Replacing t by $t + 1$ in Equation (9-7), we have

$$Z_{t+1} - \mu = \phi(Z_t - \mu) + a_{t+1} \tag{9-8}$$

Given $Z_t, Z_{t-1}, \ldots, Z_1$, we take conditional expectations of both sides of Equation (9-8) and obtain

$$\hat{Z}_t(1) - \mu = \phi[E(Z_t \mid Z_t, Z_{t-1}, \ldots, Z_1) - \mu] + E(a_{t+1} \mid Z_t, Z_{t-1}, \ldots, Z_1) \tag{9-9}$$

Now from the properties of conditional expectation [Equation (F-3)], we have that

$$E(Z_t \mid Z_t, Z_{t-1}, \ldots, Z_1) = Z_t \tag{9-10}$$

Also, since a_{t+1} is independent of $Z_t, Z_{t-1}, \ldots, Z_1$, we obtain from Equation (F-6)

$$E(a_{t+1} \mid Z_t, Z_{t-1}, \ldots, Z_1) = E(a_{t+1}) = 0 \tag{9-11}$$

Thus Equation (9-9) can be written as

$$\hat{Z}_t(1) = \mu + \phi(Z_t - \mu) \tag{9-12}$$

In words, a proportion ϕ of the current deviation from the process mean is added to the process mean to forecast the next process value.

Now consider a general lead time ℓ. Replacing t by $t + \ell$ in Equation (9-7) and taking conditional expectations of both sides produces

$$\hat{Z}_t(\ell) = \mu + \phi[\hat{Z}_t(\ell - 1) - \mu], \qquad \ell \geq 1 \tag{9-13}$$

since $E(Z_{t+\ell-1} \mid Z_t, Z_{t-1}, \ldots, Z_1) = \hat{Z}_t(\ell - 1)$ and, for $\ell \geq 1$, $a_{t+\ell}$ is independent of $Z_t, Z_{t-1}, \ldots, Z_1$.

Equation (9-13), which is recursive in ℓ, shows how the forecast for any lead time ℓ can be built up from the forecasts for shorter lead times by starting with the initial forecast $\hat{Z}_t(1)$ computed using Equation (9-12). The forecast $\hat{Z}_t(2)$ is then obtained from $\hat{Z}_t(2) = \mu + \phi[\hat{Z}_t(1) - \mu]$, then $\hat{Z}_t(3)$ from $\hat{Z}_t(2)$, and so on, until the desired $\hat{Z}_t(\ell)$ is found. Equation (9-13) and its generalizations for other ARIMA models are most convenient for actually computing the forecasts. Equation (9-13) is sometimes called the **difference-equation form** of the forecast.

However, Equation (9-13) can also be solved to yield a more explicit expression for the forecasts in terms of the observed history of the series. Iterating backwards on ℓ in (9-13), we have

$$\begin{aligned}
\hat{Z}_t(\ell) &= \phi[\hat{Z}_t(\ell - 1) - \mu] + \mu \\
&= \phi(\phi[\hat{Z}_t(\ell - 2) - \mu]) + \mu \\
&\;\;\vdots \\
&= \phi^{\ell-1}[\hat{Z}_t(1) - \mu] + \mu
\end{aligned}$$

or

$$\hat{Z}_t(\ell) = \mu + \phi^\ell(Z_t - \mu), \qquad \ell \geq 1 \tag{9-14}$$

The current deviation from the mean is discounted by a factor ϕ^ℓ, whose magnitude decreases with increasing lead time ℓ. The discounted deviation is then added to the process mean to produce the lead ℓ forecast.

As a numerical example, suppose we have an AR(1) model with $\phi = 0.7$, $\mu = 10.2$, and our current observed value is 10.6. We would forecast 1 time period ahead as

$$\begin{aligned}
\hat{Z}_t(1) &= 10.2 + 0.7(10.6 - 10.2) \\
&= 10.2 + 0.28 \\
&= 10.48
\end{aligned}$$

For lead time 2, we have from Equation (9-13)

$$\begin{aligned}
\hat{Z}_t(2) &= 10.2 + 0.7(10.48 - 10.2) \\
&= 10.2 + 0.196 \\
&= 10.396
\end{aligned}$$

Alternatively, we can use Equation (9-14):

$$\begin{aligned}
\hat{Z}_t(2) &= 10.2 + (0.7)^2(10.6 - 10.2) \\
&= 10.396
\end{aligned}$$

At lead 5 we have

$$\begin{aligned}
\hat{Z}_t(5) &= 10.2 + (0.7)^5(10.6 - 10.2) \\
&= 10.267228
\end{aligned}$$

and by lead 10 the forecast is

$$\hat{Z}_t(10) = 10.21129901$$

which is very nearly μ.

In general, since $|\phi| < 1$, we have simply

$$\hat{Z}_t(\ell) \approx \mu \quad \text{for large } \ell \tag{9-15}$$

Later we shall see that Equation (9-15) holds for *all stationary* ARMA models.

Consider now the **one-step-ahead forecast error,** $e_t(1)$. From Equations (9-8) and (9-12) we have

$$e_t(1) = Z_{t+1} - \hat{Z}_t(1)$$
$$= \phi(Z_t - \mu) + \mu + a_{t+1} - [\phi(Z_t - \mu) + \mu]$$

or

$$e_t(1) = a_{t+1} \qquad\qquad \text{(9-16)}$$

The white noise process $\{a_t\}$ can now be reinterpreted as a sequence of one-step-ahead forecast errors. We shall see that Equation (9-16) persists for completely general ARIMA models. Note also that Equation (9-16) implies that the forecast error $e_t(1)$ is independent of the history of the process $Z_t, Z_{t-1}, \dots$ up to time t. If this were not so, the dependence could be exploited to improve our forecast.

Equation (9-16) also implies that our one-step-ahead forecast error variance is given by

$$\text{Var}\,[e_t(1)] = \sigma_a^2 \qquad\qquad \text{(9-17)}$$

To investigate the properties of the forecast errors for longer leads, it is convenient to express the AR(1) model in general linear process, or MA(∞), form. From Equation (4-16) we recall that

$$Z_t = \mu + a_t + \phi a_{t-1} + \phi^2 a_{t-2} + \cdots \qquad\qquad \text{(9-18)}$$

Then (9-14) and (9-18) together yield

$$e_t(\ell) = Z_{t+\ell} - \mu - \phi^\ell(Z_t - \mu)$$
$$= a_{t+\ell} + \phi a_{t+\ell-1} + \cdots + \phi^{\ell-1}a_{t+1} + \phi^\ell a_t$$
$$+ \cdots - \phi^\ell(a_t + \phi a_{t-1} + \cdots)$$

so that

$$e_t(\ell) = a_{t+\ell} + \phi a_{t+\ell-1} + \phi^2 a_{t+\ell-2} + \cdots + \phi^{\ell-1}a_{t+1} \qquad\qquad \text{(9-18)}$$

which can also be written as

$$e_t(\ell) = \sum_{j=0}^{\ell-1} \psi_j a_{t+\ell-j} \qquad\qquad \text{(9-19)}$$

Equation (9-19) will be shown to hold for general ARIMA models (see Equations (9-44) and (9-51)).

Note that $E[e_t(\ell)] = 0$; thus the forecasts are **unbiased.** Furthermore, from Equation (9-19) we have

$$\text{Var}\,[e_t(\ell)] = \sigma_a^2 \sum_{j=0}^{\ell-1} \psi_j^2 \qquad\qquad \text{(9-20)}$$

We see that the forecast error variance increases as the lead ℓ increases. Contrast this with the result given in Equation (9-5).

In particular, for the AR(1) case

$$\text{Var}\,[e_t(\ell)] = \sigma_a^2 \frac{1 - \phi^{2\ell}}{1 - \phi^2}$$

(9-21)

which we obtain by summing a finite geometric series.
For long lead times, we see that

$$\text{Var}\,[e_t(\ell)] \approx \frac{\sigma_a^2}{1 - \phi^2} \quad \text{for large } \ell$$

(9-22)

or, by Equation (4-12),

$$\text{Var}\,[e_t(\ell)] \approx \text{Var}\,(Z_t) = \gamma_0 \quad \text{for large } \ell$$

(9-23)

Equation (9-23) will be shown to be valid for *all stationary* ARMA processes.

MA(1)

To illustrate how to solve the problems that arise in forecasting moving average or mixed models, consider the MA(1) case with a nonzero mean:

$$Z_t = \mu + a_t - \theta a_{t-1}$$

Again replacing t by $t + 1$ and taking conditional expectations of both sides, we have

$$\hat{Z}_t(1) = \mu - \theta E(a_t \,|\, Z_t, Z_{t-1}, \ldots, Z_1)$$

(9-24)

However, for an invertible model, Equation (4-44) shows that a_t is a *function* of $Z_t, Z_{t-1}, \ldots$. Thus by Equation (F-3) we have

$$E(a_t \,|\, Z_t, Z_{t-1}, \ldots, Z_1) = a_t$$

(9-25)

[In fact, an approximation is involved in this equation, since we are conditioning on only $Z_t, Z_{t-1}, \ldots, Z_1$ and not on the *infinite* history of the process. However, if, as in practice, t is large and the model is invertible, the error in the approximation will be very small. If the model is not invertible—for example, if we have overdifferenced the data—then Equation (9-25) is not even approximately valid (see Harvey, 1981c, p. 161).]

Using (9-25) in (9-24), we thus have the one-step-ahead forecast for the invertible MA(1) model:

$$\hat{Z}_t(1) = \mu - \theta a_t$$

(9-26)

The computation of a_t will be a by-product of estimating the model (see Equation (7-42)).

Notice once more that

$$
\begin{aligned}
e_t(1) &= Z_{t+1} - \hat{Z}_t(1) \\
&= \mu + a_{t+1} - \theta a_t - (\mu - \theta a_t) \\
&= a_{t+1}
\end{aligned}
$$

as in Equation (9-16), and thus Equation (9-17) also obtains.

For longer lead times we have

$$\hat{Z}_t(\ell) = \mu + E(a_{t+\ell} \mid Z_t, Z_{t-1}, \ldots, Z_1) - \theta E(a_{t+\ell-1} \mid Z_t, Z_{t-1}, \ldots, Z_1)$$

But for $\ell > 1$ both $a_{t+\ell}$ and $a_{t+\ell-1}$ are independent of $Z_t, Z_{t-1}, \ldots, Z_1$. Consequently, these conditional expected values are zero, and we have

$$\hat{Z}_t(\ell) = \mu \quad \text{for } \ell > 1 \tag{9-27}$$

Notice here that Equation (9-15) holds *exactly* for the MA(1) case when $\ell > 1$. Since for this model we trivially have $\psi_1 = -\theta$ and $\psi_j = 0$ for $j > 1$, Equations (9-19) and (9-20) also hold.

THE RANDOM WALK WITH DRIFT

To illustrate forecasting with nonstationary ARIMA series, consider the random walk with drift defined by

$$Z_t = Z_{t-1} + \theta_0 + a_t \quad \text{for } t \geq -m \tag{9-28}$$

Here

$$\hat{Z}_t(1) = E(Z_t \mid Z_t, Z_{t-1}, \ldots, Z_1) + \theta_0 + E(a_{t+1} \mid Z_t, Z_{t-1}, \ldots, Z_1)$$

so that

$$\hat{Z}_t(1) = Z_t + \theta_0 \tag{9-29}$$

Similarly, the difference-equation form for the lead ℓ forecast is

$$\hat{Z}_t(\ell) = \hat{Z}_t(\ell - 1) + \theta_0, \quad \ell \geq 1 \tag{9-30}$$

Iterating backwards on ℓ yields the explicit expression

$$\hat{Z}_t(\ell) = Z_t + \ell\theta_0, \quad \ell \geq 1 \tag{9-31}$$

In contrast to Equation (9-15), if $\theta_0 \neq 0$, the forecast does not converge for long leads ℓ but follows a straight line with slope θ_0 for all ℓ.

Note that the presence or absence of the constant term θ_0 significantly alters the forecast. For this reason constant terms should not be included in nonstationary ARIMA models unless the evidence is clear that the mean of the differenced series is significantly different from zero. Equation (3-3) for the variance of the sample mean will help assess this significance.

However, as we have seen in the AR(1) and MA(1) cases, the one-step-ahead forecast error is

$$e_t(1) = Z_{t+1} - \hat{Z}_t(1) = a_{t+1}$$

Also,

$$
\begin{aligned}
e_t(\ell) &= Z_{t+\ell} - \hat{Z}_t(\ell) \\
&= (Z_t + \ell\theta_0 + a_{t+1} + \cdots + a_{t+\ell}) - (Z_t + \ell\theta_0) \\
&= \sum_{j=0}^{\ell-1} a_{t+\ell-j}
\end{aligned}
$$

which agrees with Equation (9-19), since in this model $\psi_j = 1$ for all j. (See Equation (5-16) with $\theta = 0$.)

From Equation (9-20) we have that

$$\text{Var}\,[e_t(\ell)] = \sigma_a^2 \sum_{j=0}^{\ell-1} 1^2$$

that is,

$$\text{Var}\,[e_t(\ell)] = \ell\sigma_a^2, \qquad \ell \geq 1 \tag{9-32}$$

In contrast to the stationary case, here $\text{Var}\,[e_t(\ell)]$ grows without limit as the forecast lead time ℓ increases. We shall see that this property is characteristic of the forecast error variance for *all nonstationary* ARIMA processes.

RESULTS FOR THE GENERAL STATIONARY ARMA CASE

For the general stationary, invertible ARMA(p, q) model, the **difference-equation form** for computing forecasts is given by

$$\hat{Z}_t(\ell) = \phi_1\hat{Z}_t(\ell - 1) + \phi_2\hat{Z}_t(\ell - 2) + \cdots + \phi_p\hat{Z}_t(\ell - p) + \theta_0$$
$$- \theta_1 E(a_{t+\ell-1} \,|\, Z_t, Z_{t-1}, \ldots, Z_1) - \theta_2 E(a_{t+\ell-2} \,|\, Z_t, Z_{t-1}, \ldots, Z_1)$$
$$- \cdots - \theta_q E(a_{t+\ell-q} \,|\, Z_t, Z_{t-1}, \ldots, Z_1) \tag{9-33}$$

where

$$E(a_{t+j} \,|\, Z_t, Z_{t-1}, \ldots, Z_1) = \begin{cases} 0 & \text{for } j \geq 1 \\ a_{t+j} & \text{for } j \leq 0 \end{cases} \tag{9-34}$$

We note that $\hat{Z}_t(j)$ is a true forecast for $j > 0$, but

$$\hat{Z}_t(j) = Z_{t+j} \quad \text{for } -(p - 1) \leq j \leq 0 \tag{9-35}$$

As in Equation (9-25), Equation (9-34) for $j \leq 0$ involves some minor approximation. For an invertible model, Equation (4-48) shows that, using the π-weights, a_t can be expressed as a linear combination of the infinite sequence $Z_t, Z_{t-1}, \ldots$. However, the π-weights die out exponentially fast, and the approximation assumes that π_j is negligible for $j > t - q$.

As an example, consider an ARMA(1, 1) model. We have

$$\hat{Z}_t(1) = \phi Z_t + \theta_0 - \theta a_t \tag{9-36}$$

with

$$\hat{Z}_t(2) = \phi\hat{Z}_t(1) + \theta_0$$

and, more generally,

$$\hat{Z}_t(\ell) = \phi\hat{Z}_t(\ell - 1) + \theta_0 \quad \text{for } \ell \geq 2 \tag{9-37}$$

using Equation (9-36) to get the recursion started.

Equations (9-36) and (9-37) can also be solved by iteration to get the alternative explicit expression

$$\hat{Z}_t(\ell) = \mu + \phi^\ell(Z_t - \mu) - \phi^{\ell-1}\theta a_t, \quad \text{for } \ell \geq 1 \qquad \text{(9-38)}$$

As Equations (9-33) and (9-34) indicate, the noise terms $a_t, a_{t-1}, \ldots, a_{t-(q-1)}$ appear directly in the computation of the forecasts for leads $\ell = 1, 2, \ldots, q$. However, for $\ell > q$ the autoregressive portion of the difference equation takes over, and we have

$$\hat{Z}_t(\ell) = \phi_1\hat{Z}_t(\ell - 1) + \phi_2\hat{Z}_t(\ell - 2) + \cdots + \phi_p\hat{Z}_t(\ell - p) + \theta_0 \quad \text{for } \ell > q \qquad \text{(9-39)}$$

Thus the general nature of the forecast for long lead times will be determined by the autoregressive parameters $\phi_1, \phi_2, \ldots, \phi_p$ (and the constant term θ_0, which is related to the mean of the process).

Recalling from Equation (5-26) that $\theta_0 = \mu(1 - \phi_1 - \phi_2 - \cdots - \phi_p)$, we can rewrite (9-39) in terms of deviations from μ as

$$\hat{Z}_t(\ell) - \mu = \phi_1[\hat{Z}_t(\ell - 1) - \mu] + \phi_2[\hat{Z}_t(\ell - 2) - \mu]$$
$$+ \cdots + \phi_p[\hat{Z}_t(\ell - p) - \mu] \quad \text{for } \ell > q \qquad \text{(9-40)}$$

As a function of lead time ℓ, $\hat{Z}_t(\ell) - \mu$ satisfies the same recursive relationship as does the autocorrelation function ρ_k of the process (see Equations (4-32) and (4-42)). Thus, as in Sections 4.3 and 4.4, the roots of the AR characteristic equation will determine the general behavior of $\hat{Z}_t(\ell) - \mu$ for $\ell > q$. In particular, $\hat{Z}_t(\ell) - \mu$ can be expressed as a linear combination of exponentially decaying terms in ℓ (corresponding to the *real* roots) and damped sine wave terms (corresponding to the *complex* roots).

Thus for any *stationary* ARMA model, $\hat{Z}_t(\ell) - \mu$ decays to zero as ℓ increases, and the long-lead forecast is simply the process mean μ as given in Equation (9-15). This agrees with common sense, since for stationary ARMA models the dependence dies out as the time span between observations increases, and this dependence is the only reason we can improve on the "naive" forecast of using μ alone.

As a numerical example of forecasting for an AR(2) model, consider the model estimated in Exhibit 7.2 for the U.S. quarterly unemployment series. We have $\hat{\phi}_1 = 1.5630$, $\hat{\phi}_2 = -0.6583$, $\hat{\phi}_0 = 0.48430$, and thus $\hat{\mu} = 5.0810$. These values for ϕ_1 and ϕ_2 correspond to complex roots for the AR characteristic equation, so we should expect oscillatory behavior around μ for the forecasts. In the AR(2) case, Equations (9-33), (9-34), and (9-35) become

$$\hat{Z}_t(1) = \phi_1 Z_t \quad + \phi_2 Z_{t-1} \quad + \theta_0$$
$$\hat{Z}_t(2) = \phi_1\hat{Z}_t(1) \quad + \phi_2 Z_t \quad + \theta_0$$

and

$$\hat{Z}_t(\ell) = \phi_1\hat{Z}_t(\ell - 1) + \phi_2\hat{Z}_t(\ell - 2) + \theta_0 \quad \text{for } \ell > 2$$

Exhibit 9.1 shows a plot of $\hat{Z}_t(\ell)$ versus lead time for 30 quarters into the future for the estimated unemployment series model with $Z_t = 6.20$ and $Z_{t-1} = 6.63$. Notice that for $\ell \geq 15$, $\hat{Z}_t(\ell) \approx \hat{\mu} = 5.0810$.

EXHIBIT 9.1 Forecasts for AR(2) Model of U.S. Unemployment Rate Series

```
MTB > tsplot 4 'Forecast';
SUBC> origin is 121.

Forecast-
         - 1
         -
 5.700+
         -
         -
         -
         -
         - 2
 5.400+
         -
         -
         -
         - 3
         -
 5.100+                    3412341234123412
         -      4          12
         -              34
         -      1     2
         -      2341
 4.800+
          +---+---+---+---+---+---+---+
             128     136     144     152
```

For the general ARMA model, we can use Equation (9-33) to calculate the one-step-ahead forecast error:

$$e_t(1) = Z_{t+1} - \hat{Z}_t(1)$$

$$= \phi_1 Z_t + \phi_2 Z_{t-1} + \cdots + \phi_p Z_{t-p+1} + \theta_0 + a_{t+1} - \theta_1 a_t - \cdots - \theta_q a_{t-q+1}$$

$$- (\phi_1 Z_t + \phi_2 Z_{t-1} + \cdots + \phi_p Z_{t-p+1} + \theta_0 - \theta_1 a_t$$

$$- \theta_2 a_{t-1} - \cdots - \theta_q a_{t-q+1})$$

$$= a_{t+1}$$

so that Equation (9-16) and thus (9-17) are valid quite generally.

To argue the validity of Equation (9-19) for $e_t(\ell)$ in the present generality, we need to consider a new representation for ARIMA processes. Appendix H shows that any ARIMA model can be written in **truncated linear process** form as

$$Z_{t+\ell} = C_t(\ell) + I_t(\ell), \qquad \ell \geq 1 \tag{9-41}$$

where, for our present purposes, we need only know that

$$C_t(\ell) \quad \text{is a certain function of} \quad Z_t, Z_{t-1}, \ldots \tag{9-42}$$

and

$$I_t(\ell) = a_{t+\ell} + \psi_1 a_{t+\ell-1} + \psi_2 a_{t+\ell-2} + \cdots + \psi_{\ell-1} a_{t+1} \quad \text{for } \ell \geq 1 \tag{9-43}$$

Furthermore, for invertible models with t reasonably large, $C_t(\ell)$ is approximately a function of the *finite* history $Z_t, Z_{t-1}, \ldots, Z_1$.

Thus, using Equations (F-3) and (F-6), we have

$$\hat{Z}_t(\ell) = E(C_t(\ell) \mid Z_t, Z_{t-1}, \ldots, Z_1) + E[I_t(\ell) \mid Z_t, Z_{t-1}, \ldots, Z_1]$$

$$= C_t(\ell)$$

Finally,

$$\begin{aligned}
e_t(\ell) &= Z_{t+\ell} - \hat{Z}_t(\ell) \\
&= C_t(\ell) + I_t(\ell) - C_t(\ell) \\
&= I_t(\ell) \\
&= \sum_{j=0}^{\ell-1} \psi_j a_{t+\ell-j}
\end{aligned} \qquad \text{(9-44)}$$

Thus, for a general invertible ARIMA process,

$$E[e_t(\ell)] = 0, \qquad \ell \geq 1 \qquad \text{(9-45)}$$

and

$$\text{Var}\,[e_t(\ell)] = \sigma_a^2 \sum_{j=0}^{\ell-1} \psi_j^2, \qquad \ell \geq 1 \qquad \text{(9-46)}$$

From Equations (4-4) and (9-46), we see that for long lead times in stationary ARMA models we have

$$\text{Var}\,[e_t(\ell)] \approx \sigma_a^2 \sum_{j=0}^{\infty} \psi_j^2$$

or

$$\text{Var}\,[e_t(\ell)] \approx \gamma_0 \quad \text{for large } \ell \qquad \text{(9-47)}$$

NONSTATIONARY ARIMA MODELS

As the random walk shows, forecasting for nonstationary ARIMA models is quite similar to forecasting for stationary ARMA models, but there are some striking differences. Recall from Equation (5-12) that an ARIMA$(p, 1, q)$ model can be written as a nonstationary ARMA$(p + 1, q)$ model. We shall write this as

$$\begin{aligned}
Z_t = \varphi_1 Z_{t-1} &+ \varphi_2 Z_{t-2} + \cdots + \varphi_{p+1} Z_{t-p-1} + \theta_0 + a_t \\
&- \theta_1 a_{t-1} - \theta_2 a_{t-2} - \cdots - \theta_q a_{t-q} \quad \text{for } t > -m
\end{aligned} \qquad \text{(9-48)}$$

where the φ coefficients are directly related to the block ϕ coefficients. In particular,

$$\begin{aligned}
\varphi_1 &= 1 + \phi_1 \\
\varphi_j &= \phi_j - \phi_{j-1}, \qquad j = 2, 3, \ldots, p \\
\varphi_{p+1} &= -\phi_p
\end{aligned} \qquad \text{(9-49)}$$

and

For a general order of differencing d, we would have $p + d$ φ coefficients.

From this representation we can immediately extend Equations (9-33), (9-34), and (9-35) to cover the nonstationary cases by replacing p by $p + d$ and ϕ_j by φ_j.

As an example of the necessary calculations, consider the ARIMA$(1, 1, 1)$ case. Here

$$Z_t - Z_{t-1} = \phi(Z_{t-1} - Z_{t-2}) + \theta_0 + a_t - \theta a_{t-1}$$

so that

$$Z_t = (1 + \phi)Z_{t-1} - \phi Z_{t-2} + \theta_0 + a_t - \theta a_{t-1}$$

Thus

and

$$\hat{Z}_t(1) = (1 + \phi)Z_t - \phi Z_{t-1} + \theta_0 - \theta a_t$$
$$\hat{Z}_t(2) = (1 + \phi)\hat{Z}_t(1) - \phi Z_t + \theta_0$$
$$\hat{Z}_t(\ell) = (1 + \phi)\hat{Z}_t(\ell - 1) - \phi\hat{Z}_t(\ell - 2) + \theta_0 \quad \text{for } \ell > 2$$

(9-50)

For the general invertible ARIMA model, the truncated linear process representation given in Equations (9-41), (9-42), and (9-43) and the calculations following these equations show that we can write

and so

$$e_t(\ell) = \sum_{j=0}^{\ell-1} \psi_j a_{t+\ell-j}, \quad \ell \geq 1$$

(9-51)

and

$$E[e_t(\ell)] = 0, \quad \ell \geq 1$$

(9-52)

$$\text{Var}\,[e_t(\ell)] = \sigma_a^2 \sum_{j=0}^{\ell-1} \psi_j^2, \quad \ell \geq 1$$

(9-53)

However, for nonstationary models the ψ_j-weights do *not* decay to zero as j increases. For example, for the random walk model, $\psi_j = 1$ for all j; for the IMA(1, 1) model, $\psi_j = 1 - \theta$ for $j \geq 1$; for the IMA(2, 2) case, $\psi_j = 1 + \theta_2 + (1 - \theta_1 - \theta_2)j$ for $j \geq 1$; and for the ARI(1, 1) model, $\psi_j = (1 - \phi^{j+1})/(1 - \phi)$ (see Chapter 5).

Thus, for any nonstationary model, Equation (9-53) shows that the forecast error variance will grow without bound as the lead time ℓ increases. This fact should not be too surprising, since with nonstationary series the distant future is quite uncertain.

9.4 PREDICTION LIMITS

As in all statistical endeavors, in addition to forecasting or predicting the unknown $Z_{t+\ell}$, we would like to assess the precision of our prediction.

DETERMINISTIC TRENDS

For the deterministic trend model with a white noise stochastic component $\{X_t\}$, we recall that

$$\hat{Z}_t(\ell) = \mu_{t+\ell}$$

and

$$\text{Var}\,[e_t(\ell)] = \text{Var}\,(X_{t+\ell}) = \gamma_0$$

If the stochastic component is normally distributed, then the forecast error $e_t(\ell) = Z_{t+\ell} - \hat{Z}_t(\ell) = X_{t+\ell}$ is also normally distributed.

Thus for a given confidence level $1 - \alpha$, we could use a standard normal table (or the Minitab INVCDF command with subcommand NORMAL) to find the $(1 - \alpha/2)100$ percentage point, $z(1 - \alpha/2)$ say, and claim that

$$P\left[-z\left(1 - \frac{\alpha}{2}\right) < \frac{Z_{t+\ell} - \hat{Z}_t(\ell)}{\sqrt{\text{Var}\,[e_t(\ell)]}} < z\left(1 - \frac{\alpha}{2}\right)\right] = 1 - \alpha$$

or, equivalently,

$$P\left\{\hat{Z}_t(\ell) - z\left(1 - \frac{\alpha}{2}\right)\sqrt{\text{Var}\,[e_t(\ell)]} < Z_{t+\ell} < \hat{Z}_t(\ell) + z\left(1 - \frac{\alpha}{2}\right)\sqrt{\text{Var}\,[e_t(\ell)]}\right\}$$

$$= 1 - \alpha$$

Thus we may be $(1 - \alpha)100\%$ confident that the future observation $Z_{t+\ell}$ will be contained within the prediction limits

$$\hat{Z}_t(\ell) \pm z\left(1 - \frac{\alpha}{2}\right)\sqrt{\text{Var}\,[e_t(\ell)]} \tag{9-54}$$

For the average monthly temperature series with cosine trend, we had predicted the January average as 19.6° F and the June average as 68.3° F, each with an (estimated) standard deviation of forecast error of 3.7° F. Thus 95% prediction limits for January and June are, respectively,

$$19.6 \pm 1.96(3.7) = 19.6 \pm 7.25 = 12.35° \quad \text{to} \quad 26.85° \text{ F}$$

and

$$68.3 \pm 1.96(3.7) = 68.3 \pm 7.25 = 61.05° \quad \text{to} \quad 75.55° \text{ F}$$

Exhibit 9.2 shows a plot of the forecasts and upper and lower 95% prediction limits over a full year, from January through December.

EXHIBIT 9.2 Forecasts and Prediction Limits for Average Monthly Temperatures, Dubuque, Iowa

```
MTB > mplot 'Forecast' vs 'Lead', 'Upper' vs 'Lead', 'Lower' vs 'Lead'

       80.0+                              B
          -                         B          B
  Forecast-                              A
          -                         A          A   B
          -                    B         C   C
       60.0+                   A   C                A
          -                                         C   B
          -              B   C                      A
          -              A
          -                                    C   B
       40.0+        B   C
          -
          -         A                              A
          -      B                                     B
          -   B      C                             C
       20.0+      A                                    A
          -   A
          -      C                                     C
          -   C
            --------+--------+--------+--------+--------+--------+-Lead
                 2.00     4.00     6.00     8.00    10.00    12.00
```

Readers who are familiar with standard regression analysis will recall that since the forecast involves *estimated* parameters, the correct forecast error variance is given by $\gamma_0[1 + (1/n) + c_{n,\ell}]$, where $c_{n,\ell}$ is a certain function of the sample size n and of the lead time ℓ. However, it can be shown that for the types of trends that we are considering (namely, cosines and polynomials) and for large sample sizes n, the $1/n$ and $c_{n,\ell}$ are both negligible relative to 1. For example, with a cosine trend of period 12 over $N = n/12$ years, we have that $c_{n,\ell} = 2/n$; thus the correct forecast error variance is $\gamma_0[1 + (3/n)]$ rather than our approximate γ_0. For a linear time trend model, it can be shown that $c_{n,\ell} = 3(n + 2\ell - 1)^2/[n(n^2 - 1)] \approx 3/n$ for moderate lead ℓ and large n. Thus, again our approximation seems justified.

ARIMA MODELS

If the white noise terms a_t in a general ARIMA series each arise from a normal distribution, then from Equation (9-51) the forecast error $e_t(\ell)$ will also have a normal distribution, and the steps leading up to Equation (9-54) remain valid. However, in contrast to the deterministic trend model, recall that in the present case

$$\text{Var}\,[e_t(\ell)] = \sigma_a^2 \sum_{j=0}^{\ell-1} \psi_j^2$$

In practice, σ_a will be unknown and must be estimated from the observed time series. The necessary ψ-weights are, of course, also unknown, since they are certain functions of the unknown ϕ's and θ's. We estimate the ψ's from the estimated ϕ's and θ's. For large sample sizes, these estimations will have little effect on the actual prediction limits given above.

As a numerical example, consider the U.S. unemployment rate series once more. Here we estimated an AR(2) model with the results (see Exhibit 7.2),

$$\hat{\phi}_1 = 1.5630, \quad \hat{\phi}_2 = -0.6583, \quad \hat{\sigma}_a = \sqrt{0.1300} = 0.361, \quad \text{and} \quad \hat{\mu} = 5.0810$$

Using Equation (4-28), we then have

$$\hat{\psi}_1 = \hat{\phi}_1 = 1.5630$$
$$\hat{\psi}_2 = \hat{\phi}_1\hat{\psi}_1 + \hat{\phi}_2 = (1.5630)(1.5630) + (-0.6583) = 1.7845$$
$$\hat{\psi}_3 = \hat{\phi}_1\hat{\psi}_2 + \hat{\phi}_2\hat{\psi}_1 = (1.5630)(1.7845) + (-0.6583)(1.5630) = 1.760$$

and so forth.

We can then calculate the estimated forecast error standard deviation as

$$\sqrt{\text{Var}\,[e_t(\ell)]} = \hat{\sigma}_a \left[\sum_{j=0}^{\ell-1} \hat{\psi}_j^2 \right]^{1/2}$$

for any lead time ℓ and thus produce prediction limits on our forecasts. Exhibit 9.3 shows the same forecasts as did Exhibit 9.1, but now we also plot upper and lower 95% prediction limits to illustrate our confidence in those forecasts. Notice that after about 15 quarters the prediction limits remain essentially constant.

EXHIBIT 9.3 Forecasts and Prediction Limits for U.S. Unemployment Rates

```
MTB > mtsplot 4 'Unemp', 'Forecast', 'Lower' & 'Upper';
SUBC> tstart at 113;
SUBC> origin is 122 for 'Forecast', 'Lower' & 'Upper'.


Unemp     -
          -
          -    34                    ZZZZZZZZZZZZZZZZZZZZZZ
      7.50+ 12  1         ZZZZZZZ
          -       23    ZZ
          -        4 Z
          -        1X
          -         X
      5.00+        Y  XXXXXXXXXXXXXXXXXXXXXXXXXXXXXX
          -
          -        Y
          -         Y
          -          Y
      2.50+          Y       YYYYYYYYYYYYYYYYYY
          -           YYYYYYYY
          -
          -
          -
          -
      0.00+
          +---+---+---+---+---+---+---+---+---+
            120     128     136     144     152


X = Forecast  Y = Lower    Z = Upper
```

9.5 FORECASTING AND PREDICTION LIMITS WITH MINITAB

Judicious use of various Minitab commands can ease the computational burden of obtaining forecasts and prediction limits.

DETERMINISTIC TRENDS

In Chapter 3 we saw how to use the Minitab REGRESS command to estimate the parameters of typical trend models. Once the trends are estimated, the forecasts are just extrapolated trends that can be carried out by hand. If desired, we can also use Minitab to do all of the calculations.

Suppose we use the REGRESS command to estimate the linear trend model

$$Z_t = \beta_0 + \beta_1 t + X_t$$

Using the subcommand

COEFFICIENTS, put into **C**

we can store the estimates of β_0 and β_1 in any column, say C5. If we then use

SET into **C6**

$$n + 1:n + \ell$$

END

to obtain the appropriate time index into the future, our forecasts can be obtained from

$$\text{LET } C7 = C5(1) + C5(2) * C6$$

Prediction limits can then be formed as

$$\text{LET } C8 = C7 + 1.96 * s$$

and

$$\text{LET } C9 = C7 - 1.96 * s$$

where s is the estimated standard deviation obtained from the REGRESS command.*

With a periodic trend such as the cosine, the forecasts can be obtained by picking off the appropriate fitted values. For example, with monthly data beginning in January, if we have estimated a cosine trend and stored the fitted

---

EXHIBIT 9.4 Forecasting the U.S. Unemployment Rate Series

```
MTB > arima (2,0,0) 'Unemp';
SUBC> forecast 30 quarters, store in 'Forecast', 'Lower', 'Upper'.

Estimates at each iteration
Iteration      SSE      Parameters
    0        166.026    0.100     0.100    4.166
    1        135.995    0.250     0.025    3.774
    2        109.731    0.400    -0.052    3.392
    3         86.851    0.550    -0.129    3.014
    4         67.240    0.700    -0.208    2.639
    5         50.835    0.850    -0.286    2.266
    6         37.579    1.000    -0.365    1.895
    7         26.675    1.150    -0.435    1.476
    8         19.237    1.300    -0.500    1.032
    9         16.164    1.450    -0.591    0.728
   10         15.441    1.558    -0.657    0.505
   11         15.437    1.563    -0.659    0.486
   12         15.437    1.563    -0.658    0.484
Relative change in each estimate less than  0.0010

Final Estimates of Parameters
Type       Estimate    St. Dev.   t-ratio
AR   1      1.5630      0.0700     22.32
AR   2     -0.6583      0.0702     -9.37
Constant   0.48430     0.03280    14.76
Mean       5.0810      0.3442

No. of obs.:  121
Residuals:    SS = 15.3345  (backforecasts excluded)
              MS =  0.1300  DF = 118

Modified Box-Pierce chisquare statistic
Lag               12           24           36           48
Chisquare   18.0(DF=10)  33.0(DF=22)  44.1(DF=34)  50.1(DF=46)
```

---

*These prediction limits are valid only for large *n*. The regression subcommand PREDICT will compute prediction limits valid for all *n*, but unfortunately, PREDICT only prints the limits and does not store them for plotting.

values in C4, then

COPY C4 into **C5;**

USE rows **1:12.**

will put the forecast for one year into C5, beginning with January.

ARIMA MODELS

For ARIMA models, forecasting and prediction limits in Minitab are carried out via a subcommand to the ARIMA command. The format is as follows:

FORECAST [origin = **K**] to lead **K** [put forecast in **C,**

[put prediction limits in **C** and **C**]]

If the forecast origin is not specified, forecasts are made from the end of the series. The prediction limits are calculated with 95% confidence and they, along with the forecasts, are always printed. Once the forecasts and prediction limits are stored, a variety of plots, such as those shown in Exhibits 9.1, 9.2, and 9.3, are possible.

Exhibit 9.4 displays the Minitab output for the FORECAST subcommand

EXHIBIT 9.4 (*Continued*)

```
Forecasts from period 121
                        95 Percent Limits
Period     Forecast      Lower        Upper       Actual
  122       5.81027     5.10356      6.51697
  123       5.48419     4.17290      6.79548
  124       5.23110     3.41173      7.05046
  125       5.05018     2.84612      7.25423
  126       4.93401     2.46431      7.40370
  127       4.87154     2.23516      7.50793
  128       4.85038     2.12047      7.58029
  129       4.85843     2.08312      7.63375
  130       4.88494     2.09180      7.67808
  131       4.92108     2.12316      7.71899
  132       4.96010     2.16178      7.75842
  133       4.99731     2.19880      7.79583
  134       5.02978     2.23005      7.82951
  135       5.05603     2.25423      7.85783
  136       5.07509     2.27150      7.87980
  137       5.08913     2.28298      7.89529
  138       5.09720     2.28953      7.90488
  139       5.10097     2.29232      7.90961
  140       5.10154     2.29236      7.91072
  141       5.09995     2.29053      7.90937
  142       5.09710     2.28759      7.90660
  143       5.09368     2.28416      7.90320
  144       5.09021     2.28070      7.89973
  145       5.08705     2.27753      7.89658
  146       5.08439     2.27486      7.89393
  147       5.08231     2.27276      7.89186
  148       5.08081     2.27124      7.89038
  149       5.07984     2.27025      7.88943
  150       5.07931     2.26971      7.88890
  151       5.07911     2.26951      7.88871
```

EXHIBIT 9.5 Forecasting the Railroad Bond Yield Series

```
MTB > arima (1,1,0) 'RRBonds', 'Resid', 'Fitted';
SUBC> forecast 30 months, store in 'Forecast', 'Lower' & 'Upper'.

Estimates at each iteration
Iteration       SSE      Parameters
    0         10153.3     0.100
    1          9129.8     0.250
    2          8605.5     0.400
    3          8530.1     0.478
    4          8529.9     0.482
    5          8529.9     0.483
Relative change in each estimate less than  0.0010

Final Estimates of Parameters
Type      Estimate    St. Dev.  t-ratio
AR   1     0.4825      0.0876     5.51

Differencing: 1 regular difference
No. of obs.:  Original series 102, after differencing 101
Residuals:    SS = 8526.97  (backforecasts excluded)
              MS =   85.27  DF = 100

Modified Box-Pierce chisquare statistic
Lag              12           24           36           48
Chisquare   6.4(DF=11)  18.9(DF=23)  27.7(DF=35)  37.8(DF=47)


Forecasts from period 102
                        95 Percent Limits
Period    Forecast     Lower       Upper        Actual
 103      811.035     792.932     829.137
 104      810.569     778.196     842.941
 105      810.344     765.485     855.202
 106      810.235     754.494     865.975
 107      810.182     744.854     875.509
 108      810.156     736.259     884.054
 109      810.144     728.475     891.813
 110      810.138     721.331     898.944
 111      810.134     714.704     905.564
 112      810.133     708.500     911.765
 113      810.132     702.651     917.612
 114      810.131     697.102     923.160
 115      810.130     691.812     928.449
 116      810.130     686.748     933.512
 117      810.130     681.884     938.375
 118      810.129     677.198     943.060
 119      810.129     672.671     947.586
 120      810.128     668.289     951.967
 121      810.128     664.038     956.217
 122      810.127     659.907     960.347
 123      810.127     655.887     964.366
 124      810.126     651.969     968.284
 125      810.126     648.145     972.106
 126      810.125     644.410     975.840
 127      810.125     640.757     979.492
 128      810.125     637.182     983.067
 129      810.124     633.679     986.570
 130      810.124     630.244     990.004
 131      810.124     626.873     993.374
 132      810.123     623.563     996.683
```

EXHIBIT 9.6 Forecasts and Prediction Limits for the Railroad Bond Yield Series

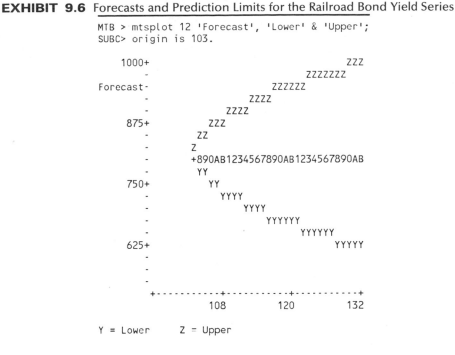

```
MTB > mtsplot 12 'Forecast', 'Lower' & 'Upper';
SUBC> origin is 103.

   1000+                                        ZZZ
       -                                    ZZZZZZZ
Forecast-                              ZZZZZZ
       -                           ZZZZ
       -                       ZZZZ
    875+                    ZZZ
       -                  ZZ
       -                Z
       -                +890AB1234567890AB1234567890AB
       -                 YY
    750+                 YY
       -                   YYYY
       -                     YYYY
       -                       YYYYYY
       -                          YYYYYY
    625+                             YYYYY
       -
       -
       -
        +-----------+-----------+-----------+
                  108         120         132
```

Y = Lower   Z = Upper

applied to the U.S. unemployment rate series. Actual values are printed only if the forecast origin is earlier than the end of the given series.

Exhibits 9.5 and 9.6 show forecasting results for the nonstationary ARI(1, 1) model applied to the data for the railroad bond yields. Note here that the prediction limits continue to spread out as the forecast lead increases.

9.6 UPDATING ARIMA FORECASTS

Suppose we are forecasting a time series that is available on a monthly basis. Our last observation is, say, for February, and we forecast values for March, April, and May. As time goes by, the actual value for March becomes known. With this new value in hand, we would like to update or revise (and, one hopes, improve) our forecasts for April and May. Of course, we could compute new forecasts from scratch. However, there is a simpler way.

For a general forecast origin t and lead time ℓ, our original forecast is denoted $\hat{Z}_t(\ell + 1)$. Once the observation at time $t + 1$ is available, we would like to update our forecast as $\hat{Z}_{t+1}(\ell)$. Equations (9-4) and (9-43) yield

$$Z_{t+\ell+1} = C_t(\ell + 1) + a_{t+\ell+1} + \psi_1 a_{t+\ell} \psi_2 a_{t+\ell-1} + \cdots + \psi_\ell a_{t+1}$$

Since $C_t(\ell + 1)$ and a_{t+1} are functions of $Z_{t+1}, Z_t, \ldots,$ whereas $a_{t+\ell+1}, a_{t+\ell}, \ldots, a_{t+2}$ are independent of $Z_{t+1}, Z_t, \ldots,$ we quickly obtain the expression

$$\hat{Z}_{t+1}(\ell) = C_t(\ell + 1) + \psi_\ell a_{t+1}$$

However, $\hat{Z}_t(\ell + 1) = C_t(\ell + 1)$, and, of course, $a_{t+1} = Z_{t+1} - \hat{Z}_t(1)$. Thus we have the general **updating equation**

$$\hat{Z}_{t+1}(\ell) = \hat{Z}_t(\ell + 1) + \psi_\ell[Z_{t+1} - \hat{Z}_t(1)] \tag{9-55}$$

Notice that $Z_{t+1} - \hat{Z}_t(1)$ is the *actual forecast error* at time $t + 1$ once Z_{t+1} has been observed.

As a numerical example, consider again the AR(2) model for the U.S. unemployment rate series. From Exhibit 9.4 with forecast origin 121, our two-quarter-ahead forecast for $t = 123$ is seen to be 5.48419. Suppose now that the observation at $t = 122$ turns out to be 5.9. With $\psi_1 = 1.5630$, our updated forecast for $t = 123$ (now one quarter ahead) is

$$\hat{Z}_{t+1}(1) = \hat{Z}_t(2) + 1.5630(5.9 - 5.81027)$$
$$= 5.48419 + 0.14025$$
$$= 5.62444$$

9.7 FORECAST WEIGHTS AND EXPONENTIALLY WEIGHTED MOVING AVERAGES

For ARIMA models without moving average terms, it is clear how the forecasts are explicitly determined from the observed series $Z_t, Z_{t-1}, \ldots, Z_1$. However, for any model with $q > 0$, the noise terms $a_t, a_{t-1}, \ldots, a_{t-(q-1)}$ appear in the forecasts, and the nature of the forecast explicitly in terms of $Z_t, Z_{t-1}, \ldots, Z_1$ is hidden. To bring out this aspect of the forecasts, we return to the inverted form of any invertible ARIMA process, namely,

$$Z_t = \sum_{j=1}^{\infty} \pi_j Z_{t-j} + a_t$$

(see Equation (4-48). Thus we can write

$$Z_{t+1} = \sum_{j=1}^{\infty} \pi_j Z_{t+1-j} + a_{t+1}$$

Taking conditional expectations of both sides, given $Z_t, Z_{t-1}, \ldots, Z_1$, we obtain a new representation of the one-step-ahead forecast:

$$\hat{Z}_t(1) = \sum_{j=1}^{\infty} \pi_j Z_{t+1-j} \tag{9-56}$$

(We are assuming here that t is sufficiently large and/or that the π-weights die out sufficiently quickly so that $\pi_t, \pi_{t+1}, \ldots$ are all negligible.)

For an invertible ARIMA model, the π-weights can be calculated recursively from the expressions

$$\pi_j = \begin{cases} \sum_{i=1}^{\min(j,q)} \theta_i \pi_{j-i} + \varphi_j, & 1 \le j \le p + d \\ \sum_{i=1}^{\min(j,q)} \theta_i \pi_{j-i}, & j > p + d \end{cases} \tag{9-57}$$

with initial value $\pi_0 = -1$. (Compare with Equations (4-41) for the ψ-weights.)
Consider in particular the nonstationary IMA(1, 1) model

$$Z_t = Z_{t-1} + a_t - \theta a_{t-1}$$

Here $p = 0$, $d = 1$, $q = 1$, and $\varphi_1 = 1$; thus

$$\pi_1 = \theta \pi_0 + 1 = 1 - \theta$$
$$\pi_2 = \theta \pi_1 = \theta(1 - \theta)$$

and generally,

$$\pi_j = \theta \pi_{j-1} \quad \text{for } j > 1$$

Thus we have explicitly

$$\pi_j = (1 - \theta)\theta^{j-1} \quad \text{for } j \ge 1 \tag{9-58}$$

so that we can write

$$\hat{Z}_t(1) = (1 - \theta)Z_t + (1 - \theta)\theta Z_{t-1} + (1 - \theta)\theta^2 Z_{t-2} + \cdots \tag{9-59}$$

In this case the π-weights decrease *exponentially*, and furthermore,

$$\sum_{j=1}^{\infty} \pi_j = (1 - \theta) \sum_{j=1}^{\infty} \theta^{j-1} = \frac{1 - \theta}{1 - \theta} = 1$$

Thus $\hat{Z}_t(1)$ is called an **exponentially weighted moving average,** or **EWMA.**
Simple algebra shows that we can also write

$$\hat{Z}_t(1) = (1 - \theta)Z_t + \theta \hat{Z}_{t-1}(1) \tag{9-60}$$

and

$$\hat{Z}_t(1) = \hat{Z}_{t-1}(1) + (1 - \theta)[Z_t - \hat{Z}_{t-1}(1)] \tag{9-61}$$

Equations (9-60) and (9-61) show how to update forecasts from origin $t - 1$ to origin t, and they express the result either as a linear combination of the new observation and the old forecast or in terms of the old forecast and the last observed forecast error.

Using the EWMA to forecast time series has been advocated, mostly on an ad hoc basis, for a number of years. See Brown (1962) and Montgomery and Johnson (1976).

The parameter $1 - \theta$ is called the **smoothing constant** in the EWMA literature, and its selection (estimation) is sometimes quite arbitrary.

From the ARIMA model-building approach, we let the data indicate whether an IMA(1, 1) model is appropriate for the series under consideration. *If so* we then estimate θ in an efficient manner and compute an EWMA forecast that we are confident is the minimum mean square error forecast. A comprehensive treatment of exponential smoothing methods and their relationships with ARIMA models is given in Abraham and Ledolter (1983).

9.8 FORECASTING TRANSFORMED SERIES

DIFFERENCING

Suppose we are interested in forecasting a series whose model involves a first difference to achieve stationarity. Two methods of forecasting can be considered:

1. forecasting the original nonstationary series, for example, by using the difference-equation form of Equation (9-33), or

2. forecasting the stationary differenced series $W_t = Z_t - Z_{t-1}$ and then "undoing" the difference to obtain the forecast for the original series.

We shall show that both methods lead to identical forecasts. This follows essentially because differencing is a *linear* operation and because the conditional expectation of a linear combination is the same linear combination of the conditional expectations.

Consider in particular the IMA(1, 1) model. Basing our work on the original nonstationary series, we forecast as

$$\hat{Z}_t(1) = Z_t - \theta a_t \qquad \text{(9-62)}$$

and

$$\hat{Z}_t(\ell) = \hat{Z}_t(\ell - 1) \quad \text{for } \ell > 1 \qquad \text{(9-63)}$$

Consider now the differenced stationary MA(1) series $W_t = Z_t - Z_{t-1}$. We would forecast $W_{t+\ell}$ as

$$\hat{W}_t(1) = -\theta a_t \qquad \text{(9-64)}$$

and

$$\hat{W}_t(\ell) = 0 \quad \text{for } \ell > 1 \qquad \text{(9-65)}$$

However, $\hat{W}_t(1) = \hat{Z}_t(1) - Z_t$; thus $\hat{W}_t(1) = -\theta a_t$ is equivalent to

$$\hat{Z}_t(1) = Z_t - \theta a_t$$

as before. Similarly, $\hat{W}_t(\ell) = \hat{Z}_t(\ell) - \hat{Z}_t(\ell - 1)$, and Equation (9-65) becomes Equation (9-63), as we have claimed.

The same result would apply to any model involving differences of any order and indeed to any type of *linear* transformation with constant coefficients. (Certain linear transformations other than differencing may be applicable to seasonal time series. See Chapter 10).

LOG TRANSFORMATIONS

As we have seen earlier, it is frequently appropriate to model the logarithms of the original series—a nonlinear transformation. Let Y_t be the original series and let $Z_t = \log(Y_t)$. It can be shown that we always have

$$E(Y_{t+\ell} \mid Y_t, Y_{t-1}, \ldots, Y_1) \geq \exp[E(Z_{t+\ell} \mid Z_t, Z_{t-1}, \ldots, Z_1)] \qquad \text{(9-66)}$$

with equality holding only in trivial cases. Thus the naive forecast $\exp[\hat{Z}_t(\ell)]$ is *not* the minimum mean square error forecast of $Y_{t+\ell}$. To evaluate the minimum mean square error forecast in original terms, we shall find the following fact useful: If X has a normal distribution with mean μ and variance σ^2, then

$$E[\exp(X)] = \exp\left(\mu + \frac{\sigma^2}{2}\right)$$

(This follows, for example, from the moment-generating function for X.) In our application

$$\mu = E(Z_{t+\ell} \mid Z_t, Z_{t-1}, \ldots, Z_1)$$

and

$$\begin{aligned}
\sigma^2 &= \text{Var}\,(Z_{t+\ell} \mid Z_t, Z_{t-1}, \ldots, Z_1) \\
&= \text{Var}\,[e_t(\ell) + C_t(\ell) \mid Z_t, Z_{t-1}, \ldots, Z_1] \quad \text{(from Equations (9-41), (9-42), (9-43),} \\
&\qquad\qquad\qquad\qquad\qquad\qquad\qquad\qquad\qquad\qquad \text{and (9-44))} \\
&= \text{Var}\,[e_t(\ell)] \mid Zt, Z_{t-1}, \ldots, Z_1) \quad \text{(since } C_t(\ell) \text{ is a function of } Z_t, Z_{t-1}, \ldots, Z_1) \\
&= \text{Var}\,[e_t(\ell)] \qquad\qquad\qquad\qquad \text{(since } e_t(\ell) \text{ is independent of } Z_t, Z_{t-1}, \ldots, Z_1)
\end{aligned}$$

Thus the minimum mean square error forecast in the original series is given by

$$\exp\{\hat{Z}_t(\ell) + \tfrac{1}{2}\text{Var}\,[e_t(\ell)]\} \qquad \text{(9-67)}$$

Throughout our discussion of forecasting, we have assumed that minimizing the mean square forecast error is the criterion of choice. For normally distributed data, this is probably a very good criterion. However, if Z_t has a normal distribution, then $Y_t = \exp(Z_t)$ will have a lognormal distribution for which a different criterion may be desirable. In particular, since the lognormal distribution is asymmetric and has a very long right tail, a criterion based on the mean absolute value may be more appropriate. For this "loss function," the optimal forecast is the **median** of the distribution of $Y_{t+\ell}$ conditional on $Y_t, Y_{t-1}, \ldots, Y_1$. Since the log transformation preserves medians and since, for a normal distribution, the mean and median are identical, the naive forecast $\exp[\hat{Z}_t(\ell)]$ will be the optimal forecast of $Y_{t+\ell}$ in the sense that it minimizes the mean absolute forecast error.

9.9 SUMMARY OF FORECASTING WITH CERTAIN ARIMA MODELS

Here we bring together various forecasting results for special ARIMA models.

AR(1): $Z_t = \mu + \phi(Z_{t-1} - \mu) + a_t$

$$\hat{Z}_t(\ell) = \mu + \phi[\hat{Z}_t(\ell - 1) - \mu]$$

$$= \mu + \phi^\ell(Z_t - \mu)$$

$$\approx \mu \quad \text{for large } \ell$$

$$\text{Var}\,[e_t(\ell)] = \sigma_a^2 \frac{1 - \phi^{2\ell}}{1 - \phi^2}$$

$$\approx \frac{\sigma_a^2}{1 - \phi^2} \quad \text{for large } \ell$$

$$\psi_j = \phi^j \quad \text{for } j > 0$$

MA(1): $Z_t = \mu + a_t - \theta a_{t-1}$

$$\hat{Z}_t(\ell) = \begin{cases} \mu - \theta a_t & \text{for } \ell = 1 \\ \mu & \text{for } \ell > 1 \end{cases}$$

$$\text{Var}\,[e_t(\ell)] = \begin{cases} \sigma_a^2 & \text{for } \ell = 1 \\ \sigma_a^2(1 + \theta^2) & \text{for } \ell > 1 \end{cases}$$

$$\psi_j = \begin{cases} -\theta & \text{for } \ell = 1 \\ 0 & \text{for } \ell > 1 \end{cases}$$

IMA(1, 1) WITH CONSTANT TERM: $Z_t = Z_{t-1} + \theta_0 + a_t - \theta a_{t-1}$

(The random walk with drift is the special case where $\theta = 0$.)

$$\hat{Z}_t(1) = Z_t + \theta_0 - \theta a_t$$

$$\hat{Z}_t(\ell) = Z_t + \ell\theta_0 - \theta a_t$$

$$= \ell\theta_0 + (1 - \theta) \sum_{j=0}^{\infty} \theta^j Z_{t-j}$$

$$\text{Var}\,[e_t(\ell)] = \sigma_a^2[1 + (\ell - 1)(1 - \theta)^2]$$

$$\psi_j = 1 - \theta \quad \text{for } j > 0$$

Note that if $\theta_0 \neq 0$, the forecasts follow a straight line with slope θ_0, but if $\theta_0 = 0$, which is the usual case, then $\hat{Z}_t(\ell)$ is the same for all leads ℓ.

IMA(2, 2): $Z_t = 2Z_{t-1} - Z_{t-2} + a_t - \theta_1 a_{t-1} - \theta_2 a_{t-2}$

$$\hat{Z}_t(1) = 2Z_t - Z_{t-1} + \theta_0 - \theta_1 a_t - \theta_2 a_{t-1} \tag{9-68}$$

$$\hat{Z}_t(2) = 2\hat{Z}_t(1) - Z_t + \theta_0 - \theta_2 a_t \tag{9-69}$$

$$\hat{Z}_t(\ell) = 2\hat{Z}_t(\ell - 1) - \hat{Z}_t(\ell - 2) + \theta_0, \quad \text{for } \ell > 2 \tag{9-70}$$

$$= A + B\ell + \frac{\theta_0}{2}\ell^2, \qquad \text{for } \ell \geq 1 \tag{9-71}$$

where

and

$$A = 2\hat{Z}_t(1) - \hat{Z}_t(2) + \theta_0 \tag{9-72}$$

$$B = \hat{Z}_t(2) - \hat{Z}_t(1) - \frac{3}{2}\theta_0 \tag{9-73}$$

If $\theta_0 \neq 0$, then the forecasts follow a quadratic curve in ℓ, but if $\theta_0 = 0$, the forecasts will form a straight line with slope $\hat{Z}_t(2) - \hat{Z}_t(1)$ and will pass through the two initial forecasts $\hat{Z}_t(1)$ and $\hat{Z}_t(2)$. It can be shown that Var$[e_t(\ell)]$ is a certain cubic function of ℓ. See Box and Jenkins (1976, p. 149). We also have

$$\psi_j = 1 + \theta_2 + (1 - \theta_1 - \theta_2)j \quad \text{for } j > 0 \tag{9-74}$$

It can also be shown that forecasting a special case of the IMA(2, 2) model, namely when $\theta_1 = 2\omega$ and $\theta_2 = -\omega^2$ with $|\omega| < 1$, is equivalent to so-called double exponential smoothing with smoothing constant $1 - \omega$. See Abraham and Ledolter (1983).

CHAPTER 9 EXERCISES

9.1. For an AR(1) model with $Z_t = 12.2$, $\phi = -0.5$, and $\mu = 10.8$, find $\hat{Z}_t(1)$.

9.2. Suppose that annual sales (in millions of dollars) of the Acme Corporation, follow the AR(2) model:

$$Z_t = 5 + 1.1Z_{t-1} - 0.5Z_{t-2} + a_t \quad \text{with } \sigma_a^2 = 2$$

a. If sales for 1985, 1984, and 1983 were $10 million, $11 million, and $9 million, respectively, forecast sales for 1986 and 1987.
b. Show that $\psi_1 = 1.1$ for this model.
c. Calculate 95% prediction limits for your forecasts in part (a) for 1986.
d. If sales in 1986 are $12 million, show how to update your forecast for 1987.

9.3. Use the estimated trend of Equation (9-6) to forecast the average monthly temperature in Dubuque, Iowa, for April. Also find a 95% prediction interval for that April forecast.

9.4. Use Exhibit 3.4 to forecast milk production per cow for the month of March 1976. Recall here that $t = 168$ corresponds to December 1975.

9.5. Consider the model

$$Z_t = \beta_0 + \beta_1 t + X_t \quad \text{with} \quad X_t = \phi X_{t-1} + a_t$$

We assume that β_0, β_1, and ϕ are known. Show that the minimum mean square error forecast ℓ steps ahead can be written as

$$\hat{Z}_t(\ell) = \beta_0 + \beta_1(t + \ell) + \phi^\ell(Z_t - \beta_0 - \beta_1 t)$$

9.6. Verify Equation (9-21).

9.7. Verify Equation (9-38).

9.8. Use Minitab to forecast the Portland, Oregon, gasoline price series (see Exercises 6.9, 7.9, and 8.7). Choose a forecast origin so that some actual values can be compared with forecast values. Are the actual values within the 95% prediction intervals?

9.9. Repeat Exercise 9.8 for the Iowa nonfarm income series. (See Exercises 6.10, 7.10, and 8.8.)

APPENDIX F CONDITIONAL EXPECTATION

If X and Y have joint p.d.f. $f(x, y)$ and we denote the marginal p.d.f. of X by $f(x)$, then the **conditional p.d.f.** of Y given $X = x$ is given by

$$f(y \mid x) = \frac{f(x, y)}{f(x)}$$

The **conditional expectation of Y given $X = x$** is then defined as

$$E(Y \mid X = x) = \int_{-\infty}^{\infty} y f(y \mid x) \, dy$$

Notice this is just the *mean* of the conditional distribution and thus has all of the properties of ordinary means and expected values. For example,

$$E(aY + bZ \mid X = x) = aE(Y \mid X = x) + bE(Z \mid X = x) \tag{F-1}$$

and

$$E[h(Y) \mid X = x] = \int_{-\infty}^{\infty} h(y) f(y \mid x) \, dy \tag{F-2}$$

In addition, several new properties hold that have no analogies with unconditional expectations:

$$E[h(X) \mid X = x] = h(x) \tag{F-3}$$

that is, given $X = x$, the random variable $h(X)$ can be treated like a constant.

More generally,

$$E[h(X, Y) \mid X = x] = E[h(x, Y) \mid X = x] \tag{F-4}$$

If we set $E[Y \mid X = x] = g(x)$, then $g(X)$ will be a random variable and we can consider $E[g(X)]$. It can be shown that

$$E[g(X)] = E(Y)$$

which is sometimes written as

$$E[E(Y \mid X)] = E(Y) \tag{F-5}$$

If Y is independent of X, then

$$E(Y \mid X) = E(Y) \tag{F-6}$$

APPENDIX G MINIMUM MEAN SQUARE ERROR PREDICTION

Suppose Y is a random variable with mean μ_Y and variance σ_Y^2. If our object is to predict Y using only a constant c, what is the *best* choice for c? Clearly, we first must define *best*. A common (and convenient) criterion is to choose c to minimize the **mean square error** of prediction, that is, to minimize

$$g(c) = E(Y - c)^2$$

If we expand $g(c)$, we have

$$g(c) = E(Y^2) - 2cE(Y) + c^2$$

Since $g(c)$ is now quadratic in c and opens upward, solving $g'(c) = 0$ will produce the required minimum. We have

$$g'(c) = -2E(Y) + 2c = 0$$

so that the optimal c is

$$c = E(Y) = \mu_Y \qquad \text{(G-1)}$$

Note also that

$$\min g(c) = E(Y - \mu_Y)^2 = \sigma_Y^2 \qquad \text{(G-2)}$$

Now consider the situation where a second random variable X is available, and we wish to use the observed value of X to help predict Y. Let $\rho = \text{Corr}(X, Y)$.

We first suppose, for simplicity, that only *linear* functions $a + bX$ can be used for the prediction. The mean square error is then given by

$$g(a, b) = E(Y - a - bX)^2$$

Again expanding $g(a, b)$ we have

$$g(a, b) = E(Y^2) + a^2 + b^2 E(X^2) - 2aE(Y) + 2abE(X) - 2bE(XY)$$

This is also quadratic in a and b and opens upward. Thus we can find the point of minimum by solving simultaneously the linear equations $\partial g(a, b)/\partial a = 0$ and $\partial g(a, b)/\partial b = 0$. We have

$$\frac{\partial g}{\partial a}(a, b) = 2a - 2E(Y) + 2bE(X) = 0$$

and

$$\frac{\partial g}{\partial b}(a, b) = 2bE(X^2) + 2aE(X) - 2E(XY) = 0$$

which we rewrite as

$$a + E(X)b = E(Y) \quad \text{and} \quad E(X)a + E(X^2)b = E(XY)$$

Multiplying the first equation by $E(X)$ and subtracting yields

$$b = \frac{E(XY) - E(X)E(Y)}{E(X^2) - [E(X)]^2}$$

which, by V4 and CV3 of Appendix A, p. 23 can be written

$$b = \frac{\text{Cov}(X, Y)}{\text{Var}(X)}$$

or

$$b = \text{Corr}(X, Y)\left[\frac{\text{Var}(Y)}{\text{Var}(X)}\right]^{1/2}$$

$$= \rho\frac{\sigma_Y}{\sigma_X} \tag{G-3}$$

Then

$$a = E(Y) - bE(X)$$

$$= \mu_Y - \rho\frac{\sigma_Y}{\sigma_X}\mu_X \tag{G-4}$$

If we let $\hat{Y}$ be the minimum mean square error prediction of Y based on a linear function of X, then we can write

$$\hat{Y} = \mu_Y - \rho\left(\frac{\sigma_Y}{\sigma_X}\right)\mu_Y + \rho\left(\frac{\sigma_Y}{\sigma_X}\right)X$$

or

$$\frac{\hat{Y} - \mu_Y}{\sigma_Y} = \rho\frac{X - \mu_X}{\sigma_X} \tag{G-5}$$

In terms of standardized random variables Y' and X', we have simply

$$Y' - \rho X' \tag{G-6}$$

Also, using Equations (G-3) and (G-4) we find

$$\min g(a, b) = E\left[Y - \mu_Y + \rho\left(\frac{\sigma_Y}{\sigma_X}\right)\mu_X - \rho\left(\frac{\sigma_Y}{\sigma_X}\right)X\right]^2$$

$$= \sigma_Y^2 E\left(\frac{Y - \mu_Y}{\sigma_Y} - \rho\frac{X - \mu_X}{\sigma_X}\right)^2$$

$$= \sigma_Y^2[\text{Var}(Y') + \rho^2\text{Var}(X') - 2\rho\text{Cov}(Y', X')]$$

$$= \sigma_Y^2(1 + \rho^2 - 2\rho\rho)$$

$$= \sigma_Y^2(1 - \rho^2) \tag{G-7}$$

(Since the left-hand side is always nonnegative, Equation (G-7) provides a proof that $-1 \leq \rho \leq +1$.)

If we compare Equation (G-7) with Equation (G-2), we see that the minimum mean square error obtained when we use a linear function of X to predict Y is *reduced* by a factor of $1 - \rho^2$ compared with that obtained by ignoring X and simply using the constant μ_Y for prediction.

Let us now consider the more general problem of predicting Y with an arbitrary function of X. Once more our criterion will be to minimize the mean square error of prediction. We need to choose the function $h(X)$, say, that minimizes

$$E[Y - h(X)]^2 \tag{G-8}$$

Using Equation (F-5), we can write Equation (G-8) as

$$E[Y - h(X)]^2 = E(E\{[Y - h(X)]^2 \mid X\}) \tag{G-9}$$

Using (F-4) the inner expectation can be written as

$$E\{[Y - h(X)]^2 \mid X = x\} = E\{[Y - h(x)]^2 \mid X = x\} \tag{G-10}$$

For each value of x, $h(x)$ is a constant, and we can apply result (G-1) to the conditional distribution of Y given $X = x$. Thus, for each x, the best choice of $h(x)$ is

$$h(x) = E(Y \mid X = x)$$

Since this choice of $h(x)$ minimizes the inner expectation in Equation (G-9), it must also provide the overall minimum of Equation (G-8). Thus

$$h(X) = E(Y \mid X) \tag{G-11}$$

is the best predictor of Y of *all* functions of X.

If X and Y have a bivariate *normal* distribution, it can be shown that

$$E(Y \mid X) = \mu_Y + \rho \frac{\sigma_Y}{\sigma_X}(X - \mu_X)$$

so that solutions (G-11) and (G-5) coincide. In this case the best of *all* functions is linear.

More generally, if Y is to be predicted based on a function of several variables $X_1, X_2, \ldots, X_n$, then it can easily be argued that the minimum mean square error predictor is given by

$$E(Y \mid X_1, X_2, \ldots, X_n) \tag{G-12}$$

APPENDIX H THE TRUNCATED LINEAR PROCESS

Suppose $\{Z_t\}$ satisfies a general ARIMA(p, d, q) model with AR characteristic polynomial $\phi(x)$, MA characteristic polynomial $\theta(x)$, and constant term θ_0. Let the distinct roots of $\phi(x) = 0$ be denoted $G_i^{-1}, i = 1, 2, \ldots, r$, with G_i^{-1} repeated p_i times ($\sum p_i = p$). Then the **truncated linear process** representation for $\{Z_t\}$ is given by

$$Z_{t+\ell} = C_t(\ell) + I_t(\ell) \quad \text{for } \ell \geq 1 \tag{H-1}$$

where

$$I_t(\ell) = \begin{cases} \sum_{j=0}^{\ell-1} \psi_j a_{t+\ell-j} & \text{for } \ell \geq 1 \\ 0 & \text{for } \ell \leq 0 \end{cases} \tag{H-2}$$

$$C_t(\ell) = \sum_{i=0}^{d} A_i \ell^i + \sum_{i=1}^{r} \left(\sum_{j=0}^{p_i-1} B_{ij} \ell^j \right) G_i^{\ell} \tag{H-3}$$

and $A_i, B_{ij}, i = 1, 2, \ldots, r, j = 1, 2, \ldots, p_i$, are constant in ℓ and depend only on $Z_t, Z_{t-1}, \ldots$.

As always, the ψ-weights are defined by the identity

$$\phi(x)(1 - x)^d (1 - \psi_1 x - \psi_2 x^2 - \cdots) = \theta(x) \tag{H-4}$$

or

$$\varphi(x)(1 - \psi_1 x - \psi_2 x^2 - \cdots) = \theta(x) \tag{H-5}$$

and can be computed recursively, as in Equations (4-41).

We shall show that the representation given by Equation (H-1) is valid by arguing that, for fixed t, $C_t(\ell)$ is essentially the **complementary function** of the defining difference equation, that is,

$$C_t(\ell) - \varphi_1 C_t(\ell - 1) - \varphi_2 C_t(\ell - 2) - \cdots - \varphi_{p+d} C_t(\ell - p - d)$$
$$= \theta_0 \quad \text{for } \ell \geq 0 \tag{H-6}$$

and that $I_t(\ell)$ is a **particular solution** (without θ_0):

$$I_t(\ell) - \varphi_1 I_t(\ell - 1) - \cdots - \varphi_{p+d} I_t(\ell - p - d)$$
$$= a_{t+\ell} - \theta_1 a_{t+\ell-1} - \cdots - \theta_q a_{t+\ell-q} \quad \text{for } \ell > q \tag{H-7}$$

Since $C_t(\ell)$ contains $p + d$ arbitrary constants, the A_i and B_{ij}, summing $C_t(\ell)$ and $I_t(\ell)$ yields the general solution of the ARIMA equation. Specific values for A_i and B_{ij} will be determined by the initial conditions $Z_t, Z_{t-1}, \ldots$.

We note that A_d is not arbitrary: We have

$$A_d = \frac{\theta_0}{(1 - \phi_1 - \phi_2 - \cdots - \phi_p)d!} \tag{H-8}$$

The proof that $C_t(\ell)$ as given by Equation (H-3) is the complementary solution and satisfies Equation (H-6) is a standard result from the theory of difference equations (see, for example, Goldberg, 1958). We shall show that the particular solution $I_t(\ell)$ defined by Equation (H-2) does satisfy Equation (H-7).

For convenience of notation, we let $\varphi_j = 0$ for $j > p + d$. Consider the left-hand side of Equation (H-7). It can be written as:

$$\begin{aligned}
(\psi_0 a_{t+\ell} + \psi_1 a_{t+\ell-1} + \cdots + \psi_{\ell-1} a_{t+1}) - \varphi_1(\psi_0 a_{t+\ell-1} + \psi_1 a_{t+\ell-2} + \cdots \\
+ \psi_{\ell-2} a_{t+1}) - \cdots - \varphi_{p+d}(\psi_0 a_{t+\ell-p-d} + \psi_1 a_{t+\ell-p-d-1} + \cdots \\
+ \psi_{\ell-p-d-1} a_{t+1})
\end{aligned} \tag{H-9}$$

Now by grouping together common a_j terms and picking off their coefficients, we obtain

Coefficient of $a_{t+\ell}$: ψ_0

Coefficient of $a_{t+\ell-1}$: $\psi_1 - \varphi_1 \psi_0$

Coefficient of $a_{t+\ell-2}$: $\psi_2 - \varphi_1 \psi_1 - \varphi_2 \psi_0$

$$\vdots$$

Coefficient of a_{t+1} : $\psi_{\ell-1} - \varphi_1 \psi_{\ell-2} - \varphi_2 \psi_{\ell-3} - \cdots - \varphi_{p+d} \psi_{\ell-p-d-1}$

If $\ell > q$, we can match these coefficients to the corresponding coefficients on the right-hand side of Equation (H-7) to obtain the relationships

$$\begin{aligned}
\psi_0 &= 1 \\
\psi_1 - \varphi_1 \psi_0 &= -\theta_1 \\
\psi_2 - \varphi_1 \psi_1 - \varphi_2 \psi_0 &= -\theta_2 \\
&\vdots \\
\psi_q - \varphi_1 \psi_{q-1} - \cdots - \varphi_q \psi_0 &= -\theta_q \\
\psi_\ell - \varphi_1 \psi_{\ell-1} - \cdots - \varphi_\ell \psi_0 &= 0 \quad \text{for } \ell > q
\end{aligned} \tag{H-10}$$

However, by comparing these relationships to Equation (H-5), we see that Equations (H-10) are precisely the equations defining the ψ-weights and thus Equation (H-7) is established.

It is instructive here to illustrate the truncated linear process representation for two models, the AR(1) and the IMA(1, 1).

In the AR(1) case we know $\psi_j = \phi^j$, so we have

$$Z_{t+\ell} = \frac{\theta_0}{1 - \phi} + B_1 \phi^\ell + \sum_{j=0}^{\ell-1} \phi^j a_{t+\ell-j}, \qquad \ell \geq 1$$

Setting $\ell = 1$ yields

$$Z_{t+1} = \frac{\theta_0}{1 - \phi} + B_1 \phi + a_{t+1}$$

But we also know that

$$Z_{t+1} = \phi Z_t + \theta_0 + a_{t+1}$$

Thus we get

$$B_1 = Z_t - \frac{\theta_0}{1 - \phi}$$

and obtain

$$\hat{Z}_t(\ell) = C_t(\ell) = \frac{\theta_0}{1 - \phi} + \left(Z_t - \frac{\theta_0}{1 - \phi}\right)\phi^\ell$$

which agrees with Equation (9-14).

In the IMA(1, 1) model, $\psi_j = 1 - \theta$ for $j > 0$, so that

$$Z_{t+\ell} = A_0 + \theta_0\ell + a_{t+\ell} + (1 - \theta)\sum_{j=1}^{\ell-1} a_{t+\ell-j}$$

Again setting $\ell = 1$, we get

$$Z_{t+1} = A_0 + \theta_0 + a_{t+1}$$

But we also have

$$Z_{t+1} = Z_t + \theta_0 + a_{t+1} - \theta a_t$$

Consequently,

$$A_0 = Z_t - \theta a_t$$

$$= (1 - \theta)\sum_{j=0}^{\infty} \theta^j Z_{t-j}$$

from the inverted form of the model. Thus

$$\hat{Z}_t(\ell) = C_t(\ell) = Z_t - \theta a_t + \theta_0\ell = (1 - \theta)\sum_{j=0}^{\infty} \theta^j Z_{t-j} + \theta_0\ell$$

which agrees with our previous results for this case.

APPENDIX I BACKCASTING

In Equations (7-51) and (7-52) we saw that the unconditional sum-of-squares function for a general ARMA(p, q) model can be expressed as

$$S(\phi, \theta, \theta_0) = \sum_{t=-\infty}^{n} \hat{a}_t^2$$

where $\hat{a}_t = a_t$ for $1 \le t \le n$, but

$$\hat{a}_t = E(a_t | Z_1, Z_2, \ldots, Z_n) \quad \text{for } t \le 0$$

are **backcasts** (or backforecasts) of the white noise terms that occur before our observation period. We shall now use our knowledge of forecasting to determine how to calculate these backcasts.

To simplify the notation without losing any of the salient ideas, we shall consider an ARMA(1, 1) model:

$$Z_t = \phi Z_{t-1} + \theta_0 + a_t - \theta a_{t-1} \tag{I-1}$$

Now in any stationary time series model, the *direction* of time is mathematically irrelevant; that is, we could equally well think of time as running backwards. Thus our observed series could just as well be generated by the **backward model**

$$Z_t = \phi Z_{t+1} + \theta_0 + b_t - \theta b_{t+1} \tag{I-2}$$

where $\{b_t\}$ is a zero-mean white noise sequence with variance σ_a^2 and is such that $b_{t-\ell}$ is independent of $Z_t, Z_{t+1}, \ldots$ for $\ell > 0$. One can easily check that all of the statistical properties of the backward model are identical to those of the usual forward ARMA(1, 1) model. Equation (I-2) is then rewritten as

$$b_t = Z_t - \phi Z_{t+1} - \theta_0 + \theta b_{t+1} \tag{I-3}$$

For a particular set of values for ϕ, θ, and θ_0, we use Equation (I-3) to calculate the b-values backwards in time starting with $t = n - 1$ and setting $b_n = 0$ to get started (compare with the discussion following Equation (7-41)). The recursive calculation then proceeds until we get back to $t = 1$. If we temporarily let $\hat{Z}_t = E(Z_t | Z_1, Z_2, \ldots, Z_n)$, we can backcast as

$$\hat{Z}_0 = \phi Z_1 \quad + \theta_0 - \theta b_1 \tag{I-4}$$
$$\hat{Z}_{-1} = \phi \hat{Z}_0 \quad + \theta_0$$

and, more generally,

$$\hat{Z}_{-\ell} = \phi \hat{Z}_{-\ell+1} + \theta_0 \quad \text{for } \ell \ge 1 \tag{I-5}$$

(Compare with Equations (9-36) and (9-37).)

Using Equations (I-4) and (I-5), we backcast sufficiently far back in time until the backcasts have essentially converged to their limit, $\theta_0/(1 - \phi)$. Suppose this has occurred at time $t = -Q$. We then use the usual forward version of the ARMA(1, 1) model to generate a_t *forward* in time according to

$$a_t = Z_t - \phi Z_{t-1} - \theta_0 + \theta a_{t-1}$$

setting $a_{-Q} = 0$ to start this final recursion. With the nonnegligible a_t in hand, we can then calculate the required unconditional sum-of-squares function as given by Equation (7-51).

In the Minitab ARIMA command, a maximum of $100 - M$ backcasts are computed where M is the largest lag appearing in the model being estimated. If the backcasts have not converged by that time, a warning message is printed.

Our general expression (7-51) for the unconditional sum-of-squares function was quoted without proof. In order to make it more plausible, we shall show that Equation (7-51) agrees with our earlier Equation (7-47), derived specifically for the AR(1) case.

If we look backward in time, it is clear that Equation (9-14) implies that

$$E(Z_t \mid Z_1, Z_2, \ldots, Z_n) = \phi^{-t+1}(Z_1 - \mu) + \mu \quad \text{for } t \le 1 \tag{I-6}$$

This in turn can be used to establish that

$$\hat{a}_t = E(a_t \mid Z_1, Z_2, \ldots, Z_n) = \phi^{-t+1}(1 - \phi^2)(Z_1 - \mu) \quad \text{for } t \le 1 \tag{I-7}$$

It is also easy to see that

$$\hat{a}_t = a_t = (Z_t - \mu) - \phi(Z_{t-1} - \mu) \quad \text{for } 2 \le t \le n \tag{I-8}$$

Consequently,

$$\sum_{t=-\infty}^{n} \hat{a}_t^2 = \sum_{t=2}^{n} [(Z_t - \mu) - \phi(Z_{t-1} - \mu)]^2 + \sum_{t=-\infty}^{1} [\phi^{-t+1}(1 - \phi^2)(Z_1 - \mu)]^2$$

$$= \sum_{t=2}^{n} [(Z_t - \mu) - \phi(Z_{t-1} - \mu)]^2 + (1 - \phi^2)(Z_1 - \mu)^2$$

(by summing the geometric series)

which does agree with Equation (7-47), as required.

CHAPTER 10 SEASONAL MODELS

In Chapter 3 we saw how seasonal deterministic trends might be modeled. However, in many areas in which time series are used, particularly business and economics, the assumption of any deterministic trend is quite suspect even though cyclical tendencies are very common in such series.

Exhibit 10.2, which appears later in the chapter, displays the logarithm of monthly air-passenger-miles within the United States from January 1960 through December 1977. As expected, the series tends to be highest during the summer vacation months, but it also generally shows an increase from November to December. There is also a general upward trend throughout the full period. We might of course assume a linear time trend in addition to a cosine or seasonal means trend, but do we really believe that the overall trend should be that regular?

An alternative is to use the **stochastic seasonal models** developed in this chapter.

10.1 SEASONAL ARMA MODELS

We shall begin by studying stationary models and then consider nonstationary generalizations in Section 10.3. We let s denote the known seasonal period; for monthly series $s = 12$ and for quarterly series $s = 4$.

Consider the series generated according to

$$Z_t = a_t - \Theta a_{t-12}$$

Notice that

$$\text{Cov}(Z_t, Z_{t-1}) = \text{Cov}(a_t - \Theta a_{t-12}, a_{t-1} - \Theta a_{t-13})$$
$$= 0$$

but that

$$\text{Cov}(Z_t, Z_{t-12}) = \text{Cov}(a_t - \Theta a_{t-12}, a_{t-12} - \Theta a_{t-24})$$
$$= -\Theta \sigma_a^2$$

It is easy to see that such a series is stationary and has nonzero autocorrelation only at lag 12.

Generalizing these ideas, we define a **seasonal MA(Q)s model** of order Q by

$$Z_t = a_t - \Theta_1 a_{t-s} - \Theta_2 a_{t-2s} - \cdots - \Theta_Q a_{t-Qs} \qquad \text{(10-1)}$$

with **seasonal MA characteristic polynomial**

$$\Theta(x) = 1 - \Theta_1 x^s - \Theta_2 x^{2s} - \cdots - \Theta_Q x^{Qs} \tag{10-2}$$

It is evident that such a series is always stationary and that the autocorrelation function will be nonzero only at the seasonal lags of $s, 2s, 3s, \ldots, Qs$. In particular,

$$\rho_{ks} = \frac{-\Theta_k + \Theta_1\Theta_{k+1} + \Theta_2\Theta_{k+2} + \cdots + \Theta_{Q-k}\Theta_Q}{1 + \Theta_1^2 + \Theta_2^2 + \cdots + \Theta_Q^2} \quad \text{for } k = 1, 2, \ldots, Q \tag{10-3}$$

(Compare with Equation (4-8).) For the model to be invertible, the roots of $\Theta(x) = 0$ must all exceed 1 in absolute value.

It is useful to note that the seasonal MA(Q)s model can also be viewed as a special case of an ordinary nonseasonal MA model of order $q = Qs$ but with all θ-values equal to zero except at the seasonal lags $s, 2s, \ldots, Qs$.

Seasonal autoregressive models can also be defined. Consider

$$Z_t = \Phi Z_{t-12} + a_t \tag{10-4}$$

where $|\Phi| < 1$ and a_t is independent of $Z_{t-1}, Z_{t-2}, \ldots$. It can be shown that $|\Phi| < 1$ ensures stationarity. Thus it is easy to argue that $E(Z_t) = 0$; multiplying Equation (10-4) by Z_{t-k}, taking expectations, and dividing by γ_0 yields

$$\rho_k = \Phi\rho_{k-12} \quad \text{for } k \geq 1 \tag{10-5}$$

Clearly

$$\rho_{12} = \Phi\rho_0 = \Phi \quad \text{and} \quad \rho_{24} = \Phi\rho_{12} = \Phi^2$$

More generally,

$$\rho_{12k} = \Phi^k \quad \text{for } k = 1, 2, \ldots \tag{10-6}$$

Furthermore, setting $k = 1$ and then $k = 11$ and using $\rho_k = \rho_{-k}$ gives us

$$\rho_1 = \Phi\rho_{11} \quad \text{and} \quad \rho_{11} = \Phi\rho_1$$

which implies that $\rho_1 = \rho_{11} = 0$. Similarly, one can show that $\rho_k = 0$ except at the seasonal lags $s, 2s, 3s, \ldots$. At those lags, the autocorrelation function decays exponentially like an AR(1) model.

With this example in mind, we define a **seasonal AR(P)s model** of order P by

$$Z_t = \Phi_1 Z_{t-s} + \Phi_2 Z_{t-2s} + \cdots + \Phi_P Z_{t-Ps} + a_t \tag{10-7}$$

with **seasonal AR characteristic polynomial**

$$\Phi(x) = 1 - \Phi_1 x^s - \Phi_2 x^{2s} - \cdots - \Phi_P x^{Ps} \tag{10-8}$$

We require that a_t be independent of $Z_{t-1}, Z_{t-2}, \ldots$ and, for stationarity, that the roots of $\Phi(x) = 0$ be greater than 1 in absolute value. Again Equation (10-7) can be seen as a special AR(p) model of order $p = Ps$ with nonzero ϕ-coefficients only at the seasonal lags $s, 2s, \ldots, Ps$.

It can be shown that the autocorrelation function is nonzero only at lags $s, 2s, \ldots$, where it behaves like a combination of decaying exponentials and damped sine functions. In particular, Equations (10-4), (10-5), and (10-6) easily generalize to the seasonal AR(1)s model to give

$$\rho_{ks} = \Phi^k \quad \text{for } k = 1, 2, \ldots \tag{10-9}$$

with zero correlation at all other lags.

10.2 MULTIPLICATIVE SEASONAL ARMA MODELS

Rarely shall we need models that incorporate autocorrelation *only* at the seasonal lags. By combining the ideas of seasonal and nonseasonal ARMA models, we can develop parsimonious models that contain autocorrelation both for the seasonal lags and for neighboring observations.

Consider a model whose MA characteristic polynomial is given by

$$(1 - \theta x)(1 - \Theta x^{12})$$

Multiplying out we have $1 - \theta x - \Theta x^{12} + \theta \Theta x^{13}$. Thus the corresponding series satisfies

$$Z_t = a_t - \theta a_{t-1} - \Theta a_{t-12} + \theta \Theta a_{t-13} \tag{10-10}$$

For this model we can check that the autocorrelation function is nonzero only at lags 1, 11, 12, and 13. We find

$$\gamma_0 = (1 + \theta^2)(1 + \Theta^2)\sigma_a^2 \tag{10-11}$$

$$\rho_1 = -\frac{\theta}{1 + \theta^2}$$

$$\rho_{11} = \rho_{13} = \frac{\theta \Theta}{(1 + \theta^2)(1 + \Theta^2)} \tag{10-12}$$

and

$$\rho_{12} = -\frac{\Theta}{1 + \Theta^2}$$

Of course, we can also introduce both short-term and seasonal autocorrelation by defining an MA model of order 12 with only θ_1 and θ_{12} nonzero. We shall see in Section 10.3 that the "multiplicative" model arises quite naturally for nonstationary models that entail differencing.

In general, then, we define a **multiplicative seasonal ARMA(p, q) $\times$ (P, Q)s model*** as a model with AR characteristic polynomial $\phi(x)\Phi(x)$ and MA

---

* Using the backshift operator of Appendix E, we can write the multiplicative seasonal ARMA(p, q) $\times$ (P, Q)s model as $\phi(B)\Phi(B)Z_t = \theta(B)\Theta(B)a_t$.

characteristic polynomial $\theta(x)\Theta(x)$ where

$$\phi(x) = 1 - \phi_1 x - \phi_2 x^2 - \cdots - \phi_p x^p$$
$$\Phi(x) = 1 - \Phi_1 x^s - \Phi_2 x^{2s} - \cdots - \Phi_P x^{Ps}$$

and   **(10-13)**

$$\theta(x) = 1 - \theta_1 x - \theta_2 x^2 - \cdots - \theta_q x^q$$
$$\Theta(x) = 1 - \Theta_1 x^s - \Theta_2 x^{2s} - \cdots - \Theta_Q x^{Qs}$$

The model may also contain a constant term θ_0. Note once more that we have just a special ARMA model with AR order $p + Ps$ and MA order $q + Qs$, but the coefficients are not completely general, being determined by only $p + P + q + Q$ coefficients. If $s = 12$, $p + P + q + Q$ will be considerably smaller than $p + Ps + q + Qs$ and will allow a more parsimonious model.

As another example, suppose $P = q = 1$ and $p = Q = 0$ with $s = 12$. The model is then

$$Z_t = \Phi Z_{t-12} + a_t - \theta a_{t-1} \qquad \textbf{(10-14)}$$

Using our standard techniques, we find that

$$\gamma_1 = \Phi \gamma_{11} - \theta \sigma_a^2 \qquad \textbf{(10-15)}$$

and

$$\gamma_k = \Phi \gamma_{k-12} \quad \text{for } k \geq 2 \qquad \textbf{(10-16)}$$

After considering the equations implied by various choices for k, we arrive at

$$\gamma_0 = \frac{1 + \theta^2}{1 - \Phi^2} \sigma_a^2 \qquad \textbf{(10-17)}$$

$$\rho_{12k} = \Phi^k, \quad k = 1, 2, \ldots \qquad \textbf{(10-18)}$$

$$\rho_{12k-1} = \rho_{12k+1} = -\frac{\theta}{1 + \theta^2} \Phi^k, \quad k = 0, 1, 2, \ldots \qquad \textbf{(10-19)}$$

with all other autocorrelations being zero.

Figure 10.1 shows a graph of this autocorrelation function for the case of $\Phi = 0.75$, $\theta = \pm 0.4$, and $s = 12$. The shape of this autocorrelation is somewhat typical of the sample autocorrelation function observed for numerous seasonal

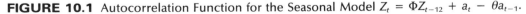

FIGURE 10.1 Autocorrelation Function for the Seasonal Model $Z_t = \Phi Z_{t-12} + a_t - \theta a_{t-1}$.

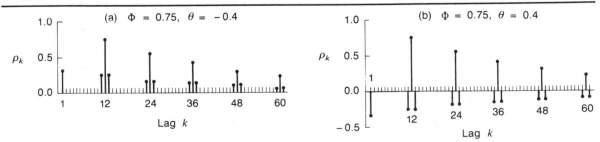

time series. The even simpler autocorrelation given by Equations (10-12) also seems to occur frequently in practice.

10.3 NONSTATIONARY SEASONAL ARIMA MODELS

An important tool in modeling nonstationary seasonal processes is the seasonal difference. The **seasonal difference** of period s for the series $\{Z_t\}$ is denoted $\nabla_s Z_t$ and is defined as

$$\nabla_s Z_t = Z_t - Z_{t-s} \tag{10-20}$$

For example, for a monthly series we consider the changes from January to January, February to February, and so forth for successive years. Note that for a series of length n, the seasonal difference series will be of length $n - s$; that is, s data values are lost due to seasonal differencing.

As an example where seasonal differencing is appropriate, consider a process generated according to

$$Z_t = S_t + a_t \tag{10-21}$$

with

$$S_t = S_{t-s} + b_t \tag{10-22}$$

where $\{a_t\}$ and $\{b_t\}$ are independent white noise series. Here $\{S_t\}$ is a "seasonal random walk," and if $\sigma_b \ll \sigma_a$, $\{S_t\}$ would model a slowly changing seasonal component.

Due to the nonstationarity of $\{S_t\}$, clearly $\{Z_t\}$ is nonstationary. However, if we seasonally difference Z_t, as given in Equation (10-21), we find

$$\begin{aligned} \nabla_s Z_t &= S_t - S_{t-s} + a_t - a_{t-s} \\ &= b_t + a_t - a_{t-s} \end{aligned} \tag{10-23}$$

An easy calculation shows that $\nabla_s Z_t$ is stationary and has the autocorrelation function of an MA(1)s model.

The model of Equations (10-21) and (10-22) could also be generalized to account for a nonseasonal, slowly changing stochastic trend. Consider

$$Z_t = M_t + S_t + a_t \tag{10-24}$$

with

$$S_t = S_{t-s} + b_t \tag{10-25}$$

and

$$M_t = M_{t-1} + c_t \tag{10-26}$$

where $\{a_t\}$, $\{b_t\}$, and $\{c_t\}$ are mutually independent white noise sequences. Here we take both a seasonal difference and an ordinary nonseasonal difference to obtain*

$$\nabla\nabla_s Z_t = \nabla(M_t - M_{t-s} + b_t + a_t - a_{t-s})$$
$$= (c_t + b_t + a_t) - (b_{t-1} + a_{t-1}) - (c_{t-s} + a_{t-s}) + a_{t-s-1} \qquad \text{(10-27)}$$

The process in Equation (10-27) is stationary and will have a nonzero autocorrelation only at lags 1, $s - 1$, s, and $s + 1$, which agrees with the correlation structure of the multiplicative seasonal model ARMA(0, 1) $\times$ (0, 1)s.

These examples lead to the definition of nonstationary seasonal models. A process $\{Z_t\}$ is said to be a **multiplicative seasonal ARIMA model** with nonseasonal orders p, d, and q, seasonal orders P, D, and Q, and seasonal period s if the differenced series

$$W_t = \nabla^d \nabla_s^D Z_t \qquad \text{(10-28)}$$

satisfies an ARMA(p, q) $\times$ (P, Q)s model.† We say that $\{Z_t\}$ is an **ARIMA(p, d, q) $\times$ (P, D, Q)s model.**

Clearly, such models represent a broad, flexible class from which to select an appropriate model for a particular series. It has been found empirically that many series can be adequately fit by these models, usually with a small number of parameters, say, three or four.

10.4 MODEL SPECIFICATION, FITTING, AND CHECKING

Model specification, fitting, and checking for seasonal models will follow the same general techniques developed in Chapters 6, 7, and 8. Here we shall simply highlight the application of these ideas specifically to seasonal models.

MODEL SPECIFICATION

As always, a careful inspection of the time plot of the series is the first step. Exhibit 10.1 shows total monthly air-passenger-miles within the United States from January 1960 through December 1977. On a scale that accommodates the range of the full series, the variation in the series early on is very small relative to the variation later, when the general level is higher. This behavior suggests that a log transformation is in order.

Exhibit 10.2 shows the corresponding logged values. We still see seasonality and upward drift, but now the variation appears comparable at

---

*It should be noted that $\nabla_s Z_t$ will in fact be stationary and $\nabla\nabla_s Z_t$ will be noninvertible. We use Equations (10-24), (10-25), and (10-26) merely to help motivate multiplicative seasonal ARIMA models.

†Using the backshift operator of Appendix E, we may write the general ARIMA(p, d, q) $\times$ (P, D, Q)s model as $\phi(B)\Phi(B)\nabla^d\nabla_s^D Z_t = \theta(B)\Theta(B)a_t$.

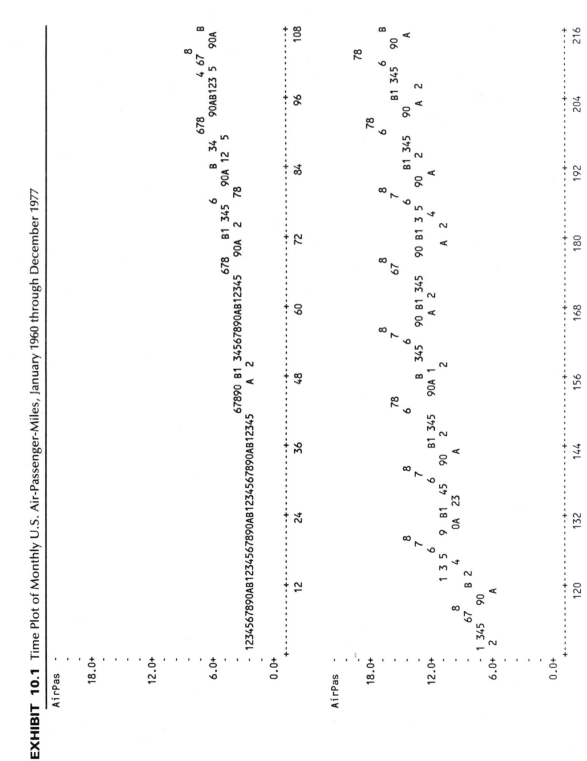

EXHIBIT 10.1 Time Plot of Monthly U.S. Air-Passenger-Miles, January 1960 through December 1977

EXHIBIT 10.2 Time Plot of Logarithms of Monthly U.S. Air-Passenger-Miles, January 1960 through December 1977

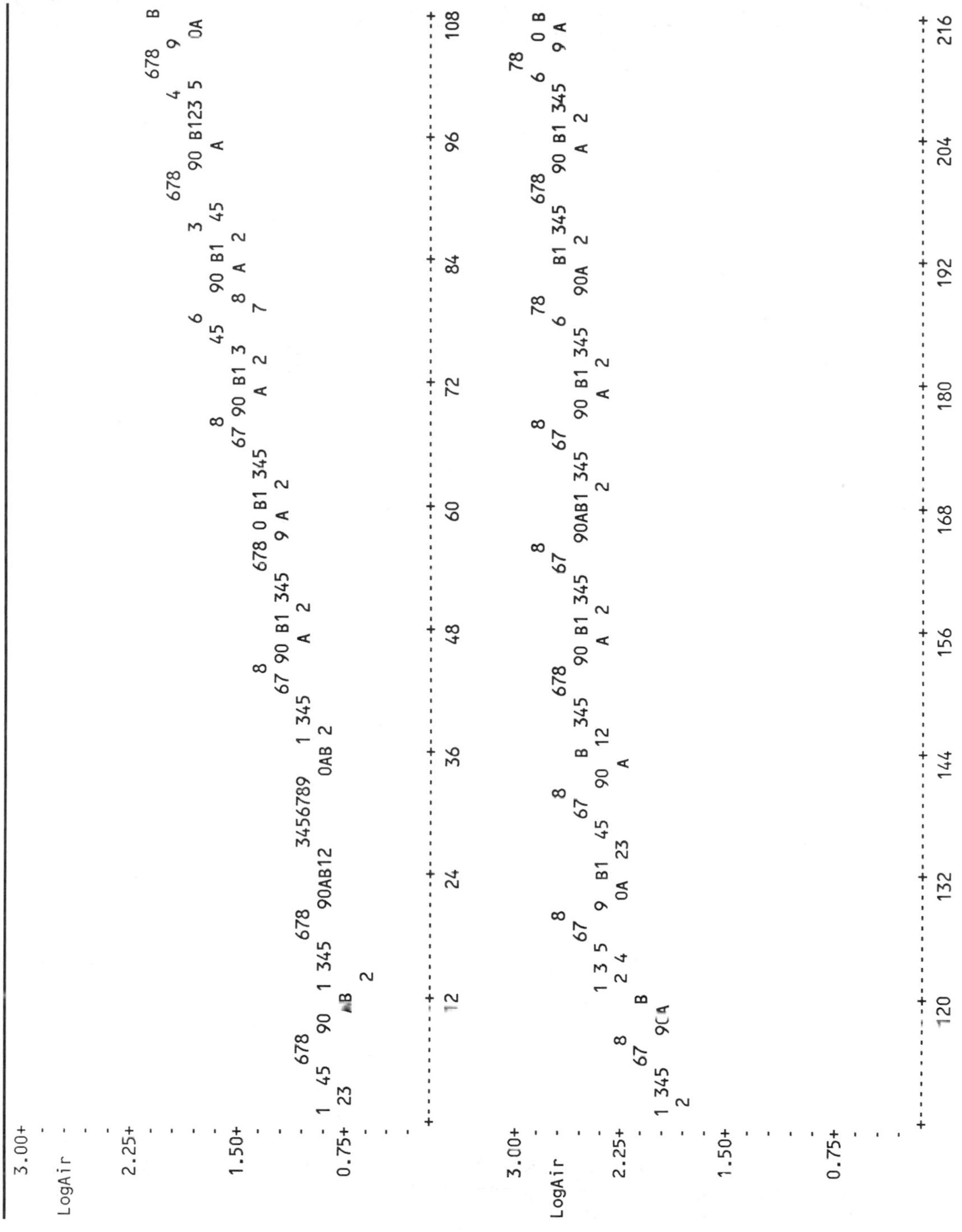

EXHIBIT 10.3 Sample Autocorrelation Function for the Logged U.S. Air-Passenger-Miles Series

```
ACF of LogAir

         -1.0 -0.8 -0.6 -0.4 -0.2  0.0  0.2  0.4  0.6  0.8  1.0
         +----+----+----+----+----+----+----+----+----+----+
    1  0.975                              XXXXXXXXXXXXXXXXXXXXXXXXX
    2  0.959                              XXXXXXXXXXXXXXXXXXXXXXXXX
    3  0.943                              XXXXXXXXXXXXXXXXXXXXXXXXX
    4  0.929                              XXXXXXXXXXXXXXXXXXXXXXXX
    5  0.917                              XXXXXXXXXXXXXXXXXXXXXXXX
    6  0.904                              XXXXXXXXXXXXXXXXXXXXXXXX
    7  0.896                              XXXXXXXXXXXXXXXXXXXXXXX
    8  0.887                              XXXXXXXXXXXXXXXXXXXXXXX
    9  0.880                              XXXXXXXXXXXXXXXXXXXXXXX
   10  0.873                              XXXXXXXXXXXXXXXXXXXXXXX
   11  0.867                              XXXXXXXXXXXXXXXXXXXXXXX
   12  0.866                              XXXXXXXXXXXXXXXXXXXXXXX
   13  0.843                              XXXXXXXXXXXXXXXXXXXXXX
   14  0.826                              XXXXXXXXXXXXXXXXXXXXXX
   15  0.810                              XXXXXXXXXXXXXXXXXXXXX
   16  0.796                              XXXXXXXXXXXXXXXXXXXXX
   17  0.782                              XXXXXXXXXXXXXXXXXXXXX
   18  0.767                              XXXXXXXXXXXXXXXXXXXX
   19  0.758                              XXXXXXXXXXXXXXXXXXXX
   20  0.747                              XXXXXXXXXXXXXXXXXXXX
   21  0.738                              XXXXXXXXXXXXXXXXXXX
   22  0.730                              XXXXXXXXXXXXXXXXXXX
   23  0.722                              XXXXXXXXXXXXXXXXXXX
   24  0.720                              XXXXXXXXXXXXXXXXXXX
   25  0.697                              XXXXXXXXXXXXXXXXXX
   26  0.681                              XXXXXXXXXXXXXXXXXX
   27  0.665                              XXXXXXXXXXXXXXXXXX
   28  0.650                              XXXXXXXXXXXXXXXXX
   29  0.634                              XXXXXXXXXXXXXXXXX
   30  0.619                              XXXXXXXXXXXXXXXX
   31  0.609                              XXXXXXXXXXXXXXXX
   32  0.596                              XXXXXXXXXXXXXXXX
   33  0.587                              XXXXXXXXXXXXXXXX
   34  0.577                              XXXXXXXXXXXXXXX
   35  0.569                              XXXXXXXXXXXXXXX
   36  0.565                              XXXXXXXXXXXXXXX
```

different levels of the series. From here on, all analysis will be based on the logarithms of the original series.

Exhibit 10.3 gives the sample autocorrelation function for the log series. the upward drift in the series is so strong that the seasonality is nearly lost in the sample autocorrelation function.

Exhibit 10.4 displays a time plot of the (nonseasonal) first-difference series. Note that all time plots are carried out with a period of 12 specified so that observations one period apart can be clearly recognized. In this exhibit, the seasonality is quite apparent: 2's, 9's, and A's tend to be low, while 3's, 6's, and B's tend to be high.

The seasonality is strikingly displayed in Exhibit 10.5, which gives the sample autocorrelation function of the first-difference series. Note the strong, positive autocorrelation at lags 12, 24, and 36, which is decaying quite slowly.

Exhibits 10.6 and 10.7 show what happens if we calculate a period 12

EXHIBIT 10.4 Time Plot of the First Difference of the Logged U.S. Air-Passenger-Miles Series

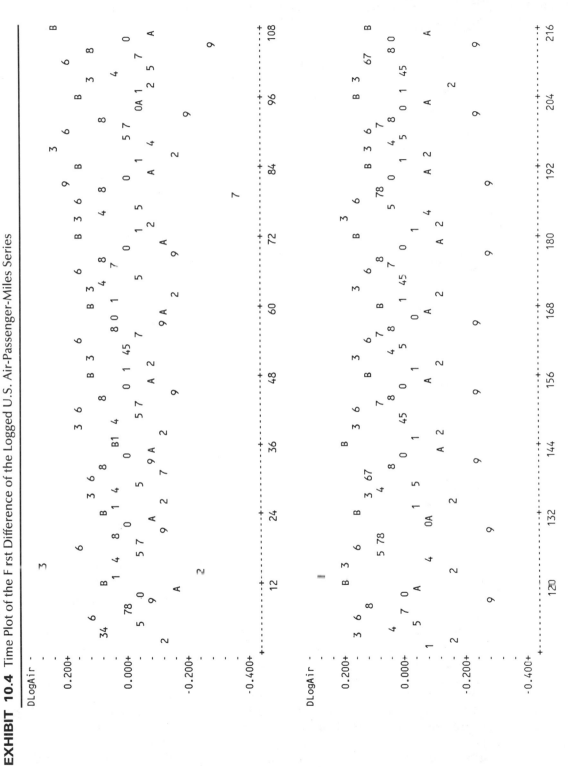

EXHIBIT 10.5 Sample Autocorrelation Function for the First Difference of the Logged U.S. Air-Passenger-Miles Series

```
ACF of DLogAir

          -1.0 -0.8 -0.6 -0.4 -0.2  0.0  0.2  0.4  0.6  0.8  1.0
          +----+----+----+----+----+----+----+----+----+----+
  1 -0.291                        XXXXXXX
  2 -0.033                             XX
  3 -0.029                             XX
  4 -0.127                           XXXX
  5  0.086                              XXX
  6 -0.172                          XXXXX
  7  0.089                              XXX
  8 -0.131                           XXXX
  9  0.012                              X
 10 -0.026                             XX
 11 -0.225                        XXXXXXX
 12  0.748                              XXXXXXXXXXXXXXXXXXX
 13 -0.260                       XXXXXXXX
 14  0.028                             XX
 15 -0.041                             XX
 16 -0.105                           XXXX
 17  0.088                              XXX
 18 -0.188                          XXXXX
 19  0.107                              XXXX
 20 -0.115                           XXXX
 21  0.005                              X
 22 -0.036                             XX
 23 -0.199                          XXXXX
 24  0.713                              XXXXXXXXXXXXXXXXXX
 25 -0.253                       XXXXXXX
 26  0.029                             XX
 27 -0.033                             XX
 28 -0.072                            XXX
 29  0.024                             XX
 30 -0.161                          XXXXX
 31  0.125                              XXXX
 32 -0.119                           XXXX
 33 -0.010                              X
 34 -0.025                             XX
 35 -0.186                          XXXXX
 36  0.656                              XXXXXXXXXXXXXXXXX
```

seasonal difference instead of the nonseasonal difference. The time plot (Exhibit 10.6) shows that most of the seasonality has been removed but that the series probably cannot be assumed stationary. The sample autocorrelation function (Exhibit 10.7) shows little seasonality but strongly indicates nonstationarity.

Finally, Exhibits 10.8 and 10.9 display the time plot and sample autocorrelation function, respectively, for the series after both nonseasonal and seasonal differencing has been applied. Both exhibits suggest that a stationary model be considered. Note that after differencing, a rather strong correlation remains at lag 12 but that the correlations at lags 24 and 36 become negligible. For this example, the effective number of observations is $n - d - Ds = 216 - 13 = 203$. Assuming a tentative order of $q = 12$, Equation (6-11) yields a standard error for r_k of 0.091. Thus the lag 17 value of 0.144 can safely be ignored. In the interest of parsimony, we might also ignore lag 2 and consider, after differencing, a simple multiplicative model as given in Equation (10-10). In

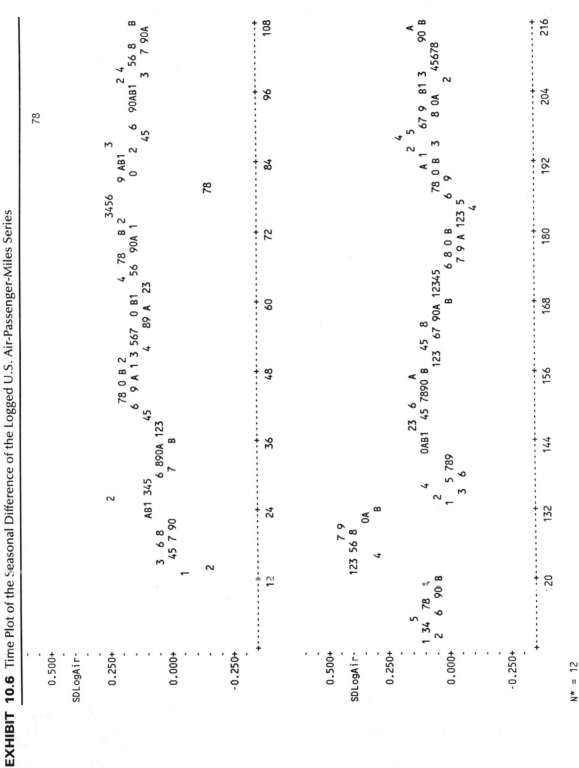

EXHIBIT 10.6 Time Plot of the Seasonal Difference of the Logged U.S. Air-Passenger-Miles Series

EXHIBIT 10.7 Sample Autocorrelation Function for the Seasonal Difference of the Logged U.S. Air-Passenger-Miles Series

```
ACF of SDLogAir

        -1.0 -0.8 -0.6 -0.4 -0.2  0.0  0.2  0.4  0.6  0.8  1.0
         +----+----+----+----+----+----+----+----+----+----+
  1  0.735                             XXXXXXXXXXXXXXXXXXX
  2  0.557                             XXXXXXXXXXXXXX
  3  0.474                             XXXXXXXXXXXX
  4  0.421                             XXXXXXXXXXX
  5  0.372                             XXXXXXXXXX
  6  0.326                             XXXXXXXXX
  7  0.243                             XXXXXX
  8  0.192                             XXXXX
  9  0.152                             XXXXX
 10  0.082                             XXX
 11 -0.021                           XX
 12 -0.163                        XXXXX
 13 -0.040                           XX
 14  0.051                             XX
 15  0.073                             XXX
 16  0.101                             XXXX
 17  0.145                             XXXXX
 18  0.110                             XXXX
 19  0.126                             XXXX
 20  0.148                             XXXXX
 21  0.161                             XXXXX
 22  0.148                             XXXXX
 23  0.178                             XXXXX
 24  0.175                             XXXXX
 25  0.147                             XXXXX
 26  0.129                             XXXX
 27  0.130                             XXXX
 28  0.107                             XXXX
 29  0.100                             XXX
 30  0.164                             XXXXX
 31  0.160                             XXXXX
 32  0.129                             XXXX
 33  0.102                             XXXX
 34  0.128                             XXXX
 35  0.109                             XXXX
 36  0.103                             XXXX
```

our general notation, then, we are entertaining the ARIMA $(0, 1, 2) \times (0, 1, 1)12$ model for further consideration, fully realizing that diagnostic checking may indicate the need for a more general model. Note that Equations (10-12) show ρ_{11} and ρ_{13} as nonzero also, but that if θ and Θ are positive and θ is small, as Exhibit 10.9 suggests, ρ_{11} and ρ_{13} will also be small and positive. This agrees with the sign and magnitude of r_{11} and r_{13}.

As a second example of specifying a seasonal model, consider once more the monthly milk production series first shown in Exhibit 1.3. In Chapter 3 we attempted to model this series using a deterministic time trend plus a seasonal means model. However, recall that the residuals from that model were far from satisfactory (see Exhibits 3.12 and 3.17). Exhibits 10.10 through 10.13 present the sample autocorrelation function of the original series, the first-difference series, the seasonal difference series, and the series obtained by

EXHIBIT 10.8 Time Plot of the Seasonal and Nonseasonal Difference of the Logged U.S. Air-Passenger-Miles Series

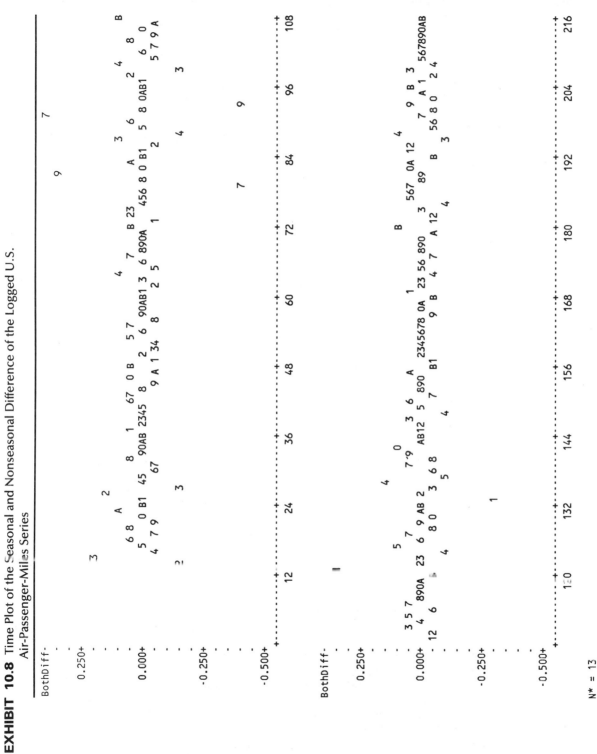

N* = 13

EXHIBIT 10.9 Sample Autocorrelation Function of the Seasonal and Nonseasonal Difference of the Logged U.S. Air-Passenger-Miles Series

```
ACF of BothDiff

              -1.0 -0.8 -0.6 -0.4 -0.2  0.0  0.2  0.4  0.6  0.8  1.0
               +----+----+----+----+----+----+----+----+----+----+
      1  -0.181                         XXXXXX
      2  -0.160                         XXXXX
      3  -0.065                          XXX
      4  -0.007                           X
      5  -0.001                           X
      6   0.068                           XXX
      7  -0.059                          XX
      8  -0.025                          XX
      9   0.061                           XXX
     10   0.074                           XXX
     11   0.073                           XXX
     12  -0.507            XXXXXXXXXXXXXX
     13   0.078                           XXX
     14   0.117                           XXXX
     15  -0.012                           X
     16  -0.031                          XX
     17   0.144                           XXXXX
     18  -0.102                        XXXX
     19  -0.009                           X
     20   0.020                           XX
     21   0.049                           XX
     22  -0.084                         XXX
     23   0.063                           XXX
     24   0.050                           XX
     25  -0.018                           X
     26  -0.036                          XX
     27   0.050                           XX
     28  -0.030                          XX
     29  -0.131                        XXXX
     30   0.135                           XXXX
     31   0.054                           XX
     32  -0.014                           X
     33  -0.098                         XXX
     34   0.083                           XXX
     35  -0.018                           X
     36  -0.015                           X
```

applying both differences. Exhibit 10.13 suggests that a multiplicative ARIMA $(0, 1, 1) \times (0, 1, 1)12$ model should be investigated.

MODEL FITTING

Having specified a tentative seasonal model for a particular time series, we proceed to estimate the parameters of that model as efficiently as possible. As we have remarked earlier, the multiplicative seasonal ARIMA models are just special cases of our general ARIMA models. As such, all of our work on parameter estimation in Chapter 7 carries over to the seasonal case.

From a practical point of view, the starting-value problems discussed in Chapter 7 are a little more serious with seasonal models. Consider, for example, the ARMA$(0, 0, 0) \times (0, 0, 1)12$ model:

$$Z_t = a_t - \Theta a_{t-12}$$

EXHIBIT 10.10 Sample Autocorrelation Function for the Monthly Milk Production Series

```
ACF of Milkpr

        -1.0 -0.8 -0.6 -0.4 -0.2  0.0  0.2  0.4  0.6  0.8  1.0
        +----+----+----+----+----+----+----+----+----+----+
  1  0.892                          XXXXXXXXXXXXXXXXXXXXXXXX
  2  0.778                          XXXXXXXXXXXXXXXXXXXX
  3  0.620                          XXXXXXXXXXXXXXXX
  4  0.487                          XXXXXXXXXXXXX
  5  0.428                          XXXXXXXXXXXX
  6  0.376                          XXXXXXXXXX
  7  0.415                          XXXXXXXXXXX
  8  0.454                          XXXXXXXXXXX
  9  0.562                          XXXXXXXXXXXXXX
 10  0.687                          XXXXXXXXXXXXXXXXX
 11  0.769                          XXXXXXXXXXXXXXXXXXXX
 12  0.845                          XXXXXXXXXXXXXXXXXXXXXXX
 13  0.745                          XXXXXXXXXXXXXXXXXXX
 14  0.638                          XXXXXXXXXXXXXXXX
 15  0.490                          XXXXXXXXXXXXX
 16  0.364                          XXXXXXXXXX
 17  0.306                          XXXXXXXXX
 18  0.255                          XXXXXXX
 19  0.287                          XXXXXXXX
 20  0.321                          XXXXXXXXX
 21  0.417                          XXXXXXXXXXX
 22  0.529                          XXXXXXXXXXXXXX
 23  0.603                          XXXXXXXXXXXXXXXX
 24  0.673                          XXXXXXXXXXXXXXXXX
 25  0.583                          XXXXXXXXXXXXXXX
 26  0.488                          XXXXXXXXXXXXX
 27  0.355                          XXXXXXXXXX
 28  0.240                          XXXXXXX
 29  0.188                          XXXXXX
 30  0.141                          XXXXX
 31  0.169                          XXXXX
 32  0.196                          XXXXX
 33  0.281                          XXXXXXX
 34  0.383                          XXXXXXXXXX
 35  0.451                          XXXXXXXXXXX
 36  0.516                          XXXXXXXXXXXXX
```

In inverted form this becomes

$$a_t = Z_t + \Theta Z_{t-12} + \Theta^2 Z_{t-24} + \cdots$$

and even though $|\Theta| < 1$, we might have to go far into the past before we could consider Θ^k negligible. For this reason, *conditional* least-squares estimates should probably not be used for seasonal models, and we may have to backcast rather far into the past to obtain the unconditional least-squares estimates. Recall that Minitab gives a warning if the backcasts are not converging.

The Minitab command to estimate multiplicative seasonal models has the format

ARIMA p = **K**, d = **K**, q = **K**, P = **K**, D = **K**, Q = **K**, s = **K** series in **C**

Additional columns may be specified for storing the residuals, fitted values, and estimated parameters in that order. The subcommands CONSTANT, NOCONSTANT, STARTING values, and FORECAST may also be used. If $d > 0$ or $D > 0$, no constant is assumed, but if $d = D = 0$, a constant term will be

EXHIBIT 10.11 Sample Autocorrelation Function for the First Difference of the Milk Production Series

```
ACF of DiffMilk

        -1.0 -0.8 -0.6 -0.4 -0.2  0.0  0.2  0.4  0.6  0.8  1.0
        +----+----+----+----+----+----+----+----+----+----+
 1   0.023                        XX
 2   0.256                        XXXXXX
 3  -0.114                    XXXX
 4  -0.359               XXXXXXXXXX
 5  -0.058                      XX
 6  -0.502          XXXXXXXXXXXXXX
 7  -0.033                      XX
 8  -0.347               XXXXXXXXXX
 9  -0.082                     XXX
10   0.233                        XXXXXX
11   0.028                        XX
12   0.911                        XXXXXXXXXXXXXXXXXXXXXXXXX
13   0.031                        XX
14   0.238                        XXXXXX
15  -0.103                    XXXX
16  -0.330               XXXXXXXXXX
17  -0.058                      XX
18  -0.464          XXXXXXXXXXXXXX
19  -0.031                      XX
20  -0.320               XXXXXXXXXX
21  -0.079                     XXX
22   0.208                        XXXXX
23   0.028                        XX
24   0.833                        XXXXXXXXXXXXXXXXXXXXXX
25   0.029                        XX
26   0.222                        XXXXXX
27  -0.086                     XXX
28  -0.300               XXXXXXXXX
29  -0.056                      XX
30  -0.422          XXXXXXXXXXXX
31  -0.023                      XX
32  -0.297               XXXXXXXX
33  -0.079                     XXX
34   0.188                        XXXXX
35   0.022                        XX
36   0.762                        XXXXXXXXXXXXXXXXXXX
```

estimated unless instructed otherwise. If starting values are to be used, they should be given in the following order: AR, seasonal AR, MA, seasonal MA, and constant. The starting value for the constant term is always optional.

Exhibit 10.14 shows the printout when the ARIMA command is applied to the logged air-passenger-miles series.

DIAGNOSTIC CHECKING

Model criticism for seasonal models is no different than for nonseasonal models except that we pay special attention to the seasonal lags. We look at the normality of the residuals by evaluating their histogram and normal scores. We check on the randomness of the residuals by plotting them over time, computing their sample autocorrelation function, the modified Box–Pierce

EXHIBIT 10.12 Sample Autocorrelation Function for the Seasonal Difference of the Milk Production Series

```
ACF of SDifMilk

          -1.0 -0.8 -0.6 -0.4 -0.2  0.0  0.2  0.4  0.6  0.8  1.0
          +----+----+----+----+----+----+----+----+----+----+
   1   0.856                         XXXXXXXXXXXXXXXXXXXXXX
   2   0.773                         XXXXXXXXXXXXXXXXXXXX
   3   0.689                         XXXXXXXXXXXXXXXXXX
   4   0.576                         XXXXXXXXXXXXXXX
   5   0.493                         XXXXXXXXXXXXX
   6   0.438                         XXXXXXXXXXXX
   7   0.391                         XXXXXXXXXXX
   8   0.306                         XXXXXXXXX
   9   0.213                         XXXXXX
  10   0.130                         XXXX
  11   0.021                         XX
  12  -0.079                      XXX
  13  -0.054                       XX
  14  -0.077                      XXX
  15  -0.088                      XXX
  16  -0.087                      XXX
  17  -0.095                      XXX
  18  -0.137                     XXXX
  19  -0.176                    XXXXX
  20  -0.180                   XXXXXX
  21  -0.188                   XXXXXX
  22  -0.190                   XXXXXX
  23  -0.171                    XXXXX
  24  -0.175                    XXXXX
  25  -0.173                    XXXXX
  26  -0.181                   XXXXXX
  27  -0.184                   XXXXXX
  28  -0.171                    XXXXX
  29  -0.169                    XXXXX
  30  -0.143                    XXXXX
  31  -0.121                     XXXX
  32  -0.108                     XXXX
  33  -0.099                      XXX
  34  -0.096                      XXX
  35  -0.081                      XXX
  36  -0.099                      XXX
```

statistic, and the runs test. Overfitting with models more general than the model believed adequate will also be employed.

   We illustrate these ideas on the air-passenger-miles and milk production data.

AIR PASSENGER SERIES

We earlier specified an ARIMA$(0, 1, 1) \times (0, 1, 1)12$ model based on the logs of the series. Exhibit 10.14 shows the results of such a fit. However, Exhibit 10.15 gives the sample autocorrelation function of the residuals from this model. Since $\pm 2/\sqrt{203} = \pm 0.14$, we see that $r_2 = -0.207$ is significantly different from zero. So is r_3, but in the interest of parsimony we shall expand our model slowly.

EXHIBIT 10.13 Sample Autocorrelation Function for the Seasonal and Nonseasonal Difference of the Milk Production Series

```
ACF of BothDiff

              -1.0 -0.8 -0.6 -0.4 -0.2  0.0  0.2  0.4  0.6  0.8  1.0
              +----+----+----+----+----+----+----+----+----+----+
     1 -0.212                         XXXXXX
     2  0.014                              X
     3  0.120                              XXXX
     4 -0.100                          XXXX
     5 -0.091                           XXX
     6 -0.030                            XX
     7  0.127                              XXXX
     8  0.024                             XX
     9 -0.043                            XX
    10  0.095                             XXX
    11 -0.034                            XX
    12 -0.437              XXXXXXXXXXXXX
    13  0.187                              XXXXXX
    14 -0.066                           XXX
    15 -0.049                            XX
    16  0.029                             XX
    17  0.115                             XXXX
    18 -0.006                             X
    19 -0.112                         XXXX
    20  0.022                             XX
    21 -0.025                            XX
    22 -0.065                           XXX
    23  0.082                             XXX
    24 -0.019                             X
    25  0.027                             XX
    26 -0.000                             X
    27 -0.049                            XX
    28  0.026                             XX
    29 -0.080                           XXX
    30  0.012                             X
    31  0.021                             XX
    32  0.012                             X
    33  0.022                             XX
    34 -0.022                            XX
    35  0.090                             XXX
    36 -0.066                           XXX
```

Consider the more general model ARIMA$(0, 1, 2) \times (0, 1, 1)12$. Exhibit 10.16 displays the results of the estimation. Note that we would reject the hypothesis $\theta_2 = 0$ at the usual significance levels because the t-ratio is 3.39. Furthermore, Exhibit 10.17 attests to the improvement in the model, since no correlations exceed 0.14.

For seasonal models the modified Box–Pierce statistic should be compared to chi-squared critical values with $K - p - q - P - Q$ degrees of freedom. Here $p + q + P + Q$ is the number of parameters estimated, and n in Equation (8-12) is the effective number of observations after differencing. The maximum lag K should be chosen somewhat larger for seasonal models; for monthly series, $K = 36$ or $K = 48$ is usually a good choice.

For the air-passenger-miles data, $K = 36$ produces a modified Box–Pierce statistic of 18.4 with $36 - 2 - 1 = 33$ degrees of freedom; this shows no inadequacy in the model. The runs test also confirms the randomness

EXHIBIT 10.14 Estimation of the ARIMA$(0, 1, 1) \times (0, 1, 1)$ 12 Model for the Logged U.S. Air-Passenger-Miles Series

```
MTB > arima (0,1,1)X(0,1,1)12 'LogPas', 'Resid' & 'Fitted'

Estimates at each iteration
Iteration      SSE      Parameters
      0     1.15980    0.100    0.100
      1     1.01506    0.132    0.250
      2     0.89961    0.164    0.400
      3     0.81675    0.194    0.550
      4     0.76305    0.233    0.700
      5     0.75453    0.259    0.739
      6     0.75451    0.275    0.732
      7     0.75451    0.275    0.732
Relative change in each estimate less than  0.0010

Final Estimates of Parameters
Type      Estimate    St. Dev.   t-ratio
MA   1      0.2753      0.0684      4.03
SMA 12      0.7317      0.0526     13.92

Differencing: 1 regular, 1 seasonal of order 12
No. of obs.:  Original series 216, after differencing 203
Residuals:    SS = 0.737759  (backforecasts excluded)
              MS = 0.003670  DF = 201

Modified Box-Pierce chisquare statistic
Lag             12            24           36            48
Chisquare   19.2(DF=10)  25.5(DF=22)  31.7(DF=34)  50.8(DF=46)
```

hypothesis. However, the histogram of the residuals, Exhibit 10.18, reveals two extreme residuals. Since MS = 0.003418, the two standardized residuals are about $-0.34/\sqrt{0.003418} = -5.82$ and $0.29/\sqrt{0.003418} = 4.94$, both unusually large for a normal distribution. The normal-scores test and corresponding plot given in Exhibit 10.18 also point out the nonnormality. The time plot of the residuals in Exhibit 10.19 reveals the unusual residuals at $t = 79$ and $t = 121$. We shall return to this problem of outliers in Chapter 11.

MILK PRODUCTION SERIES

Exhibit 10.20 presents the estimated ARIMA$(0, 1, 1) \times (0, 1, 1)$12 model for the milk production series. The sample autocorrelation function shown in Exhibit 10.21 displays the lack of autocorrelation in the residuals. The modified Box–Pierce statistic based on 36 correlations comes out to be 19.37—quite insignificant. Also, the runs test does not reject randomness at the 5% level.

Normality of the residuals is somewhat better in this example. Exhibit 10.22 displays the histogram and the normal-scores test and plot. However, we would reject normality at the 1% level: Two large residuals are the main source of difficulty.

Putting the normality question aside, how does this stochastic seasonal model compare with our earlier deterministic seasonal model? First note that the stochastic model contains just 2 parameters, θ and Θ, whereas the

EXHIBIT 10.15 Sample Autocorrelation Function of Residuals from the ARIMA(0, 1, 1) × (0, 1, 1) 12 Model for the U.S. Air-Passenger-Miles Series

```
ACF of Resid

         -1.0 -0.8 -0.6 -0.4 -0.2  0.0  0.2  0.4  0.6  0.8  1.0
         +----+----+----+----+----+----+----+----+----+----+
  1   0.034                           XX
  2  -0.207                       XXXXXX
  3  -0.134                         XXXX
  4  -0.063                          XXX
  5   0.039                           XX
  6   0.111                           XXXX
  7  -0.027                           XX
  8  -0.055                           XX
  9   0.031                           XX
 10  -0.041                           XX
 11  -0.002                            X
 12  -0.076                          XXX
 13  -0.023                           XX
 14   0.033                           XX
 15  -0.020                            X
 16  -0.023                           XX
 17   0.116                           XXXX
 18  -0.014                            X
 19  -0.003                            X
 20   0.009                            X
 21   0.022                           XX
 22  -0.077                          XXX
 23   0.067                           XXX
 24   0.021                           XX
 25  -0.023                           XX
 26   0.019                            X
 27   0.048                           XX
 28   0.001                            X
 29  -0.082                          XXX
 30   0.039                           XX
 31   0.078                           XXX
 32   0.024                           XX
 33  -0.084                          XXX
 34   0.010                            X
 35  -0.000                            X
 36  -0.019                            X
```

deterministic model uses 13 parameters—one for each month plus the linear time trend. From Exhibits 3.4 and 10.20, we see that the parsimonious stochastic model has a residual mean square of only 52.02 compared with the deterministic model's residual mean square of 261—a substantial difference indeed.

10.5 FORECASTING SEASONAL MODELS

Computing forecasts with seasonal models is, as expected, most easily carried out recursively using the difference equation form for the model, as in Equations (9-33), (9-34), and (9-48). For example, consider the ARIMA(0, 1, 1) ×

EXHIBIT 10.16 Estimation of the ARIMA$(0, 1, 2) \times (0, 1, 1)$ 12 Model for the Logged U.S. Air-Passenger-Miles Series

```
MTB > arima (0,1,2)X(0,1,1)12 'LogPas' 'Resid' 'Fitted'

Estimates at each iteration
Iteration      SSE      Parameters
    0        1.12650   0.100   0.100   0.100
    1        0.97218   0.142   0.132   0.250
    2        0.85757   0.178   0.163   0.400
    3        0.77358   0.210   0.194   0.550
    4        0.74499   0.258   0.233   0.608
    5        0.72743   0.274   0.244   0.758
    6        0.71400   0.284   0.232   0.720
    7        0.71113   0.284   0.234   0.737
Unable to reduce sum of squares any further

Final Estimates of Parameters
Type      Estimate   St. Dev.   t-ratio
MA   1      0.2840    0.0688      4.13
MA   2      0.2343    0.0690      3.39
SMA 12      0.7371    0.0533     13.83

Differencing: 1 regular, 1 seasonal of order 12
No. of obs.:  Original series 216, after differencing 203
Residuals:    SS = 0.692880  (backforecasts excluded)
              MS = 0.003464  DF = 200

Modified Box-Pierce chisquare statistic
Lag              12          24          36          48
Chisquare   6.6(DF= 9)  13.8(DF=21)  18.4(DF=33)  32.6(DF=45)

MTB > cdf 18.4;
SUBC> chisq 33.
    0.018903
```

$(1, 0, 1)$12 model

$$Z_t - Z_{t-1} = \Phi(Z_{t-12} - Z_{t-13}) + a_t - \theta a_{t-1} - \Theta a_{t-12} + \theta\Theta a_{t-13} \tag{10-29}$$

which we rewrite as

$$Z_t = Z_{t-1} + \Phi Z_{t-12} - \Phi Z_{t-13} + a_t - \theta a_{t-1} - \Theta a_{t-12} + \theta\Theta a_{t-13} \tag{10-30}$$

The one-step-ahead forecast is then seen to be

$$\hat{Z}_t(1) = Z_t + \Phi Z_{t-11} - \Phi Z_{t-12} - \theta a_t - \Theta a_{t-11} + \theta\Theta a_{t-12} \tag{10-31}$$

Then

$$\hat{Z}_t(2) = \hat{Z}_t(1) + \Phi Z_{t-10} - \Phi Z_{t-11} - \Theta a_{t-10} + \theta\Theta a_{t-11} \tag{10-32}$$

and so forth. The noise terms $a_{t-13}, a_{t-12}, \ldots, a_t$ will enter into the forecasts for $\ell = 1, 2, \ldots, 13$, but for $\ell > 13$ the autoregressive part of the model takes over and we have

$$\hat{Z}_t(\ell) = \hat{Z}_t(\ell - 1) + \Phi\hat{Z}_t(\ell - 12) - \Phi\hat{Z}_t(\ell - 13) \quad \text{for } \ell > 13 \tag{10-33}$$

To understand the general nature of the forecasts, we consider some special cases.

EXHIBIT 10.17 Sample Autocorrelation Function of Residuals from the ARIMA$(0, 1, 2) \times (0, 1, 1)$ 12
Model for the Logged U.S. Air-Passenger-Miles Series

```
ACF of Resid

            -1.0 -0.8 -0.6 -0.4 -0.2  0.0  0.2  0.4  0.6  0.8  1.0
            +----+----+----+----+----+----+----+----+----+----+
  1   0.008                               X
  2   0.008                               X
  3  -0.063                              XXX
  4  -0.028                              XX
  5   0.021                               XX
  6   0.091                               XXX
  7  -0.025                              XX
  8  -0.054                              XX
  9   0.024                               XX
 10  -0.076                             XXX
 11  -0.005                              X
 12  -0.085                             XXX
 13  -0.029                             XX
 14   0.014                              X
 15  -0.005                              X
 16  -0.025                             XX
 17   0.123                               XXXX
 18  -0.021                             XX
 19   0.029                               XX
 20   0.000                              X
 21   0.046                               XX
 22  -0.074                             XXX
 23   0.078                               XXX
 24   0.018                              X
 25  -0.002                              X
 26   0.027                               XX
 27   0.039                               XX
 28   0.021                               XX
 29  -0.061                             XXX
 30   0.049                               XX
 31   0.053                               XX
 32   0.037                               XX
 33  -0.071                             XXX
 34   0.025                               XX
 35  -0.004                              X
 36   0.011                              X
```

SEASONAL AR(1)12

The seasonal AR(1) 12 model is

$$Z_t = \Phi Z_{t-12} + a_t \tag{10-34}$$

Clearly we have

$$\hat{Z}_t(\ell) = \Phi \hat{Z}_t(\ell - 12) \tag{10-35}$$

However, by iterating back on ℓ, we can also write

$$\hat{Z}_t(\ell) = \Phi^{k+1} Z_{t+r-11} \tag{10-36}$$

EXHIBIT 10.18 Normality of Residuals from the ARIMA(0, 1, 2) × (0, 1, 1) 12 Model for the Logged U.S. Air-Passenger-Miles Series

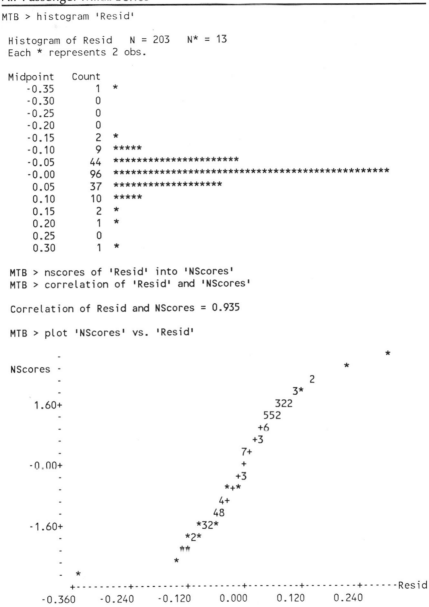

```
MTB > histogram 'Resid'

Histogram of Resid   N = 203   N* = 13
Each * represents 2 obs.

Midpoint    Count
  -0.35       1    *
  -0.30       0
  -0.25       0
  -0.20       0
  -0.15       2    *
  -0.10       9    *****
  -0.05      44    **********************
  -0.00      96    ************************************************
   0.05      37    ******************
   0.10      10    *****
   0.15       2    *
   0.20       1    *
   0.25       0
   0.30       1    *

MTB > nscores of 'Resid' into 'NScores'
MTB > correlation of 'Resid' and 'NScores'

Correlation of Resid and NScores = 0.935

MTB > plot 'NScores' vs. 'Resid'
```

where k and r are defined by $\ell = 12k + r + 1$ with $0 \leq r < 12$ and $k = 0, 1, 2, \ldots$. In other words, k is the integer part of $(\ell - 1)/12$, and $r/12$ is the fractional part of $(\ell - 1)/12$.

If our last observation is in December, then the next January is forecast as Φ times the last observed January, February as Φ times the last observed February, and so on. Two Januaries ahead is forecast as Φ^2 times the last observed January. Looking just at Januaries, the forecasts into the future will

EXHIBIT 10.19 Time Plot of Residuals from the ARIMA(0, 1, 2) × (0, 1, 1) 12 Model for the Logged U.S. Air-Passenger-Miles Series

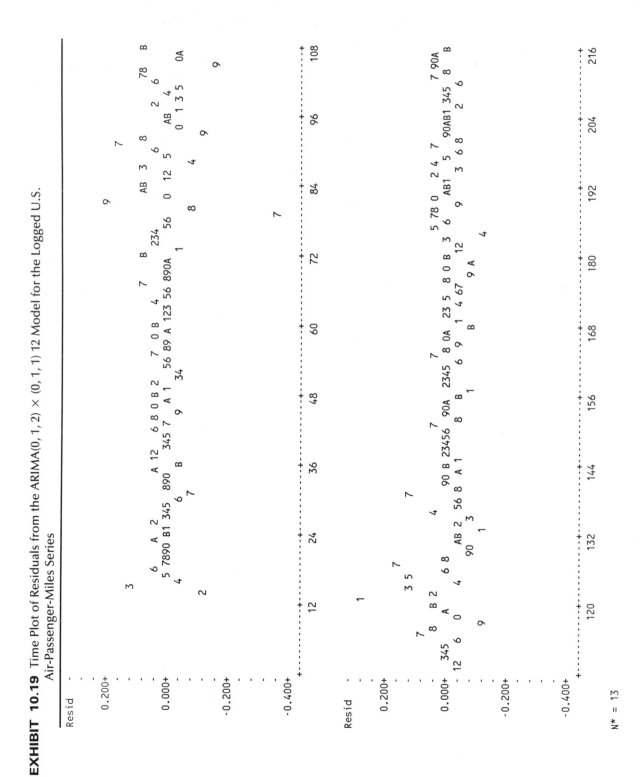

N* = 13

EXHIBIT 10.20 Estimation of the ARIMA(0, 1, 1) × (0, 1, 1) 12 Model for the Milk Production Series

```
MTB > arima (0,1,1)X(0,1,1)12 'Milkpr', 'Resid', 'Fitted'

Estimates at each iteration
Iteration       SSE      Parameters
    0        10964.9    0.100    0.100
    1         9748.0    0.151    0.250
    2         8880.2    0.188    0.400
    3         8312.5    0.214    0.550
    4         8135.7    0.230    0.679
    5         8135.1    0.227    0.674
    6         8135.1    0.227    0.673
    7         8135.1    0.227    0.673
Relative change in each estimate less than  0.0010

Final Estimates of Parameters
Type      Estimate    St. Dev.   t-ratio
MA    1     0.2271     0.0794      2.86
SMA  12     0.6728     0.0639     10.53

Differencing: 1 regular, 1 seasonal of order 12
No. of obs.:  Original series 168, after differencing 155
Residuals:    SS = 7959.79  (backforecasts excluded)
              MS =   52.02  DF = 153

Modified Box-Pierce chisquare statistic
Lag                12            24            36            48
Chisquare    9.8(DF=10)  16.2(DF=22)  19.4(DF=34)  31.8(DF=46)

MTB > cdf 19.4;
SUBC> chisq 34.
    0.021100
```

decay exponentially at a rate determined by the magnitude of Φ. All of the forecasts for each month will behave similarly but with different initial forecasts depending on the particular month under consideration.

Using Equation (9-46) and the fact that the ψ-weights are nonzero only for multiples of 12, namely

$$\psi_j = \begin{cases} \Phi^{j/12} & \text{if } j = 0, 12, 24, \ldots \\ 0 & \text{otherwise} \end{cases} \tag{10-37}$$

we have that the forecast error variance can be written as

$$\text{Var}\,[e_t(\ell)] = \frac{1 - \Phi^{2k+2}}{1 - \Phi^2}\,\sigma_a^2 \tag{10-38}$$

where again $\ell = 12k + r + 1$.

SEASONAL MA(1)12

For the seasonal MA(1)12, we have

$$Z_t = a_t - \Theta a_{t-12} + \theta_0 \tag{10-39}$$

EXHIBIT 10.21 Sample Autocorrelation Function of Residuals from the ARIMA(0, 1, 1) × (0, 1, 1) 12 Model for the Milk Production Series

```
MTB > acf 36 'Resid'

ACF of Resid
```

```
              -1.0 -0.8 -0.6 -0.4 -0.2  0.0  0.2  0.4  0.6  0.8  1.0
              +----+----+----+----+----+----+----+----+----+----+
    1  -0.023                             XX
    2   0.032                             XX
    3   0.116                             XXXX
    4  -0.087                            XXX
    5  -0.032                             XX
    6  -0.101                            XXXX
    7   0.071                             XXX
    8  -0.007                             X
    9   0.013                             X
   10   0.025                             XX
   11  -0.140                           XXXX
   12  -0.018                             X
   13   0.079                             XXX
   14  -0.068                            XXX
   15  -0.032                             XX
   16  -0.002                             X
   17   0.090                             XXX
   18  -0.062                            XXX
   19  -0.082                            XXX
   20  -0.008                             X
   21  -0.025                             XX
   22  -0.053                             XX
   23   0.043                             XX
   24  -0.012                             X
   25   0.037                             XX
   26  -0.036                             XX
   27  -0.062                            XXX
   28  -0.017                             X
   29  -0.053                             XX
   30  -0.038                             XX
   31  -0.026                             XX
   32  -0.016                             X
   33   0.006                             X
   34  -0.060                             XX
   35   0.027                             XX
   36  -0.016                             X
```

In this case we see that

$$\hat{Z}_t(1) = -\Theta a_{t-11} + \theta_0$$
$$\hat{Z}_t(2) = -\Theta a_{t-10} + \theta_0$$
$$\vdots$$
$$\hat{Z}_t(12) = -\Theta a_t \quad + \theta_0$$

and

$$\hat{Z}_t(\ell) = \quad \theta_0 \quad \text{for } \ell > 12$$

(10-40)

Here we obtain different forecasts for the months of the first year, but from then on all forecasts are given by the process mean.

For this model $\psi_0 = 1$, $\psi_{12} = -\Theta$, and $\psi_j = 0$ otherwise. Thus from

EXHIBIT 10.22 Normality of Residuals from the ARIMA(0, 1, 1) × (0, 1, 1) 12 Model for the Milk Production Series

```
MTB > hist of 'Resid'

Histogram of Resid   N = 155   N* = 13

Midpoint   Count
   -15        5   *****
   -10       17   ****************
    -5       36   ************************************
     0       48   ************************************************
     5       26   **************************
    10       16   ****************
    15        5   *****
    20        0
    25        1   *
    30        1   *

MTB > nscores of 'Resid' into 'NScores'
MTB > corr of 'Resid' & 'NScores'

Correlation of Resid and NScores = 0.983

MTB > plot 'NScores' against 'Resid'


          -                                              *
NScores   -                                        *
          -
          -                                  2
          -                             **2
  1.50+                              4**
          -                            72
          -                          453
          -                        *644
          -                       8+
 -0.00+                          99*
          -                      864
          -                    3363
          -                    75
          -                  35*
 -1.50+                    42
          -              *2  *
          -              **
          -            *
          -            *
          --+---------+---------+---------+---------+---------+-------Resid
         -20.0     -10.0      0.0      10.0      20.0     30.0
```

Equation (9-46)

$$\text{Var}\,[e_t(\ell)] = \begin{cases} \sigma_a^2, & 1 \leq \ell \leq 12 \\ (1 + \Theta^2)\sigma_a^2, & 12 < \ell \end{cases} \tag{10-41}$$

ARIMA(0, 0, 0) × (0, 1, 1)12

The ARIMA(0, 0, 0) × (0, 1, 1)12 model is

$$Z_t - Z_{t-12} = a_t - \Theta a_{t-12} \tag{10-42}$$

or

$$Z_{t+\ell} = Z_{t+\ell-12} + a_{t+\ell} - \Theta a_{t+\ell-12}$$

so that

$$\hat{Z}_t(1) = Z_{t-11} - \Theta a_{t-11}$$
$$\hat{Z}_t(2) = Z_{t-10} - \Theta a_{t-10}$$
$$\vdots$$
$$\hat{Z}_t(12) = Z_t - \Theta a_t$$

(10-43)

and then

$$\hat{Z}_t(\ell) = \hat{Z}_t(\ell - 12), \qquad \ell > 12$$

It follows that all Januaries will forecast identically, all Februaries identically, and so forth.

If we invert this model, we find that

$$Z_t = (1 - \Theta)(Z_{t-12} + \Theta Z_{t-24} + \Theta^2 Z_{t-36} + \cdots) + a_t$$

Consequently, we can write

$$\hat{Z}_t(1) = (1 - \Theta) \sum_{j=0}^{\infty} \Theta^j Z_{t-11-12j}$$
$$\hat{Z}_t(2) = (1 + \Theta) \sum_{j=0}^{\infty} \Theta^j Z_{t-10-12j}$$
$$\vdots$$
$$\hat{Z}_t(12) = (1 - \Theta) \sum_{j=0}^{\infty} \Theta^j Z_{t-12j}$$

(10-44)

From this representation we see that the forecast for each January is an exponentially weighted moving average of all observed Januaries, and similarly for each of the other months.

In this case we have $\psi_j = 1 - \Theta$ for $j = 12, 24, \ldots$, and zero otherwise. The forecast error variance is then

$$\text{Var}\,[e_t(\ell)] = \sigma_a^2[1 + k(1 - \Theta)^2]$$

(10-45)

where $\ell = 12k + r + 1, k = 0, 1, \ldots$, and $0 \leq r < 12$.

ARIMA(0, 1, 1) × (0, 1, 1)12

For the ARIMA(0, 1, 1) × (0, 1, 1)12 model

$$Z_t = Z_{t-1} + Z_{t-12} - Z_{t-13} + a_t - \theta a_{t-1} - \Theta a_{t-12} + \theta\Theta a_{t-13}$$

(10-46)

the forecasts satisfy

$$
\begin{aligned}
\hat{Z}_t(1) &= Z_t & + Z_{t-11} & & - Z_{t-12} - \theta a_t - \Theta a_{t-11} + \theta\Theta a_{t-12} \\
\hat{Z}_t(2) &= \hat{Z}_t(1) & + Z_{t-10} & & - Z_{t-11} & - \Theta a_{t-10} + \theta\Theta a_{t-11} \\
&\vdots \\
\hat{Z}_t(12) &= \hat{Z}_t(11) & + Z_t & & - Z_{t-1} & - \Theta a_t + \theta\Theta a_{t-1} \\
\hat{Z}_t(13) &= \hat{Z}_t(12) & + \hat{Z}_t(1) & & - Z_t & + \theta\Theta a_t
\end{aligned}
$$

(10-47)

and

$$\hat{Z}_t(\ell) = \hat{Z}_t(\ell - 1) + \hat{Z}_t(\ell - 12) - \hat{Z}_t(\ell - 13) \text{ for } \ell > 13$$

To understand the general pattern of these forecasts, we can use the representation

$$\hat{Z}_t(\ell) = A_1 + A_2\ell + \sum_{j=0}^{6}\left[B_{1j}\cos\frac{2\pi j\ell}{12} + B_{2j}\sin\frac{2\pi j\ell}{12}\right], \qquad \ell > 0 \qquad \text{(10-48)}$$

where A_i and B_{ij} are dependent on $Z_t, Z_{t-1}, \ldots$, or, alternatively, determined from the initial forecasts $\hat{Z}_t(1), \hat{Z}_t(2), \ldots, \hat{Z}_t(13)$.

This result follows from the general theory of difference equations and involves the roots of $(1 - x)(1 - x^{12}) = 0$ (see Appendix H and also Abraham and Box, 1978.)

Notice that Equation (10-48) reveals that the forecasts are composed of a linear trend on top of a periodic component. However, the coefficients A_i and B_{ij} are more dependent on recent data than on past data and will adapt to changes in the process as our forecast origin changes and the forecasts are updated. This is in stark contrast to fitting a deterministic linear trend plus seasonal component, where the coefficients depend rather equally on both recent and past data and remain the same for all future forecasts.

UPDATING AND PREDICTION LIMITS

Updating forecasts can be carried out for seasonal models just as we did for nonseasonal models using Equation (9-55) without change.

Prediction limits will also apply for seasonal models using Equation (9-54).

NUMERICAL EXAMPLES

Forecasts and prediction limits are computed for seasonal multiplicative ARIMA models with the FORECAST subcommand to the ARIMA command in Minitab (see Section 9.5).

Exhibit 10.23 shows the use of this subcommand with the milk production series, and Exhibit 10.24 gives a plot of the forecasts for two years together with the 95% prediction limits. Notice how the linear trend and seasonality are reproduced in the forecasts.

Exhibits 10.25 and 10.26 give similar results for the logged air-passenger-miles series.

EXHIBIT 10.23 Fitting and Forecasting the Milk Production Series

```
MTB > arima (0,1,1)X(0,1,1)12 'Milkpr', 'Resid', 'Fitted';
SUBC> forecast 36, save in 'Forecast', 'Lower' & 'Upper'.
```

```
Estimates at each iteration
Iteration       SSE       Parameters
    0        10964.9    0.100    0.100
    1         9748.0    0.151    0.250
    2         8880.2    0.188    0.400
    3         8312.5    0.214    0.550
    4         8135.7    0.230    0.679
    5         8135.1    0.227    0.674
    6         8135.1    0.227    0.673
    7         8135.1    0.227    0.673
Relative change in each estimate less than  0.0010
```

```
Final Estimates of Parameters
Type      Estimate    St. Dev.    t-ratio
MA   1     0.2271      0.0794      2.86
SMA 12     0.6728      0.0639     10.53
```

```
Differencing: 1 regular, 1 seasonal of order 12
No. of obs.:  Original series 168, after differencing 155
Residuals:    SS = 7959.79  (backforecasts excluded)
              MS =   52.02  DF = 153
```

```
Modified Box-Pierce chisquare statistic
Lag              12          24          36          48
Chisquare   9.8(DF=10)  16.2(DF=22)  19.4(DF=34)  31.8(DF=46)
```

Forecasts from period 168

| | | 95 Percent Limits | | |
Period	Forecast	Lower	Upper	Actual
169	864.00	849.86	878.14	
170	812.00	794.13	829.87	
171	922.00	901.05	942.95	
172	933.00	909.37	956.63	
173	996.00	969.97	1022.03	
174	967.00	938.77	995.23	
175	926.00	895.72	956.28	
176	888.00	855.81	920.19	
177	847.00	813.01	880.99	
178	857.00	821.29	892.71	
179	827.00	789.66	864.34	
180	873.00	834.09	911.91	
181	894.00	852.10	935.90	
182	842.00	797.66	886.34	
183	952.00	905.35	998.65	
184	963.00	914.14	1011.86	
185	1026.00	975.03	1076.97	
186	997.00	944.01	1049.99	
187	956.00	901.06	1010.94	
188	918.00	861.18	974.82	
189	877.00	818.36	935.64	
190	887.00	826.59	947.41	
191	857.00	794.87	919.13	
192	903.00	839.20	966.80	
193	924.00	857.39	990.61	
194	872.00	802.98	941.02	
195	982.00	910.65	1053.35	
196	993.00	919.40	1066.60	
197	1056.00	980.21	1131.79	
198	1027.00	949.08	1104.92	
199	986.00	906.01	1065.99	
200	948.00	865.99	1030.01	
201	907.00	823.02	990.97	
202	917.00	831.10	1002.90	
203	887.00	799.22	974.78	
204	933.00	843.38	1022.62	

EXHIBIT 10.24 Time Plot of Forecasts and Upper and Lower 95% Prediction Limits for the Milk
Production Series

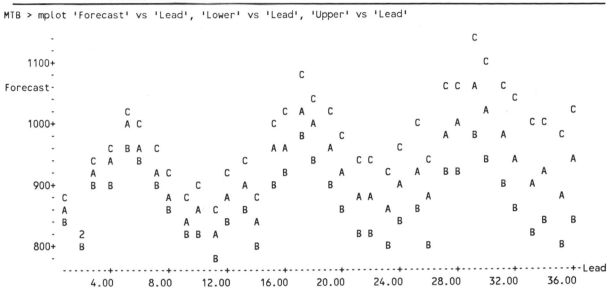

```
MTB > mplot 'Forecast' vs 'Lead', 'Lower' vs 'Lead', 'Upper' vs 'Lead'
```

CHAPTER 10 EXERCISES

10.1 Based on quarterly data, a seasonal model of the form

$$Z_t = Z_{t-4} + a_t - \theta_1 a_{t-1} - \theta_2 a_{t-2}$$

has been fit to a certain time series.

    a. Find the first four ψ-weights for this model.

    b. Suppose that $\theta_1 = 0.5$, $\theta_2 = -0.25$, and $\sigma_a = 1$ and that the data for the last four quarters are

	I	II	III	IV
Series	25	20	25	40
Residual	2	1	2	3

Find the forecasts for the next four quarters.

    c. Find 95% prediction intervals for the forecasts in b.

10.2. An AR model has AR characteristic polynomial

$$(1 - 1.6x + 0.7x^2)(1 - 0.8x^{12})$$

    a. Is the model stationary?

    b. Identify the model as a certain seasonal ARIMA model.

EXHIBIT 10.25 Fitting and Forecasting the Logged U.S. Air-Passenger-Miles Series

```
MTB > arima (0,1,2)X(0,1,1)12 'LogPas', 'Resid' & 'Fitted';
SUBC> forecast 24 months into 'Forecast', 'Lower' & 'Upper'.
```

Estimates at each iteration

Iteration	SSE	Parameters		
0	1.12650	0.100	0.100	0.100
1	0.97218	0.142	0.132	0.250
2	0.85757	0.178	0.163	0.400
3	0.77358	0.210	0.194	0.550
4	0.74499	0.258	0.233	0.608
5	0.72743	0.274	0.244	0.758
6	0.71400	0.284	0.232	0.720
7	0.71113	0.284	0.234	0.737

Unable to reduce sum of squares any further

Final Estimates of Parameters

Type		Estimate	St. Dev.	t-ratio
MA	1	0.2840	0.0688	4.13
MA	2	0.2343	0.0690	3.39
SMA	12	0.7371	0.0533	13.83

Differencing: 1 regular, 1 seasonal of order 12
No. of obs.: Original series 216, after differencing 203
Residuals: SS = 0.692880 (backforecasts excluded)
             MS = 0.003464 DF = 200

Modified Box-Pierce chisquare statistic

Lag	12	24	36	48
Chisquare	6.6(DF= 9)	13.8(DF=21)	18.4(DF=33)	32.6(DF=45)

Forecasts from period 216

		95 Percent Limits		
Period	Forecast	Lower	Upper	Actual
217	2.82896	2.71357	2.94434	
218	2.67525	2.53333	2.81716	
219	2.85318	2.70077	3.00559	
220	2.84864	2.68641	3.01087	
221	2.84539	2.67390	3.01687	
222	2.94931	2.76904	3.12958	
223	3.05144	2.86279	3.24008	
224	3.08482	2.88816	3.28149	
225	2.86283	2.65847	3.06720	
226	2.89746	2.68567	3.10925	
227	2.81225	2.59329	3.03121	
228	2.95049	2.72458	3.17639	
229	2.94388	2.70219	3.18557	
230	2.79017	2.53642	3.04392	
231	2.96810	2.70482	3.23139	
232	2.96356	2.69108	3.23604	
233	2.96031	2.67893	3.24169	
234	3.06423	2.77423	3.35423	
235	3.16636	2.86798	3.46473	
236	3.19974	2.89322	3.50627	
237	2.97775	2.66330	3.29221	
238	3.01238	2.69019	3.33458	
239	2.92717	2.59742	3.25692	
240	3.06540	2.72826	3.40255	

EXHIBIT 10.26 Time Plot of Forecasts and Upper and Lower 95% Prediction Limits for the Logged U.S. Air-Passenger-Miles Series

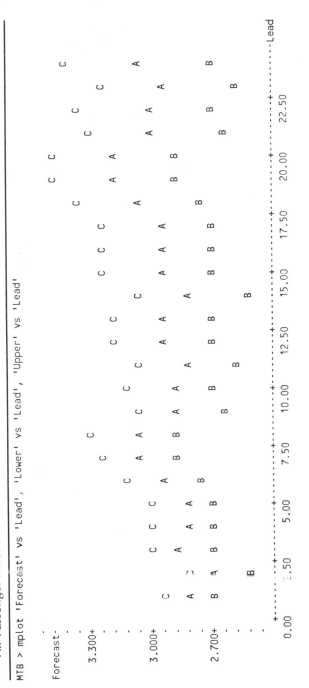

MTB > mplot 'Forecast' vs 'Lead', 'Lower' vs 'Lead', 'Upper' vs 'Lead'

10.3. Suppose that $\{Z_t\}$ satisfies

$$Z_t = a + bt + S_t + X_t$$

where S_t is deterministic and periodic with period s and $\{X_t\}$ is a seasonal ARIMA$(p, 0, q) \times (P, 1, Q)s$ series. What is the model for $W_t = Z_t - Z_{t-s}$?

10.4. For the seasonal model $Z_t = \Phi Z_{t-4} + a_t - \theta a_{t-1}$ with $|\Phi| < 1$, find γ_0 and ρ_k for $k > 0$.

10.5. Identify the following as certain multiplicative seasonal ARIMA models:

a. $Z_t = 0.5Z_{t-1} + Z_{t-4} - 0.5Z_{t-5} + a_t - 0.3a_{t-1}$

b. $Z_t = Z_{t-1} + Z_{t-12} - Z_{t-13} + a_t - 0.5a_{t-1} - 0.5a_{t-12} + 0.25a_{t-13}$

10.6. Verify Equations (10-17), (10-18), and (10-19).

10.7. Develop a seasonal ARIMA model for the Portland, Oregon, bus ridership series. Be sure to carry out the model-building process fully. Are log transformations indicated? Do your residuals pass the usual tests?

10.8. Suppose that the process $\{Z_t\}$ develops according to

$$Z_t = Z_{t-4} + a_t \quad \text{for } t > 4$$

with $Z_t = a_t$ for $t = 1, 2, 3, 4$. Find the variance function and autocorrelation function for $\{Z_t\}$.

CHAPTER 11 SPECIAL TOPICS

In this chapter we present several special topics in time series analysis. In most cases our discussion will be brief and will refer to the time series literature for further information. For the most part the various sections can be read independently.

11.1 COMBINED DETERMINISTIC TREND/ARIMA MODELS

Consider again the model

$$Z_t = \mu_t + X_t \tag{11-1}$$

where μ_t is a deterministic mean function. In ordinary regression analysis we assume that $\{X_t\}$ is white noise. However, as we have seen, there are situations where such an assumption is not realistic; in such cases we may wish to allow $\{X_t\}$ to follow some ARIMA model. We again emphasize that we should have substantial evidence supporting the assumption of a deterministic mean function before proceeding with such a model. Deterministic mean functions will rarely be appropriate in the social sciences but may find some application in the physical and biological sciences.

Recall from Section 3.4 that that for typical μ_t, that is, polynomials, linear combinations of sines and cosines, and seasonal means, the ordinary least-squares estimates of the unknown parameters in μ_t are quite acceptable for large sample sizes. Thus, a reasonable approach to fitting the model of Equation (11-1) in those cases is to estimate μ_t by standard regression methods, obtaining residuals $\hat{X}_t$, and then to consider $\{\hat{X}_t\}$ as a time series to be modeled by a suitable ARIMA process.

As an example, consider the milk production series once more. Here there are specific biological reasons to support a deterministic seasonal effect. An upward linear trend might be justified on the basis of improvements in feed, genetics, and animal health care. As we saw in Exhibit 3.17 in particular, the residuals from such a model are far from white noise. On the basis of Exhibit 3.17, we now consider fitting an AR(1) model to those residuals.

Exhibit 11.1 displays the estimation results, but we still note difficulties in the residuals. For example, the modified Box-Pierce statistic at lag 24 of 49.5 exceeds the 5% critical value of 35.17 with 23 degrees of freedom. Exhibit 11.2

EXHIBIT 11.1   AR(1) Model of Residuals for the Milk Production Series

```
MTB > arima (1,0,0) 'Resid' 'NewRes';
SUBC> noconstant.

Estimates at each iteration
Iteration        SSE      Parameters
    0         33659.5       0.100
    1         25069.6       0.250
    2         18289.4       0.400
    3         13319.1       0.550
    4         10158.5       0.700
    5          8807.7       0.850
    6          8753.0       0.885
    7          8752.8       0.887
    8          8752.8       0.887
Relative change in each estimate less than  0.0010

Final Estimates of Parameters
Type      Estimate      St. Dev.   t-ratio
AR   1      0.8870        0.0361     24.58

No. of obs.:  168
Residuals:    SS = 8750.91   (backforecasts excluded)
              MS =   52.40   DF = 167

Modified Box-Pierce chisquare statistic
Lag               12            24            36            48
Chisquare   37.7(DF=11)   49.5(DF=23)   53.4(DF=35)   61.3(DF=47)

MTB > cdf 49.5;
SUBC> chisq 23 df.
     0.998929
```

shows the sample autocorrelation function for these residuals and reveals problems at lags 1, 12, and perhaps 24.

These results suggest that we estimate a model for the residuals such as ARIMA$(1, 0, 1) \times (0, 0, 1)12$. The results are shown in Exhibit 11.3. The adequacy of the model is supported by both the modified Box–Pierce statistic and the sample autocorrelation function of the residuals (Exhibit 11.4).

How does this model, in which we estimate 16 parameters for the full model, compare with our earlier ARIMA$(0, 1, 1) \times (0, 1, 1)12$ model? From Exhibit 10.20 we see that the residual mean square for that model is 52.02. Exhibit 11.3 shows a residual mean square of 47.69. However, this is a misleading estimate since it is based on residuals from the original model, in which we estimated 13 parameters. If we adjust it as (47.69)165/152, we get 51.77, which is only slightly smaller than the 52.02 obtained in the parsimonious ARIMA$(0, 1, 1) \times (0, 1, 1)12$ model.

Ideally, we would like to estimate the parameters of the trend and the ARIMA model simultaneously. Minitab is currently not capable of such estimation, but Exhibit 11.5 shows such an estimation produced by the SAS/ETS* system. Comparing these results with Exhibits 3.4 and 11.3, we see very little difference in the final results. Another quality system capable of

* SAS/ETS ™ is a registered trademark of SAS Institute, Inc., Cary, NC, USA.

EXHIBIT 11.2 Sample Autocorrelation Function of Residuals from AR(1) Fit of Residuals for the Milk Production Series

```
MTB > acf 36 'NewRes'

ACF of NewRes

             -1.0 -0.8 -0.6 -0.4 -0.2  0.0  0.2  0.4  0.6  0.8  1.0
             +----+----+----+----+----+----+----+----+----+----+
   1 -0.189                        XXXXXX
   2  0.035                           XX
   3  0.150                           XXXXX
   4 -0.045                           XX
   5  0.086                           XXX
   6 -0.154                        XXXXX
   7  0.134                           XXXX
   8 -0.031                          XX
   9  0.086                           XXX
  10  0.039                          XX
  11 -0.145                        XXXXX
  12  0.265                           XXXXXXX
  13 -0.013                          X
  14 -0.031                         XX
  15  0.019                          X
  16 -0.015                          X
  17  0.107                           XXXX
  18 -0.137                       XXXX
  19 -0.014                          X
  20 -0.032                         XX
  21  0.039                          XX
  22 -0.071                        XXX
  23 -0.031                         XX
  24  0.142                           XXXXX
  25 -0.020                          X
  26 -0.006                          X
  27 -0.047                         XX
  28 -0.021                         XX
  29 -0.038                         XX
  30 -0.086                        XXX
  31  0.010                          X
  32 -0.066                        XXX
  33  0.015                          X
  34 -0.048                         XX
  35  0.004                          X
  36 -0.001                          X
```

simultaneous estimation is distributed by Scientific Computing Associates (P.O. Box 625, De Kalb, IL 60115). The SCA system allows estimation using either a conditional least-squares or an exact maximum likelihood criterion. Results obtained by both of these methods differed insignificantly from those given by either SAS/ETS or by our two-step Minitab method.

Forecasting with the model of Equation (11-1) is straightforward in that we just need to combine the forecast using the deterministic trend, $\mu_{t+\ell}$, with the ARIMA forecast, $\hat{X}_t(\ell)$, to get

$$\hat{Z}_t(\ell) = \mu_{t+\ell} + \hat{X}_t(\ell) \tag{11-2}$$

Of course, in practice $\mu_{t+\ell}$ will be estimated.

EXHIBIT 11.3 ARIMA$(1, 0, 1) \times (0, 0, 1)$ 12 Model of Residuals for the Milk Production Series

```
MTB > arima (1,0,1)X(0,0,1)12 'Resid' 'NextRes';
SUBC> noconstant.

Estimates at each iteration
Iteration      SSE      Parameters
       0    43308.8     0.100    0.100    0.100
       1    24589.8     0.249   -0.050    0.078
       2    20475.2     0.399    0.049    0.065
       3    15864.0     0.549    0.131    0.042
       4    11302.6     0.699    0.182   -0.010
       5     8292.5     0.849    0.193   -0.133
       6     7897.1     0.918    0.203   -0.226
       7     7894.6     0.926    0.211   -0.220
       8     7894.6     0.925    0.209   -0.222
       9     7894.6     0.926    0.209   -0.222
      10     7894.6     0.926    0.209   -0.222
Relative change in each estimate less than  0.0010

Final Estimates of Parameters
Type     Estimate    St. Dev.   t-ratio
AR    1    0.9256     0.0345     26.86
MA    1    0.2092     0.0879      2.38
SMA  12   -0.2216     0.0806     -2.75

No. of obs.:  168
Residuals:   SS = 7868.35  (backforecasts excluded)
             MS =   47.69  DF = 165

Modified Box-Pierce chisquare statistic
Lag              12            24            36            48
Chisquare   10.2(DF= 9)   22.2(DF=21)   25.4(DF=33)   32.1(DF=45)
```

11.2 STATE SPACE MODELS

Control theory engineers have developed and successfully used so-called **state space** models and **Kalman filtering** over the last decade or so. Kalman (1960) originated the ideas. Meinhold and Singpurwalla (1983) give the derivations of the required equations from a Bayesian point of view using multivariate normal distribution theory. We base our exposition on the paper of Jones (1980).

Consider a general stationary and invertible ARMA(p, q) process $\{Z_t\}$. Put

$$m = \max(p, q + 1) \tag{11-3}$$

and define the **state** of the process at time t as the column vector $\mathbf{Z}(t)$ of length m whose jth element is the forecast $\hat{Z}_t(j)$, $j = 0, 1, \ldots, m - 1$, based on Z_t, $Z_{t-1}, \ldots$. Note that the lead element of $\mathbf{Z}(t)$ is just $\hat{Z}_t(0) = Z_t$.

Recall the updating Equation (9-55), which can be written as

$$\hat{Z}_{t+1}(\ell) = \hat{Z}_t(\ell + 1) + \psi_\ell a_{t+1} \tag{11-4}$$

EXHIBIT 11.4 Sample Autocorrelation Function of Residuals in ARIMA Fit of Milk Production Series Residuals

```
MTB > acf 36 'NextRes'

ACF of NextRes

              -1.0 -0.8 -0.6 -0.4 -0.2  0.0  0.2  0.4  0.6  0.8  1.0
              +----+----+----+----+----+----+----+----+----+----+
   1  -0.012                              X
   2   0.034                              XX
   3   0.119                              XXXX
   4  -0.052                             XX
   5  -0.017                              X
   6  -0.117                           XXXX
   7   0.105                              XXXX
   8  -0.003                              X
   9   0.044                              XX
  10   0.025                              XX
  11  -0.106                           XXXX
  12   0.022                              XX
  13   0.021                              XX
  14  -0.032                             XX
  15  -0.001                              X
  16   0.023                              XX
  17   0.105                              XXXX
  18  -0.102                           XXXX
  19  -0.074                            XXX
  20  -0.036                             XX
  21   0.007                              X
  22  -0.084                            XXX
  23  -0.001                              X
  24   0.156                              XXXXX
  25   0.003                              X
  26  -0.000                              X
  27  -0.047                             XX
  28  -0.030                             XX
  29  -0.066                            XXX
  30  -0.063                            XXX
  31   0.018                              X
  32  -0.031                             XX
  33   0.012                              X
  34  -0.017                              X
  35   0.024                              XX
  36  -0.040                             XX
```

We shall use this expression directly for $\ell = 0, 1, \ldots, m - 2$. For $\ell = m - 1$ we have

$$\hat{Z}_{t+1}(m - 1) = \hat{Z}_t(m) + \psi_{m-1}a_{t+1}$$
$$= \phi_1\hat{Z}_t(m - 1) + \phi_2\hat{Z}_t(m - 2) + \cdots + \phi_p\hat{Z}_t(m - p) + \psi_{m-1}a_{t+1} \quad \text{(11-5)}$$

where the last expression comes from Equation (9-40) with $\mu = 0$.

The matrix formulation of Equations (11-4) and (11-5) relating $\mathbf{Z}(t + 1)$ to $\mathbf{Z}(t)$ and a_{t+1}, called the **equations of state** (or **Akaike Markovian representation**)

EXHIBIT 11.5 Simultaneous Estimation of Combined Model for the Milk Production Series with SAS/ETS

```
2            DATA;
3              INPUT MILKPR TIME I1-I12 ;
4              CARDS;

               (Data in here.)

173            ;
174            PROC ARIMA;
175              IDENTIFY VAR=MILKPR NOPRINT
176              CROSSCOR=(TIME I1 I2 I3 I4 I5 I6 I7 I8 I9 I10 I11 I12);
177              ESTIMATE P=1 Q=(1)(12) NOCONSTANT
178              INPUT=(TIME I1 I2 I3 I4 I5 I6 I7 I8 I9 I10 I11 I12);
```

```
ARIMA: LEAST SQUARES ESTIMATION
ESTIMATES DID NOT IMPROVE AFTER HIGH RIDGE. ESTIMATES MIGHT NOT HAVE CONVERGED.

PARAMETER    ESTIMATE    STD ERROR    T RATIO  LAG  VARIABLE

MA1,1         0.213806   0.0914514      2.34    1   MILKPR
MA2,1        -0.210831   0.0843051     -2.50   12   MILKPR
AR1,1         0.926507   0.0357081     25.95    1   MILKPR
NUM1          1.6979     0.0847206     20.04    0   TIME
NUM2        591.256      6.41556       92.16    0   I1
NUM3        551.866      6.60257       83.58    0   I2
NUM4        644.375      6.74972       95.47    0   I3
NUM5        659.223      6.86237       96.06    0   I4
NUM6        720.422      6.94259      103.77    0   I5
NUM7        692.014      6.99282       98.96    0   I6
NUM8        642.232      7.01282       91.58    0   I7
NUM9        600.049      7.00325       85.68    0   I8
NUM10       557.721      6.96328       80.09    0   I9
NUM11       561.447      6.8916        81.47    0   I10
NUM12       530.593      6.78624       78.19    0   I11
NUM13       567.308      6.64353       85.39    0   I12

VARIANCE  ESTIMATE =   51.9852
STD ERROR ESTIMATE =    7.21007
NUMBER OF RESIDUALS=   168


AUTOCORRELATION CHECK OF RESIDUALS

TO    CHI                          AUTOCORRELATIONS
LAG   SQUARE DF    PROB
  6    5.62  3   0.131     -0.012  0.033  0.117 -0.053 -0.017 -0.119
 12   10.13  9   0.340      0.102 -0.005  0.046  0.021 -0.104  0.035
 18   14.56 15   0.484      0.020 -0.030  0.001  0.023  0.103 -0.104
 24   22.36 21   0.379     -0.074 -0.039  0.007 -0.089  0.001  0.157
 30   24.58 27   0.598     -0.001  0.006 -0.045 -0.028 -0.064 -0.063
```

is given as

where

$$\mathbf{Z}(t + 1) = \mathbf{F}\mathbf{Z}(t) + \mathbf{G}a_{t+1} \qquad (11\text{-}6)$$

$$\mathbf{F} = \begin{bmatrix} 0 & 1 & 0 & 0 & \cdots & 0 \\ 0 & 0 & 1 & 0 & \cdots & 0 \\ 0 & 0 & 0 & 1 & \cdots & 0 \\ & & & \ddots & & \\ 0 & 0 & 0 & 0 & & 1 \\ \phi_m & \phi_{m-1} & \cdot & \cdot & \cdot & \phi_1 \end{bmatrix} \qquad (11\text{-}7)$$

and

$$\mathbf{G} = \begin{bmatrix} 1 \\ \psi_1 \\ \cdot \\ \cdot \\ \cdot \\ \psi_{m-1} \end{bmatrix} \qquad (11\text{-}8)$$

with $\phi_j = 0$ for $j > p$. Note that the simplicity of Equation (11-6) is obtained at the expense of having to deal with vector-valued processes.

Because the state space formulation also usually allows for measurement error, we do not observe Z_t but only observe Y_t through the **observational equation**

$$Y_t = \mathbf{H}\mathbf{Z}(t) + b_t \qquad (11\text{-}9)$$

where $\mathbf{H} = (1, 0, 0, \ldots, 0)$ and $\{b_t\}$ is another zero-mean white noise process independent of $\{a_t\}$. The special case of *no* measurement error is obtained by setting $b_t = 0$ in Equation (11-9). Equivalently, this case is obtained by taking $\sigma_b^2 = 0$ in subsequent equations.

More general state space models allow $\mathbf{F}$, $\mathbf{G}$, and $\mathbf{H}$ to be more general, possibly also depending on time.

EVALUATION OF THE LIKELIHOOD FUNCTION AND KALMAN FILTERING

First a definition: The **covariance matrix** for a vector of random variables $\mathbf{X}$ of dimension $n \times 1$ is defined to be an $n \times n$ matrix whose ijth entry is the covariance between the ith and jth components of $\mathbf{X}$.

If $\mathbf{Y} = \mathbf{A}\mathbf{X} + \mathbf{B}$, then it is easily shown that the covariance matrix for $\mathbf{Y}$ is $\mathbf{AVA'}$, where $\mathbf{V}$ is the covariance matrix for $\mathbf{X}$ and the prime denotes the transpose matrix.

Getting back to the Kalman filter, we let $\mathbf{Z}(t + 1 \mid t)$ denote the $m \times 1$ vector whose jth component is

$$E[\hat{Z}_{t+1}(j) \mid Y_t, Y_{t-1}, \ldots, Y_1] \quad \text{for } j = 0, 1, \ldots, m - 1$$

Similarly, let $\mathbf{Z}(t \mid t)$ be the vector whose jth component is

$$E[\hat{Z}_t(j) \mid Y_t, Y_{t-1}, \ldots, Y_1) \quad \text{for } j = 0, 1, \ldots, m - 1$$

Then, since a_{t+1} is independent of $Z_t, Z_{t-1}, \ldots$, and hence also of $Y_t, Y_{t-1}, \ldots$, we see from Equation (11-6) that

$$\mathbf{Z}(t + 1 \mid t) = \mathbf{F}\mathbf{Z}(t \mid t) \tag{11-10}$$

Also letting $\mathbf{P}(t + 1 \mid t)$ be the covariance matrix for the "forecast error" $\mathbf{Z}(t + 1) - \mathbf{Z}(t + 1 \mid t)$ and $\mathbf{P}(t \mid t)$ be the covariance matrix for the "forecast error" $\mathbf{Z}(t) - \mathbf{Z}(t \mid t)$, we have from Equation (11-6) that

$$\mathbf{P}(t + 1 \mid t) = \mathbf{F}\mathbf{P}(t \mid t)\mathbf{F}' + \sigma_a^2 \mathbf{G}\mathbf{G}' \tag{11-11}$$

From Equation (11-9), then, with $t + 1$ replacing t,

$$Y(t + 1 \mid t) = \mathbf{H}\mathbf{Z}(t + 1 \mid t) \tag{11-12}$$

where $Y(t + 1 \mid t) = E(Y_{t+1} \mid Y_t, Y_{t-1}, \ldots, Y_1)$.

It can now be shown that the following relationships hold (see, for example, Harvey, 1981c):

$$\mathbf{Z}(t + 1 \mid t + 1) = \mathbf{Z}(t + 1 \mid t) + \mathbf{k}(t + 1)[Y_{t+1} - Y(t + 1 \mid t)] \tag{11-13}$$

where

$$\mathbf{k}(t + 1) = \mathbf{P}(t + 1 \mid t)\mathbf{H}'[\mathbf{H}\mathbf{P}(t + 1 \mid t)\mathbf{H}' + \sigma_b^2]^{-1} \tag{11-14}$$

and

$$\mathbf{P}(t + 1 \mid t + 1) = \mathbf{P}(t + 1 \mid t) - \mathbf{k}(t + 1)\mathbf{H}\mathbf{P}(t + 1 \mid t) \tag{11-15}$$

Collectively, Equations (11-10) through (11-15) are referred to as the **Kalman filter equations**.

The quantity

$$e_{t+1} = Y_{t+1} - Y(t + 1 \mid t) \tag{11-16}$$

in Equation (11-13) is the prediction error and is independent of (or at least uncorrelated with) the past observations $Y_t, Y_{t-1}, \ldots$. Since we are allowing for measurement error, e_{t+1} is not, in general, the same as a_{t+1}.

From Equations (11-16) and (11-9) we have

$$v_{t+1} = \text{Var}(e_{t+1}) = \mathbf{H}\mathbf{P}(t + 1 \mid t)\mathbf{H}' + \sigma_b^2 \tag{11-17}$$

Now consider the likelihood function for the observed series $Y_1, Y_2, \ldots, Y_n$. From the definition of the conditional probability density function, we can write

$$f(Y_1, Y_2, \ldots, Y_n) = f(Y_n \mid Y_{n-1}, Y_{n-2}, \ldots, Y_1)f(Y_{n-1}, Y_{n-2}, \ldots, Y_1)$$

or, by taking logs,

$$\log f(Y_1, Y_2, \ldots, Y_n) = \log f(Y_{n-1}, Y_{n-2}, \ldots, Y_1)$$
$$+ \log f(Y_n \mid Y_{n-1}, Y_{n-2}, \ldots, Y_1) \tag{11-18}$$

Assume now that we are dealing with normal distributions, that is, that $\{a_t\}$ and $\{b_t\}$ are normal white noise processes. Then it is known that the distribution of Y_n conditional on $Y_{n-1}, Y_{n-2}, \ldots, Y_1$ is also a normal distribution with mean $Y(n \mid n - 1)$ and variance v_n. The second term on the right-hand side of Equation (11-18) can then be written

$$\log f(Y_n \mid Y_{n-1}, Y_{n-2}, \ldots, Y_1) = -\tfrac{1}{2} \log 2\pi - \tfrac{1}{2} \log v_n - \frac{\tfrac{1}{2}[Y_n - Y(n \mid n - 1)]^2}{v_n}$$

Furthermore, the first term on the right-hand side of Equation (11-18) can be decomposed similarly again and again until we have

$$\log f(Y_1, Y_2, \ldots, Y_n) = \sum_{t=2}^{n} \log f(Y_t \mid Y_{t-1}, Y_{t-2}, \ldots, Y_1) + \log f(Y_1) \quad \textbf{(11-19)}$$

which then becomes the **prediction error decomposition** of the likelihood, namely,

$$\log f(Y_1, Y_2, \ldots, Y_n) = -\frac{n}{2} \log 2\pi - \frac{1}{2} \sum_{t=1}^{n} \log v_t - \frac{1}{2} \sum_{t=1}^{n} \frac{[Y_t - Y(t \mid t - 1)]^2}{v_t} \quad \textbf{(11-20)}$$

with $Y(1 \mid 0) = 0$ and $v_1 = \text{Var}(Y_1)$.

The overall strategy for computing the likelihood for a given set of parameter values is to use the Kalman filter equations to generate recursively the prediction errors and their variances and then use the prediction error decomposition of the likelihood function. Only one point remains: We need initial values $\mathbf{Z}(0 \mid 0)$ and $\mathbf{P}(0 \mid 0)$ to get the recursions started.

THE INITIAL-STATE COVARIANCE MATRIX

The initial-state vector $\mathbf{Z}(0 \mid 0)$ will be a vector of zeroes for a zero-mean process, and $\mathbf{P}(0 \mid 0)$ is the covariance matrix for $\mathbf{Z}(0) - \mathbf{Z}(0 \mid 0) = \mathbf{Z}(0)$. Now because $\mathbf{Z}(0) = [Z_0, \hat{Z}_0(1), \ldots, \hat{Z}_0(m - 1)]'$, it is necessary for us to evaluate

$$\text{Cov}[\hat{Z}_0(i), \hat{Z}_0(j)] \quad \text{for } i, j = 0, 1, \ldots, m - 1$$

From the truncated linear process form, Equation (9-41) with $C_i(\ell) = \hat{Z}_i(\ell)$, we may write, for $j > 0$

$$Z_j = \hat{Z}_0(j) + \sum_{k=-j}^{-1} \psi_{j+k} a_{-k} \quad \textbf{(11-21)}$$

Multiplying Equation (11-21) by Z_0 and taking expected values yields

$$\gamma_j = E(Z_0 Z_j) = E[\hat{Z}_0(0)\hat{Z}_0(j)] \quad \text{for } j \geq 0 \quad \textbf{(11-22)}$$

Now multiply Equation (11-21) by itself with j replaced by i and take expected values. Recalling that the a's are independent of past Z's and assuming

$0 < i \leq j$, we obtain

$$\gamma_{j-1} = \text{Cov}\,[\hat{Z}_0(i), \hat{Z}_0(j)] + \sigma_a^2 \sum_{k=0}^{i-1} \psi_k \psi_{k+j-i} \tag{11-23}$$

Combining Equations (11-22) and (11-23) we have the required elements of $P(0\,|\,0)$ as

$$\text{Cov}\,[\hat{Z}_0(i), \hat{Z}_0(j)] = \begin{cases} \gamma_j, & 0 = i \leq j \leq m-1 \\ \gamma_{j-i} - \sigma_a^2 \displaystyle\sum_{k=0}^{i-1} \psi_k \psi_{k+j-i}, & 1 \leq i \leq j \leq m-1 \end{cases} \tag{11-24}$$

where the ψ-weights are obtained from the recursion (4-41) and γ_k, the autocovariance function for the process, is obtained as in Appendix D, p. 82.

The variance σ_a^2 can be removed from the problem by dividing σ_b^2 by σ_a^2. The prediction error variance v_t is then replaced by $\sigma_a^2 v_t$ in the log likelihood of Equation (11-20), and we set $\sigma_a^2 = 1$ in Equation (11-11). Dropping unneeded constants, we get for the new log likelihood

$$\lambda = \sum_{t=1}^{n} \left\{ \log\,(\sigma_a^2 v_t) + \frac{[Y_t - Y(t\,|\,t-1)]^2}{\sigma_a^2 v_t} \right\} \tag{11-25}$$

which can be minimized analytically with respect to σ_a^2. We get

$$\sigma_a^2 = \frac{1}{n}\sum_{t=1}^{n} \frac{[Y_t - Y(t\,|\,t-1)]^2}{v_t} \tag{11-26}$$

Substituting this back into Equation (11-25), we find that now

$$\lambda = \sum_{t=1}^{n} \log v_t + n \log \sum_{t=1}^{n} \frac{[Y_t - Y(t\,|\,t-1)]^2}{v_t} \tag{11-27}$$

which must be minimized numerically with respect to $\phi_1, \phi_2, \ldots, \phi_p$, $\theta_1, \theta_2, \ldots, \theta_q$, and σ_b^2. Having done so, we return to Equation (11-26) to estimate σ_a^2. The function defined by Equation (11-27) is sometimes called the **concentrated log likelihood function**.

11.3 MISSING DATA

Throughout all of our previous discussions, we have assumed that our time series are observed at each time point equally spaced from $t = 1$ to $t = n$. However, sometimes one or more data values are missing. If the missing values are very early in the series, we might reasonably drop the early data entirely and consider the series to have started *after* the last missing value. What if the missing values occur elsewhere? For example, Dunsmuir and Robinson (1981b) consider a series of length 798 but with 135 values missing that are scattered throughout the history of the series. In working with missing values, we assume that they appear at random in the series; that is, there is no pattern to where they are missing and they are not missing because, for example, the value would have been too large for our recording instrument.

Our first concern should be how to specify a tentative model for such a series. Can meaningful estimates of the autocorrelation and partial autocorrelation functions be made? A little thought suggests that we can still estimate the autocorrelations with the sample autocorrelation function of Equation (6-1) by simply omitting from the sum of products (and sum of squares) any term for which at least one factor is missing. For example, for the series 1, 2, 1, *, and 0, where "*" means missing, we have

$$\bar{Z} = 1$$

$$(1 - 1)^2 + (2 - 1)^2 + (1 - 1)^2 + (0 - 1)^2 = 2$$

and

$$r_1 = \frac{(1 - 1)(2 - 1) + (2 - 1)(1 - 1)}{2} = 0$$

The properties of such estimates have been considered by Marshall (1980) and Dunsmuir and Robinson (1981a). Once the sample autocorrelations have been obtained, the sample partial autocorrelations follow from the Yule–Walker equations (6-20) or from the Levinson–Durbin recursions (6-21). Unfortunately, Minitab's ACF and PACF commands have not been set up to handle missing data (except at the beginning or end of a series, such as after differencing).

Exhibit 11.6 gives a Minitab macro for computing the sample autocorrelation function with general missing data. As an example, suppose that the 10th

EXHIBIT 11.6 Minitab Macro to Compute the Sample Autocorrelation Function with Missing Data

```
Macro ACFM


Note: Computes ACF with missing data (Uses C50-52 & K50-52)
Note
Note: Enter data column number and maximum lag wanted.
Note: Type END to terminate input.
Noecho
set 'terminal' c50
let k50=c50(1)          # data column number
let k51=c50(2)          # max lag
let c50=ck50-mean(ck50) # center data
let c51=c50
let k52=1
erase c52
exec 'ACFM.1' k51
let c50=c52/ssq(c50)
print c50
Note: Sample ACF is in C50.
erase k50-52 c51 c52
Echo


    Submacro ACFM.1

  lag c51 c51
  let c52(k52)=sum(c50*c51)
  let k52=k52+1
```

EXHIBIT 11.7 Sample Autocorrelation Function for the Railroad Bond Yield Series with Z_{10} and Z_{50} Missing

```
MTB > Note: Computes ACF with missing data (Uses C50-52 & K50-52)
MTB > Note
MTB > Note: Enter data column number and maximum lag wanted.
MTB > Note: Type END to terminate input.
DATA> 1 20
DATA> end
C50
   0.919512    0.875754    0.821726    0.770185    0.722610    0.663467    0.598340
   0.533299    0.463462    0.435092    0.359343    0.287460    0.211176    0.141486
   0.079798    0.017839   -0.042942   -0.100471   -0.150290   -0.194705

MTB > Note: Sample ACF is in C50.
MTB > tsplot c50

        C50
         - 12
  0.80 +    34
         -      5
         -       6
         -        78
         -         9
  0.40 +          0
         -          12
         -           3
         -            4
         -             5
 -0.00 +              6
         -               78
         -                90
         +---------+---------+
                  10        20
```

and 50th observations of the railroad bond data are missing. Exhibit 11.7 displays the sample autocorrelation from the altered series, which should be compared with that for the original series (see Exhibit 6.19). Of course, the numbers are not identical, but the overall picture is the same.

The series needs to be differenced to achieve stationarity. Exhibit 11.8 shows the sample autocorrelation function for the first-difference series. Note that for every interior missing value in the original series, there will be *two* missing values in the differenced series. Nevertheless, the autocorrelation function for the differenced missing data series compares quite favorably with our earlier results, reported in Exhibit 6.20.

Having specified a model, we need to get good parameter estimates. If a pure AR model seems appropriate, we could solve the Yule–Walker equations via the Levinson–Durbin recursions to estimate the ϕ's. However, if the model contains any moving average components, the method-of-moments estimates are usually quite poor.

Full maximum likelihood estimates may be found using state space models and Kalman filtering. If the observation Y_{t+1} is missing, the recursion proceeds through Equations (11-10), (11-11), and (11-12). Equations (11-13),

EXHIBIT 11.8 Sample Autocorrelation Function for the First Difference of the Railroad Bond Yield Series with Missing Data

```
MTB > exec'acfm'
MTB > Note: Computes ACF with missing data (Uses C50-52 & K50-52)
MTB > Note
MTB > Note: Enter data column number and maximum lag wanted.
MTB > Note: Type END to terminate input.
DATA> 2 20
DATA> end
C50
    0.465214    0.181939    0.120745    0.100921    0.130032    0.090674    0.036617
    0.083857    0.200334    0.129676    0.074313    0.041032   -0.056533   -0.075833
   -0.128409   -0.193781   -0.095731   -0.050281   -0.055385   -0.075046

MTB > Note: Sample ACF is in C50.
MTB > tsplot c50


        C50
         - 1
         -
         -
         -
   0.25  +
         -   2       9
         -     5   0
         -   34 6 8
         -     7    12
  -0.00  +
         -           3     89
         -           4   7 0
         -             5
         -             6
  -0.25  +
          +---------+---------+
                   10        20
```

(11-14), and (11-15) are replaced by

$$\mathbf{Z}(t + 1 \mid t + 1) = \mathbf{Z}(t + 1 \mid t) \qquad \text{(11-28)}$$

and

$$\mathbf{P}(t + 1 \mid t + 1) = \mathbf{P}(t + 1 \mid t) \qquad \text{(11-29)}$$

The corresponding term in the concentrated log likelihood of Equation (11-27) is omitted. This approach was used effectively by Jones (1980), Jones and Tryon (1983) and Tryon and Jones (1983). A related method, based on the **EM algorithm**, is given in Shumway and Stoffer (1982). Dunsmuir (1981) gives a good overview of missing data problems in time series.

INTERPOLATION

When data are missing within a time series record, it may be of interest to predict their values from the available data. This is the problem of **interpolation**. Interpolation would be useful, for example, if a given data value was clearly erroneous, perhaps due to a recording error. In other circumstances, a data value may be "unusual" due to a strike or national holiday. It is then useful to interpolate the series over the period in question to account for the effect of the unusual circumstance and perhaps to predict the effect of similar circumstances in the future. Brubacher and Wilson (1976) use interpolation to estimate the effect of holidays on electricity demand.

Suppose that Z_t is the only missing observation. Then from our general theory it follows that the minimum mean square error prediction of Z_t is given by

$$E(Z_t \mid Z_{t+j}, j = \pm1, \pm2, \dots) \tag{11-30}$$

However, the two-sided nature of this predictor makes it rather difficult to compute even for ARIMA models. Abraham (1981) takes a simpler yet effective approach. Using only observations *before* time t, let $\hat{Z}_t$ be the forecast of Z_t. Then using only observations *after* time t, let $\tilde{Z}_t$ be the backcast of Z_t. Both $\hat{Z}_t$ and $\tilde{Z}_t$ are readily obtained using the methods of Chapter 9. Finally, we interpolate Z_t with that linear combination of $\hat{Z}_t$ and $\tilde{Z}_t$ that has the minimum mean square error. In general, the coefficients of the best linear combination will depend upon the parameters of the ARIMA model for the series. Abraham showed that for the AR(1) case, each forecast should be weighted with $1/(1 + \phi^2)$; for the MA(1) model, the sum of the two forecasts is best; and for any first difference model, the forecasts should be averaged.

As an example, suppose again that the tenth observation of the railroad bond yield series is missing. We have estimated an ARI(1, 1) model for this series, and from Exhibit 7.3 we have $\hat{\phi} = 0.4825$. From Chapter 9 we know that $\hat{Z}_t = (1 + \hat{\phi})Z_{t-1} - \hat{\phi}Z_{t-2}$. Since $Z_8 = 654$ and $Z_9 = 649$, our forecast is $\hat{Z}_{10} = 1.4825(649) - 0.4825(654) = 646.5875$. Also, we have $Z_{11} = 659$ and $Z_{12} = 672$; thus our backcast is $\tilde{Z}_{10} = 1.4825(659) - 0.4825(672) = 652.7275$. Finally, our interpolated value is the average of these two, namely, 649.6575, which can be compared with the actual value of 651.

11.4 CALENDAR PROBLEMS—HOLIDAY AND TRADING DAY VARIATIONS

Most monthly time series that represent a total of some variable, such as retail sales, contain calendar effects due to different lengths of months, trading day variation, and holiday variation. **Trading day variation** refers to variation caused by months having different numbers of Mondays, Tuesdays, and so on. An example where this could have an effect is a business that always does bookkeeping on Mondays. A month with five Mondays would tend to show higher sales than a month with only four. **Holiday variation** refers to fluctuations in economic activity due to annual changes in the month in which a certain holiday falls. For example, Easter can fall on dates in both March and April and has a substantial effect on certain retail sales series. Holiday effects are to be distinguished from seasonal effects, which are attributable to the *same* month every year and are handled by seasonal models. Our discussion here relies heavily on the work of Bell and Hillmer (1983).

The basic idea is to model the observed series $\{Z_t\}$ as

$$Z_t = \mu_t + X_t \qquad \text{(11-31)}$$

where μ_t is a deterministic component designed to account for the trading day variation (and/or holiday variation) and $\{X_t\}$ is an ARIMA process of some sort to account for the remaining variation in $\{Z_t\}$.

If β_i, $i = 1, 2, \ldots, 7$, represent the average levels of activity on each day of the week for the series in question, then the effect attributable to the number of each such days in month t can be written

$$\mu_t = \sum_{i=1}^{7} \beta_i D_{i,t} \qquad \text{(11-32)}$$

where $D_{i,t}$ is the known number of day i's in month t. The β's, of course, are unknown parameters.

Specifying a model for the stochastic component $\{X_t\}$ can usually be done by looking at the sample autocorrelations and partial autocorrelations of the observed series $\{Z_t\}$. This appears to work well, since the structure for $\{X_t\}$ tends to dominate the trading day effect. Suppose that both nonseasonal and seasonal differencing is required to achieve stationarity. We then consider

$$\nabla\nabla_{12}Z_t = \sum_{j=1}^{7} \beta_j \nabla\nabla_{12}D_{i,t} + \nabla\nabla_{12}X_t \qquad \text{(11-33)}$$

We can then tentatively estimate the β's by regressing $\nabla\nabla_{12}Z_t$ on the seven predictors $\nabla\nabla_{12}D_{i,t}$, $i = 1, 2, \ldots, 7$. The residuals from such a fit will then be modeled as an ARMA model. Ideally, one should then estimate all of the parameters (β's, ϕ's, and θ's) simultaneously to obtain our final model.

Bell and Hillmer (1983) consider the example of monthly retail sales of lumber and building materials from January 1967 to September 1979. They compare models that do and do not account for trading day variation and show that the model with trading day effects is substantially better.

Holiday effects are somewhat more difficult to model; we refer the reader to Bell and Hillmer (1983) for more information.

11.5 OUTLIERS

Exhibit 11.9 shows the first 60 values of the Iowa City, Iowa, monthly bus ridership time series. Careful examination of the plot shows an observation at time $t = 48$ that is quite inconsistent with the rest of the series. Such an observation is usually called an **outlier**. Exhibit 11.10 displays the seasonal difference series; here the anomaly shows up both at $t = 48$ and $t = 60$.

Such outliers may be due to recording errors, errors in the original

EXHIBIT 11.9 Iowa City Bus Ridership Series with Outlier at $t = 48$

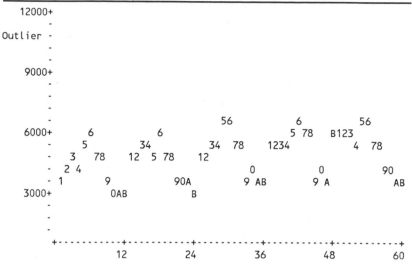

measurements, calculation errors in aggregating daily data into monthly data, digit transformations, or other errors when entering numbers into the computer. In other cases, the unusual values may be due to strikes, political unrest, or other external events. For the bus ridership series of Exhibit 11.9, the error was easily traced back and corrected. If such wild observations are not either corrected or adjusted for in some way, the subsequent analysis can be greatly affected.

For example, with the bus ridership data we take logs and difference both seasonally and nonseasonally. If the outlier is not corrected, we find a lag 12

EXHIBIT 11.10 Seasonal Difference of Bus Ridership Series with Outlier

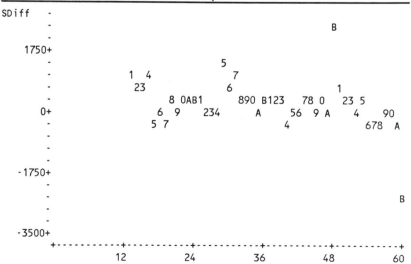

sample autocorrelation of -0.412 whereas with corrected data the corresponding correlation is -0.277.

Such a difference will clearly affect parameter estimates. Exhibit 11.11 displays the model fitting based on the logged data both with and without the outlier. Note that the seasonal MA parameter estimates are quite similar and the nonseasonal MA parameter estimates are somewhat different, but that the noise variance estimates are very different. With the outlier present, the estimated noise variance is nearly twice as large as with the correct data. This will seriously affect the assessment of the precision of the forecast (see Equations (9-53) and (9-54)).

In this example the outlier was quite obvious once the time plot was viewed. Unfortunately, this will not always be the case. The problem of detecting outliers in time series was first considered in Fox (1972), where two statistical models for time series outliers were suggested.

In the **additive outlier** (AO) model, we assume that the underlying process $\{Z_t\}$ satisfies an ordinary ARIMA model, say, but that we only observe

$$Y_t = Z_t + N_t \qquad \text{(11-34)}$$

where N_t is zero except for those times t where an outlier occurs. At an outlier,

EXHIBIT 11.11 Parameter Estimation for Bus Ridership Series: (a) Without Outlier; (b) With Outlier

(a)   arima (0,1,1)X(0,1,1)12 'Riders'

```
Estimates at each iteration
Iteration       SSE      Parameters
    0        0.684953    0.100    0.100
    1        0.600016    0.250    0.237
    2        0.552786    0.379    0.387
    3        0.511942    0.465    0.503
    4        0.486620    0.509    0.640
    5        0.467175    0.539    0.756
    6        0.453804    0.564    0.850
    7        0.452403    0.590    0.865
    8        0.452180    0.604    0.867
    9        0.452131    0.610    0.868
   10        0.452120    0.613    0.868
   11        0.452117    0.615    0.868
   12        0.452117    0.615    0.868
   13        0.452116    0.616    0.868
Relative change in each estimate less than   0.0010

Final Estimates of Parameters
Type      Estimate    St. Dev.   t-ratio
MA    1    0.6157      0.0742      8.30
SMA  12    0.8683      0.0700     12.41

Differencing: 1 regular, 1 seasonal of order 12
No. of obs.:  Original series 136, after differencing 123
Residuals:    SS = 0.397471  (backforecasts excluded)
              MS = 0.003285  DF = 121

Modified Box-Pierce chisquare statistic
Lag              12            24            36            48
Chisquare   20.8(DF=10)   30.4(DF=22)   38.8(DF=34)   52.9(DF=46)
```

EXHIBIT 11.11 (Continued)

(b)
```
MTB > arima (0,1,1)X(0,1,1)12 'Outlier'

Estimates at each iteration
Iteration       SSE      Parameters
    0        1.49377    0.100    0.100
    1        1.25328    0.231    0.250
    2        1.08158    0.371    0.400
    3        0.99217    0.521    0.476
    4        0.91295    0.583    0.626
    5        0.84251    0.622    0.776
    6        0.80265    0.651    0.926
    7        0.79464    0.679    0.905
    8        0.79420    0.691    0.902
    9        0.79414    0.696    0.902
   10        0.79413    0.698    0.902
   11        0.79413    0.699    0.902
   12        0.79413    0.699    0.902
Relative change in each estimate less than  0.0010

Final Estimates of Parameters
Type      Estimate    St. Dev.   t-ratio
MA    1     0.6994      0.0657     10.65
SMA 12     0.9019      0.0650     13.88

Differencing: 1 regular, 1 seasonal of order 12
No. of obs.:  Original series 136, after differencing 123
Residuals:    SS = 0.725422  (backforecasts excluded)
              MS = 0.005995  DF = 121

Modified Box-Pierce chisquare statistic
Lag              12            24           36           48
Chisquare   11.9(DF=10)   17.7(DF=22)  34.1(DF=34)  48.3(DF=46)
```

N_t could be either stochastic or deterministic, say, $N_t = \delta \neq 0$. This might be an appropriate model for recording errors, for example.

In the **innovations outlier** (IO) model, the observed series $\{Z_t\}$ satisfies the ARIMA equation, but the noise (innovation) $\{a_t\}$ in the model has an unusual distribution, at least for some t. For example, a_t might be normal with variance σ_a^2 and mean zero for nonoutlier t-values but mean $\delta \neq 0$ for outlier t-values. Alternatively, $\{a_t\}$ might be independent and identically distributed but with a nonnormal distribution for which extreme innovations are more likely than with a normal distribution. This model might be appropriate where a strike or other external shock affected a series.

To illustrate the effects of these two kinds of outliers in a known model, we have simulated an AR(1) process with zero mean, $\phi = 0.8$, and normal white noise with $\sigma_a^2 = 1$. The Minitab macros of Appendix B, p. 76 were used to generate a series of length 50. Exhibit 11.12(a) gives the time plot of the series without outliers. To simulate an AO-type outlier, we took the observation at $t = 41$ and added 3 to it. The resulting series is plotted in Exhibit 11.12(b). It is doubtful that the outlier would be detected by viewing the time plot.

The IO case was obtained by using the same white noise sequence but altering a_{41} to be 3.5, a rather extreme innovation from a normal distribution with mean zero and variance 1. A plot of the resulting time series is displayed in

Exhibit 11.12(c). Once more the outlier is not at all obvious. Note also that an IO outlier will affect the observed series for time points after 41 since, for example,

$$Z_{42} = \phi Z_{41} + a_{42}$$
$$= \phi(\phi Z_{40} + a_{41}) + a_{42}$$
$$= \phi^2 Z_{40} + \phi a_{41} + a_{42}$$

For a stationary ARIMA process, the effect will eventually die out, but for nonstationary models the effect persists in the observed series forever.

Based on their sample autocorrelation functions, all three data sets indicate that an AR(1) model is appropriate. The results of the ARIMA command applied to each case are summarized in Table 11.1. Again the most striking

EXHIBIT 11.12 Simulation of Outliers in an AR(1) Model: (a) No Outlier; (b) AO at $t = 41$; (c) IO at $t = 41$

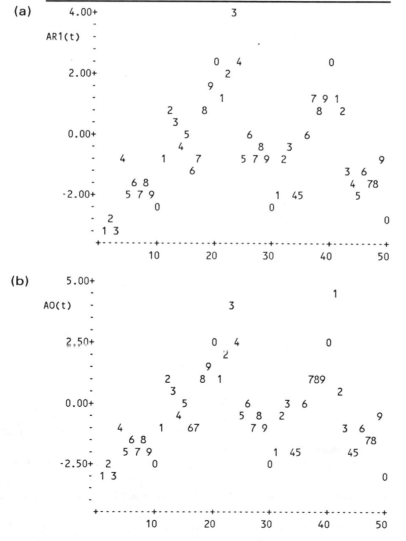

EXHIBIT 11.12 (Continued)

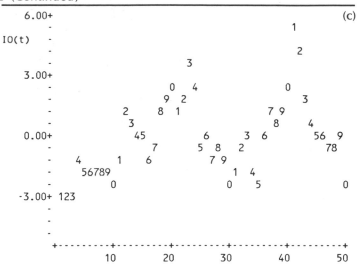

feature of the table is the increased estimate of the noise variance in the series with outliers.

Considerable research is currently being done on the time series outlier problem and definitive answers are not yet available. A good summary of this research through 1980 is given in Martin (1981), which also contains a wealth of references on the subject.

TABLE 11.1 Comparison of Parameter Estimates with and without Outliers

Data Set	ϕ	Est. SD	θ_0	Est. SD	σ_a^2
No outliers	0.79	0.10	−0.19	0.16	1.19
AO	0.76	0.10	−0.19	0.16	1.47
IO	0.80	0.10	−0.11	0.17	1.46

Note: The correct model is AR(1) with $\phi = 0.8$, $\theta_0 = 0$, and $\sigma_a^2 = 1$.

11.6 NONNORMAL PROCESSES

Some of our results for ARIMA models assume that the noise a_t is normally distributed. This assumption was used, for example, in all of our calculations for likelihoods and prediction limits. In this section we shall look at some of the implications of nonnormal innovations.

We first note that the derivations of the autocorrelation functions for ARMA models given in Chapter 4 do *not* depend on normality. Furthermore, the large-sample distribution theory for the sample autocorrelations and sample partial autocorrelations given in Chapter 6 does *not* depend on normally distributed white noise. Thus our specification techniques should still pick out appropriate ARIMA models with nonnormal noise. However, our estimation methods may or may not be acceptable. Least-squares estimation is well-known to be sensitive to extreme observations and will likely produce poor estimates if the noise distribution has "heavier tails" than a normal distribution. See Denby and Martin (1979), Martin (1980, 1981), and Birch and Martin (1981).

For some nonnormal noise distributions, least-squares estimation may be quite acceptable. Our residual analysis will then demonstrate the nonnormality of the residuals. Exhibit 11.13 gives the time plot of a series generated according to an AR(1) scheme with $\phi = 0.6$ but with *exponentially* distributed noise with p.d.f. $f(a) = \exp(-a)$, $a > 0$. Here $E(a_t) = 1 = \text{Var}(a_t)$. (The simulation was carried out using a minor modification of the AR(1) macro given in Appendix B, p. 76.) Exhibit 11.14 shows the histogram and sample autocorrelation function

EXHIBIT 11.13 Time Plot of an AR(1) Series with $\phi = 0.6$ and Exponentially Distributed White Noise

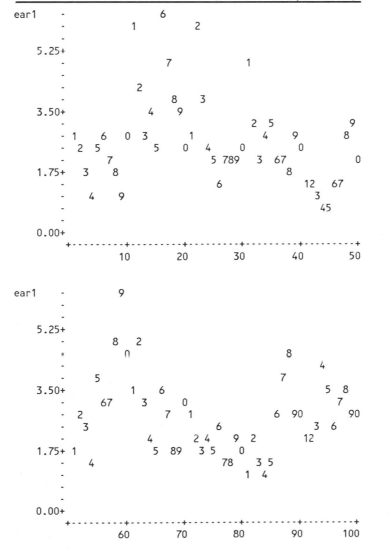

EXHIBIT 11.14 Histogram and Sample Autocorrelation Function for Exponentially Distributed White Noise

```
Histogram of exp   N = 151

Midpoint    Count
     0.0      28   ****************************
     0.5      43   *******************************************
     1.0      33   *********************************
     1.5      15   ***************
     2.0      16   ****************
     2.5       6   ******
     3.0       4   ****
     3.5       3   ***
     4.0       0
     4.5       2   **
     5.0       1   *


MTB > acf 'exp'

ACF of exp

           -1.0 -0.8 -0.6 -0.4 -0.2  0.0  0.2  0.4  0.6  0.8  1.0
            +----+----+----+----+----+----+----+----+----+----+
    1   0.022                           XX
    2  -0.098                          XXX
    3   0.056                           XX
    4   0.009                            X
    5   0.028                           XX
    6   0.179                           XXXXX
    7  -0.112                         XXXX
    8  -0.072                          XXX
    9   0.003                            X
   10  -0.136                         XXXX
   11   0.095                            XXX
   12   0.017                            X
   13  -0.124                         XXXX
   14  -0.056                           XX
   15  -0.040                           XX
   16  -0.118                         XXXX
   17   0.097                            XXX
   18  -0.080                          XXX
   19  -0.058                           XX
   20   0.016                            X
   21  -0.049                           XX
   22  -0.115                         XXXX
```

of the exponential white noise used for the simulation. The sample autocorrelation function and partial autocorrelation function of the resulting Z_t series are shown in Exhibit 11.15, which clearly indicates the AR(1) nature of the series.

Exhibit 11.16 gives the AR(1) model-fitting results. The estimates $\hat{\phi} = 0.55$, $\hat{\theta}_0 = 1.20$, and $\hat{\sigma}_a^2 = 0.996$ are quite good relative to their theoretical values of 0.6, 1, and 1, respectively. Exhibit 11.17 presents the sample autocorrelation function and histogram of the residuals from this fit and reveals the independence and nonnormality of the noise quite clearly.

Even when we are dealing with nonnormal innovations, the observed series Z_t may be nearly normal. This will be especially true in autoregressive

EXHIBIT 11.15 Sample Autocorrelation Function and Partial Autocorrelation Function for AR(1) Series with Exponential White Noise

```
ACF of ear1

            -1.0 -0.8 -0.6 -0.4 -0.2  0.0  0.2  0.4  0.6  0.8  1.0
             +----+----+----+----+----+----+----+----+----+----+
    1   0.550                         XXXXXXXXXXXXXX
    2   0.247                         XXXXXXX
    3   0.220                         XXXXXXX
    4   0.191                         XXXXX
    5   0.199                         XXXXX
    6   0.198                         XXXXX
    7   0.024                         XX
    8  -0.039                         XX
    9  -0.053                         XX
   10  -0.082                        XXX
   11  -0.027                         XX
   12  -0.129                       XXXX
   13  -0.259                    XXXXXXX
   14  -0.289                   XXXXXXXX
   15  -0.246                    XXXXXXX
   16  -0.232                    XXXXXXX
   17  -0.157                      XXXXX
   18  -0.166                      XXXXX
   19  -0.215                     XXXXXX
   20  -0.154                      XXXXX


MTB > pacf 'ear1'

PACF of ear1

            -1.0 -0.8 -0.6 -0.4 -0.2  0.0  0.2  0.4  0.6  0.8  1.0
             +----+----+----+----+----+----+----+----+----+----+
    1   0.550                         XXXXXXXXXXXXXX
    2  -0.080                        XXX
    3   0.169                         XXXXX
    4   0.012                         X
    5   0.113                         XXXX
    6   0.040                         XX
    7  -0.185                      XXXXX
    8   0.007                         X
    9  -0.085                        XXX
   10  -0.037                         XX
   11   0.060                         XXX
   12  -0.197                      XXXXXX
   13  -0.091                        XXX
   14  -0.129                       XXXX
   15  -0.014                         X
   16  -0.067                        XXX
   17   0.043                         XX
   18  -0.020                         X
   19  -0.071                        XXX
   20   0.047                         XX
```

models where the observed series is a linear combination of many noise values and a "central limit theorem effect" takes place. A theoretical basis for this effect is given in Mallows (1967). However, under certain conditions Z_t may follow an AR(1) model and still have an exponential distribution (a distribution far from normal). It is shown in Gaver and Lewis (1980) that this is possible for

EXHIBIT 11.16 Model Fitting for AR(1) Series with Exponential White Noise

```
MTB > arima (1,0,0) 'ear1' 'Resid'

Estimates at each iteration
Iteration       SSE      Parameters
   0          126.734   0.100   2.489
   1          110.563   0.250   2.061
   2          100.841   0.400   1.634
   3           97.574   0.543   1.227
   4           97.561   0.550   1.201
   5           97.561   0.550   1.199
Relative change in each estimate less than  0.0010

Final Estimates of Parameters
Type       Estimate     St. Dev.  t-ratio
AR    1      0.5501      0.0844      6.52
Constant     1.19934     0.09951    12.05
Mean         2.6656      0.2212

No. of obs.:  100
Residuals:    SS = 97.5599  (backforecasts excluded)
              MS =  0.9955  DF = 98

Modified Box-Pierce chisquare statistic
Lag                12          24           36           48
Chisquare    11.0(DF=11)  26.3(DF=23)  36.6(DF=35)  47.8(DF=47)
```

any ϕ such that $0 \leq \phi < 1$ (see Exercise 11.2). In Lusk and Wright (1982) the distribution for Z_t from an MA(1) model with exponential noise is given (see Exercise 11.1). For θ near 0 the distribution is, of course, nearly exponential itself and quite nonnormal. As $\theta \rightarrow 1$, however, the distribution approaches that of a bilateral exponential, that is, with p.d.f. $f(z) = \frac{1}{2} \exp(-|z|)$, $-\infty < z < \infty$. This is a symmetric distribution difficult to distinguish from normality based on typical samples. Moment properties of ARMA models with nonnormal noise are investigated in Davies, Spedding, and Watson (1980).

CHAPTER 11 EXERCISES

11.1. (Mathematical statistics required) Suppose $Z_t = a_t - \theta a_{t-1}$ where the $\{a_t\}$ are independent and identically distributed, each with p.d.f. $f(a) = \exp(-a)$, $a > 0$.

a. If $0 < \theta < 1$, show that the p.d.f. for Z_t is

$$f(z) = \begin{cases} \dfrac{1}{1 + \theta} e^{-z} & z > 0 \\[2mm] \dfrac{1}{1 + \theta} e^{-z/\theta} & z < 0. \end{cases}$$

b. If $-1 < \theta < 0$, show that the p.d.f. for Z_t is

$$f(z) = \frac{1}{1 - \theta}(e^{-z} - e^{-z/\theta}), \qquad z > 0$$

EXHIBIT 11.17 Residual Analysis for AR(1) Fit with Exponential White Noise

```
ACF of Resid

                -1.0 -0.8 -0.6 -0.4 -0.2  0.0  0.2  0.4  0.6  0.8  1.0
                +----+----+----+----+----+----+----+----+----+----+
    1    0.044                              XX
    2   -0.147                           XXXXX
    3    0.067                              XXX
    4    0.026                              XX
    5    0.065                              XXX
    6    0.194                              XXXXXX
    7   -0.079                            XXX
    8   -0.050                             XX
    9   -0.004                             X
   10   -0.091                            XXX
   11    0.117                              XXXX
   12   -0.016                             X
   13   -0.154                           XXXXX
   14   -0.140                           XXXX
   15   -0.050                             XX
   16   -0.115                            XXXX
   17    0.020                              X
   18   -0.017                             X
   19   -0.148                           XXXXX
   20    0.014                              X

MTB > hist 'Resid'

Histogram of Resid   N = 100

Midpoint    Count
    -1.0      21   *********************
    -0.5      29   *****************************
     0.0      23   ***********************
     0.5      10   **********
     1.0       6   ******
     1.5       5   *****
     2.0       1   *
     2.5       2   **
     3.0       1   *
     3.5       1   *
     4.0       1   *
```

11.2. (Mathematical statistics required) Suppose $Z_t = \phi Z_{t-1} + a_t$ where $0 \le \phi < 1$. Suppose further that $a_t = I_t E_t$ where $\{I_t\}$ is a sequence of Bernoulli random variables with success probability ϕ, and $\{E_t\}$ is a sequence of independent exponentially distributed random variables, each with parameter λ. Assume also that the two sequences $\{I_t\}$ and $\{E_t\}$ are independent. Under these conditions, show that Z_t has an exponential distribution with parameter λ.

APPENDIX J MINITAB PRIMER

INTRODUCTION

Minitab is an easy-to-use statistical computing system that is available for many computers, from mainframes to micros. For information on availability, contact:

Minitab, Inc.
3081 Enterprise Drive
State College, Pennsylvania 16801
Telephone: (814) 238-3280

An elementary introduction to Minitab is given in the *Minitab Handbook*, 2nd ed. (1985), published by Wadsworth, Inc., 7625 Empire Drive, Florence, Kentucky 41042 (telephone: 800-354-9706). More detailed information can be found in the *Minitab Reference Manual*; and essential information is given on a pocket-sized *Minitab Quick Reference Card*. Both of these may be available at your local computer center. If not, they can be ordered directly from Minitab, Inc.

Minitab also maintains an on-line HELP facility, which gives up-to-date information about the system. Exhibit J.1 gives an example of the help available. The reader should return to this exhibit after reading a few sections of this primer.

Minitab operates on data stored in a computer worksheet. On this worksheet, data are kept in columns denoted C1, C2, . . . ; in matrices denoted M1, M2, . . . ; or in single constants designated as K1, K2, ... Minitab employs about 180 commands to perform statistical analyses on the stored data. Columns in the worksheet may have names (enclosed in single quotes, for example, 'Temp' and 'Unemp'). However, for simplicity, we will give the command syntax using generic names: C for column, K for constant, M for matrix, and E for any of the above. Also, in this primer, portions of a Minitab command that are optional are enclosed in brackets. Minitab command lines always begin with a command name but may contain extra text to describe the required operation. Throughout the book, only the portions of the command description in **boldface** are actually required. In command names, only the first four characters are required. Furthermore, command names and C, K, and M can be either upper or lower case.

Accessing Minitab will vary from installation to installation. Consult your computer center personnel for details. In addition, Minitab can be used in both batch and interactive modes. This primer is written for the interactive mode, but most of the commands work the same in both modes.

EXHIBIT J.1

```
MTB > help help
The HELP facility in Minitab helps you learn about Minitab.

A.   For general information about Minitab, type

            HELP OVERVIEW

You will be given a choice of topics, including  1. Introduction
to Minitab,  2. Syntax of Commands,  3. Subcommands.

B.   To see what commands Minitab has, type

            HELP COMMANDS

You will be given a choice of categories, including  1. General
Information,  2. Input,  3. Output,  4. Editing and Manipulating
Data.


Do you wish to see more information on the HELP facility in Minitab?
When Minitab types MORE?, type Y if you do, and N if you do not.
More? yes
C.   If you want information on a specific Minitab command, and you
know the name of the command, then type HELP, followed by the command
name.  For example,

            HELP TABLE

will give you a description of the command TABLE.

Some commands have subcommands.  If you want information on a
subcommand, type HELP, followed by the command, followed by the
subcommand.  For example, ROWPERCENT is a subcommand of TABLE.
To get information on ROWPERCENT, type

            HELP TABLE ROWPERCENT

More? yes
D.   Minitab types each section of help, one page (screen) at a
time, and then asks MORE?  Type Y (or YES) if you want more, and
N (or NO) if you do not.

Minitab will type all sections, without asking MORE?, if you use
a +.  For example,

            HELP TABLE +


To leave Minitab, type

            STOP
```

ENTERING AND EDITING DATA

Entering data into the Minitab worksheet or, more generally, into the computer for storage is the least interesting part of any analysis—to say the least. However, it is unavoidable, and we should learn the most efficient methods for doing so. In Minitab, the main commands for entering data are SET and READ. SET enters data into a single column; READ enters data into multiple columns, one row at a time. Thus, in analyzing a single time series, SET will be the usual command for inputting the series values. The basic format for the SET command is:

SET the following data into column **C**

The data are then typed on the following lines, and spaces or commas are used to separate different data entries. When one line is full, press the RETURN key and continue entering data on the next line. To terminate the input, type **END** at the beginning of a data line.

EXHIBIT J.1 (*Continued*)

```
MTB > help overview
General help is available in the categories below.  Type HELP
OVERVIEW followed by the appropriate number.  For example,

          HELP OVERVIEW 1

    1    Introduction to Minitab    6    Missing Value Code
    2    Syntax of Commands         7    Files and Devices
    3    Subcommands                8    Computer and Local Details
    4    Stored Constants           9    Documentation
    5    Matrices



MTB > help overview 1
OVERVIEW 1.  Introduction to Minitab

Minitab is a general purpose statistics package.  It consists of a
worksheet (rows by columns) where data are stored, and about 150
commands.  The worksheet contains 1000 columns, denoted by
C1, C2, C3, ..., C1000.  The columns usually correspond to the
variables in your data, the rows to the observations.
More? yes
Here is a very simple example of a Minitab program.

    READ C1 C2
     40 18
     36 32
     14 10
    END
    LET C3 = C1 + C2
    PRINT C1-C3

The first command, READ, says to put the data that follow into C1
and C2.  The command END tells Minitab you are finished typing data.
At this point, the worksheet looks like

              C1      C2
    ------------------------------------------
              40      18
              36      32
              14      10
    ------------------------------------------
More? yes
```

EXHIBIT J.1 (*Continued*)

The command LET C3 = C1 + C2 tells Minitab to add C1 and C2. After
this command, the worksheet looks like

```
        C1    C2    C3
-------------------------------------------
        40    18    58
        36    32    68
        14    10    24
-------------------------------------------
```

The last command, PRINT C1-C3, says to print out the numbers in
C1 through C3. Minitab responds with the following output

```
    ROW    C1    C2    C3

     1     40    18    58
     2     36    32    68
     3     14    10    24
```
More? yes
When you are finished with this Introduction, you might try this
program yourself. Type each line, then push the carriage control
(return key). Minitab will prompt you for each line:

MTB > says Minitab expects a command, such as READ or LET or HELP.

DATA> says Minitab expects data.
More? yes
Another Example

This is a slightly more complicated example. It will give you the
general flavor of what Minitab can do. The first command reads 50
observations of temperature (in degrees Fahrenheit) and yield of a
chemical reaction from a data file called CHEM. The LOGE command
does a log transformation. Then a plot of the transformed data is
done. REGRESS regresses log (yield) on temperature and stores the
residuals in C20. Then PLOT and HISTOGRAM do a residual analysis.

```
READ 'CHEM' INTO C1 C2
LET C1 = (5/9)*(C1-32)
PLOT C2 VS C1                    (YIELD VS TEMP)
LOGE C2, PUT INTO C12            (LOG BASE E)
PLOT C12 VS C1                   (LOG(YIELD) VS TEMP)
REGRESS C12 ON 1 PRED C1, PUT RESIDS INTO C20
PLOT C20 VS C1                   (RESIDUALS VS TEMP)
HISTOGRAM C20
```

Consider the following example:

> MTB > **SET** the following time series values into **C1**
>
> DATA > **10.6 9.8 7.7 9.9**
>
> DATA > **11.2 9.5**
>
> DATA > **END**

Here, column 1 will contain 10.6, 9.8, 7.7, 9.9, 11.2, and 9.5 from top to bottom.
MTB> is the Minitab prompt, which is displayed in interactive Minitab
whenever a command is expected.

What if you make a mistake? Of course, several kinds of errors are possible. Suppose that the second number, 9.8, should have been 8.9. If we discover the error before pressing the RETURN key for that data line, we may backspace to the error and correct it immediately. (We may also have to retype the remainder of the line, depending on the particular equipment used. Experiment!) If we discover the error after the END command has been entered, we can correct the value using the LET command as:

LET C1(2) = 8.9

Here C1(2) refers to the second row of column 1.

The values in a column may be displayed at any time with the PRINT command:

PRINT E

Multiple columns, constants, or matrices may be printed using

PRINT E, E, . . . , E

Suppose we find that we left out an entry in the series. For example, suppose that the third entry was omitted, which is 8.6. We can INSERT it as:

INSERT the following data between rows **K** and **K** of column **C**

The data are typed on the following lines and terminated with END. Consider the following example:

MTB > **INSERT** between rows **2** and **3** of column **C1**

DATA > **8.6**

DATA > **END**

With this instruction, column 1 would contain 10.6, 9.8, 8.6, 7.7, 9.9, 11.2, and 9.5, in that order. INSERT may also be used to add data to the bottom of existing columns. The format is simply

INSERT at the end of **C**

Rows may be deleted using the DELETE command.

DELETE row **K** of **C**

A group of consecutive rows may be deleted with a single command by using a colon. For example,

DELETE 2:5 C1

will delete rows 2 through 5 of C1.

In interactive model building, a given series is unlikely to be analyzed completely in one Minitab session. However, when we exit Minitab (using the STOP command) the worksheet is erased. Therefore, as soon as we are sure that the data have been entered correctly and no further editing is necessary, we should save the series in a "permanent" computer file using the WRITE command. The format is

WRITE into **'filename'** the data in column **C**

The filename must be enclosed in single quotation marks, and it should be meaningful to the user of the data. In the body of this book, we have used such names as *TempDub* for the Dubuque temperature series and *Milkpr* for the milk production series.

In a subsequent Minitab session, the series can be reentered into the worksheet by

SET 'filename' into column **C**

where, of course, filename is replaced by the name of the file containing the desired series.

Multiple columns may be written using

WRITE into **'filename'** the data in columns **C, C, . . . , C**

When WRITE is used with multiple columns, the file is written with one row per line. However, when a single column is written using the WRITE command, as many numbers as will fit are placed on each line. The user need not be concerned with this, since SET will recover the column correctly in any case.

Because a file created with the WRITE command is a standard ASCII file, it can be used with other computer programs and modified with your computer's editor. Consequently, data files can also be set up outside of Minitab using any editor available for your system. Any such file can then be entered into Minitab using SET (for one column) or READ (for multiple columns).

Another way to save data is with the SAVE command. This command has the format

SAVE the worksheet in **'filename'**

This command saves all the columns currently in use, all names assigned to the columns, and any constants or matrices in use.

A worksheet stored with the SAVE command can only be reentered with the RETRIEVE command:

RETRIEVE from **'filename'**

which restores the worksheet to exactly the way it was when stored with the SAVE command.

Files created with the SAVE command are stored in a special compact format that only Minitab can use. Using the SAVE command is especially useful if your Minitab analysis must be interrupted before you are finished. Using RETRIEVE at the next session enables you to continue where you left off. Just remember that SAVE and RETRIEVE go together and SET (or READ) and WRITE go together. A file that was created with SAVE cannot be obtained with SET or READ, and a file that was created with WRITE cannot be recalled with the RETRIEVE command.

We should note that every computer system has its own idiosyncracies when handling files. Some require you to assign the files before entering Minitab, and each has its own rules for assigning filenames. Consult your computer center for details concerning your system.

Columns may be given names. If they are named, they can be referenced either by name or by number. Obviously, meaningful names are easier to remember than arbitrary numbers. Furthermore, all output involving a named column will be labeled with that name. Such names are very useful when printed output is reevaluated a week or a month later and also when the Minitab output is incorporated directly into a printed project report. The specific format for the NAME command is:

NAME for **C** is **'name'**

A name may be up to eight characters long, using any characters except the single quotation mark ('). Blanks may not be used at the beginning or end of the name. Subsequently when a name is used in place of a column number, it must be enclosed in single quotes. Examples are:

MTB < **NAME** for **C2** is **'Months'**

MTB > **NAME** for **C1** is **'Milkpr'**

MTB > **PRINT 'Milkpr'**

ENTERING PATTERNED DATA

Occasionally data need to be entered that follow a particular pattern. For example, suppose the consecutive numbers 1960 through 1984 must be put into a column as year indices. We could, of course, type them in individually, but there is a better way. The SET command has a feature that allows such patterned data to be input easily. We could use

MTB > **SET** into **C2**

DATA > **1960:1984**

DATA > **END**

In general, any list of consecutive integers may be abbreviated using the colon.

In addition, sequences of numbers with increments different from 1 may be entered using a slash. Suppose we wanted to enter only the even numbers from 1960 to 1984. We do this as follows:

MTB > **SET C2**

DATA > **1960:1984/2**

In general, the number following the slash is the increment between the numbers in the list.

Parentheses may be used to repeat lists of data. If we need to set up a column containing the quarter number for five years of quarterly data, we can write

MTB > **SET C3**

DATA > **5(1:4)**

After this command is executed, column 3 will contain 1, 2, 3, 4, 1, 2, 3, 4, . . . , 1, 2, 3, 4. In general, the number before the opening parenthesis is a repetition factor.

PLOTTING DATA

Once the time series values have been entered into the worksheet, values should be plotted against time. This is easily accomplished with the TSPLOT command:

TSPLOT [with period **K**] time series in **C**

The horizontal axis is labeled 1, 2, . . . rather than with actual time. Exhibit 2.1 shows an example of this command. Series collected monthly should be plotted with a period of 12 so that the same plotting symbols will be used for the same months. Exhibits 1.2 and 1.3 show monthly data plots; Exhibit 1.1 gives a similar plot for quarterly data.

SUBCOMMANDS

Some Minitab commands have subcommands, which allow more control. To use a subcommand, end the main command line with a semicolon (;). The prompt SUBC> will then be given. Now type the subcommand; type another semicolon after it if another subcommand is to follow. End the final subcommand with a period. If you forget the period, type a period or END on the next subcommand line.

If while in interactive Minitab you discover an error after entering a subcommand, type ABORT as the next subcommand. This cancels the entire command and brings back the MTB > prompt.

The TSPLOT command has four subcommands: INCREMENT, START, ORIGIN, and TSTART.

The subcommand ORIGIN = K specifies the observation number of the first time series value. For a monthly series in C1 that begins with March, we would specify a period of 12 and use the subcommand

ORIGIN = 3

Then all Januaries will be plotted with a 1, all Februaries with a 2, and so forth. Octobers will be plotted with 0's, Novembers with A's, and Decembers with B's. If the ORIGIN subcommand is omitted, the first observation is plotted as a 1.

The TSTART = K subcommand sets the first displayed time point on the time axis. This is especially useful for looking only at the last part of a series in conjunction with forecasting. If a series contains 112 observations, then the subcommand TSTART = 101 will only plot observations at times 101, 102, . . . , 112.

The subcommands INCREMENT and START are used to control the vertical axis. INCREMENT = K specifies the distance between the labeled marks on the axis (the tick marks), and START = K [end at K] specifies the first (and, optionally, the last) tick mark.

If the series is too long to fit across one page, the plot is broken into several pieces. A page normally holds about 80 columns, which, allowing for margins and labeling, permits a series of 60 observations across. If your equipment can handle more than this, the width may be changed with the OW (output width) command. The height of the plot is normally 21 lines tall with 2 lines for labels, but this can also be changed with the HEIGHT command:

HEIGHT of plots is **K** lines

Several time series can be plotted on the same graph using the multiple time series plot command MTSPLOT. The format is

MTSPLOT   [period **K**] series in **C**, series in **C**, . . . series in **C**

The first series is plotted with the symbols given in TSPLOT. The additional series are plotted with symbols given in the legend on the plot. If more than one series would be plotted in the same position, a + is plotted in that position. An example of this type of plot is given in Exhibit 9.3.

The INCREMENT, START, and TSTART subcommands are the same as for TSPLOT. The subcommand ORIGIN = K specifies that K is the origin for *all* series plotted. The subcommand

ORIGIN **K** for series **C**, . . . , **C, K** for series **C**, . . . , **C**, . . . , **K** for series **C**, . . . , **C**

assigns *different* origins to different series. Any series for which the origin is not specified is given an origin of 1. This is especially useful for simultaneously plotting a series, plotting predicted values for that series including forecasts, and plotting forecast limits. See Exhibit 9.3.

HISTOGRAMS

Histograms graphically display the distribution of any set of data. Exhibit J.2 shows a histogram for the monthly average temperatures over 12 years recorded in Dubuque, Iowa. The command in its simplest form is written as

HISTOGRAM of data in **C**

Minitab automatically chooses the class intervals in a reasonable manner, but the user may specify the intervals through subcommands.

The subcommands INCREMENT and START have the format

INCREMENT = **K**

which specifies the distance between the class midpoints or, equivalently, the width of the intervals and

START with midpoint at **K** [end with midpoint at **K**]

which specifies the position of the beginning (and, optionally, the ending) interval. Any data lying outside this range are not included in the histogram. An

EXHIBIT J.2

```
MTB > histogram of c1

Histogram of TempDub    N = 144

Midpoint    Count
       10        3   ***
       15        6   ******
       20       15   ***************
       25       12   ************
       30        7   *******
       35       11   ***********
       40        7   *******
       45        6   ******
       50       13   *************
       55       11   ***********
       60       14   **************
       65       10   **********
       70       25   *************************
       75        4   ****
```

example is:

> MTB > **HIST**OGRAM of **'TempDub'**;
> SUBC > **INCRE**MENT of **10** degrees**;**
> SUBC > **START** at **15** degrees.

Similar one-variable displays are produced by the commands DOTPLOT, STEM-AND-LEAF, and BOXPLOT.

SCATTER PLOTS

Scatter plots are useful in showing relationships between two variables. As an example, consider displaying the temperature data versus the month of the year for the Dubuque data. With the 12 years of data in column 1, we use

> MTB > **SET C2**
> DATA < **12(1 : 12)**
> DATA > **PLOT C1** versus **C2**

Exhibit J.3 gives the result of these commands. (Notice that the data input may be terminated with any valid Minitab command instead of with the END command.) The PLOT command has the general format

PLOT C versus C

The first column specified is plotted on the vertical axis, and the second on the horizontal axis. Each point is plotted with an asterisk (∗) except when more than one point falls on the same plotting position, in which case a count is given. If more than nine points occupy the same position, a plus sign (+) is plotted.

Minitab automatically sets up appropriate scales to fit one screen. If other

EXHIBIT J.3

```
MTB > name c2 'Month'
MTB > set c2
DATA> 12(1:12)
DATA> plot c1 vs. c2
```

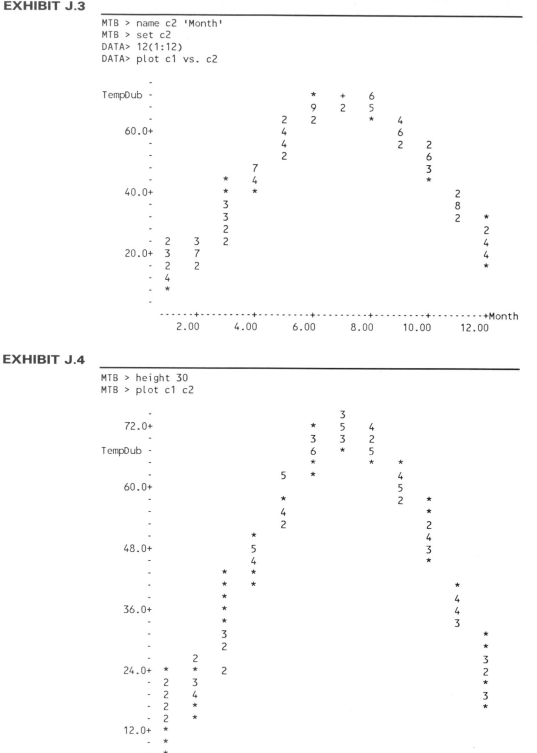

scales are desired, they may be specified with the subcommands

XINCREMENT = **K**

XSTART at **K** [go to **K**]

YINCREMENT = **K**

YSTART at **K** [go to **K**]

If your system prints output on paper, you may want to use HEIGHT and WIDTH to further control the plots. Exhibit J.4 shows the monthly temperature data plotted with increased height. This gives the advantage of more plotting positions, which gives greater detail. However, such a plot no longer fits on one screen.

Other plotting commands include LPLOT (letter plot), MPLOT (multiple plots), and TPLOT (three-dimensional plot). See the *Minitab Handbook* or the *Minitab Reference Manual*, or use HELP for details.

PRINTED OUTPUT

Printed, or "hardcopy," output is frequently needed for the final report of an analysis. In interactive Minitab, if the commands PAPER and OUTFILE are given, the output that follows goes to a printer or file, respectively, until a NOPAPER, NOOUTFILE, or another OUTFILE is given. Output will also appear on your terminal screen with these commands. Screen output may be turned off with the NOTERM subcommand. Other subcommands for PAPER or OUTFILE include OW = K and OH = K to control the output width and height, respectively.

Further Minitab commands are explained as they are needed throughout the book.

APPENDIX K DATA SETS

This appendix contains listings of all the time series used in this book. They are all set up as they would appear in computer files suitable for reading (actually "setting") by Minitab. In particular, the files are self-documented since comments following the symbol # are ignored by Minitab. This file format is constructed using any editor available for your computer rather than through Minitab.

In all cases the data are read across each row before moving to the next row. Thus the Minitab command SET will properly input the series.

```
# U.S. quarterly unemployment rates (seasonally adjusted)
# First quarter 1948 through first quarter 1978, n=121
#
3.73  3.67  3.77  3.83  4.67  5.87  6.70  6.97  6.40  5.57  4.63  4.23
3.50  3.10  3.17  3.37  3.07  2.97  3.23  2.83  2.70  2.57  2.73  3.70
5.27  5.80  5.97  5.33  4.73  4.40  4.10  4.23  4.03  4.20  4.13  4.13
3.93  4.10  4.23  4.93  6.30  7.37  7.33  6.37  5.83  5.10  5.27  5.60
5.13  5.23  5.53  6.27  6.80  7.00  6.77  6.20  5.63  5.53  5.57  5.53
5.77  5.73  5.50  5.57  5.47  5.20  5.00  5.00  4.90  4.67  4.37  4.10
3.87  3.80  3.77  3.70  3.77  3.83  3.83  3.93  3.73  3.57  3.53  3.43
3.37  3.43  3.60  3.60  4.17  4.80  5.17  5.87  5.93  5.97  5.97  5.97
5.83  5.77  5.53  5.27  5.03  4.93  4.77  4.67  5.17  5.13  5.50  6.57
8.37  8.90  8.37  8.40  7.63  7.43  7.83  7.93  7.37  7.07  6.90  6.63
6.20
```

```
# Average monthly temperatures in Dubuque, Iowa
# January 1964 through December 1975,  n=144
#
24.7  25.7  30.6  47.5  62.9  68.5  73.7  67.9  61.1  48.5  39.6  20.0 # 1964
16.1  19.1  24.2  45.4  61.3  66.5  72.1  68.4  60.2  50.9  37.4  31.1 # 1965
10.4  21.6  37.4  44.7  53.2  68.0  73.7  68.2  60.7  50.2  37.2  24.6 # 1966
21.5  14.7  35.0  48.3  54.0  68.2  69.6  65.7  60.8  49.1  33.2  26.0 # 1967
19.1  20.6  40.2  50.0  55.3  67.7  70.7  70.3  60.6  50.7  35.8  20.7 # 1968
14.0  24.1  29.4  46.6  58.6  62.2  72.1  71.7  61.9  47.6  34.2  20.4 # 1969
 8.4  19.0  31.4  48.7  61.6  68.1  72.2  70.6  62.5  52.7  36.7  23.8 # 1970
11.2  20.0  29.6  47.7  55.8  73.2  68.0  67.1  64.9  57.1  37.6  27.7 # 1971
13.4  17.2  30.8  43.7  62.3  66.4  70.2  71.6  62.1  46.0  32.7  17.3 # 1972
22.5  25.7  42.3  45.2  55.5  68.9  72.3  72.3  62.5  55.6  38.0  20.4 # 1973
17.6  20.5  34.2  49.2  54.8  63.8  74.0  67.1  57.7  50.8  36.8  25.5 # 1974
20.4  19.6  24.6  41.3  61.8  68.5  72.0  71.1  57.3  52.5  40.6  26.2 # 1975
```

```
# Milk production in pounds per cow per month
#   January 1962 through December 1975,   n=168
#
589   561   640   656   727   697   640   599   568   577   553   582  # 1962
600   566   653   673   742   716   660   617   583   587   565   598  # 1963
628   618   688   705   770   736   678   639   604   611   594   634  # 1964
658   622   709   722   782   756   702   653   615   621   602   635  # 1965
677   635   736   755   811   798   735   697   661   667   645   688  # 1966
713   667   762   784   837   817   767   722   681   687   660   698  # 1967
717   696   775   796   858   826   783   740   701   706   677   711  # 1968
734   690   785   805   871   845   801   764   725   723   690   734  # 1969
750   707   807   824   886   859   819   783   740   747   711   751  # 1970
804   756   860   878   942   913   869   834   790   800   763   800  # 1971
826   799   890   900   961   935   894   855   809   810   766   805  # 1972
821   773   883   898   957   924   881   837   784   791   760   802  # 1973
828   778   889   902   969   947   908   867   815   812   773   813  # 1974
834   782   892   903   966   937   896   858   817   827   797   843  # 1975
```

```
#   Monthly AA railroad bond yields (% X 100)
# January 1968 through June 1976 ,  n=102
#
639   643   640   653   667   667   663   654   649   651   659   672  # 1968
670   675   692   702   706   710   722   729   740   755   763   788  # 1969
818   826   821   819   827   848   881   879   878   878   868   856  # 1970
844   824   820   819   813   815   822   818   815   792   769   775  # 1971
771   773   780   779   774   772   775   770   766   771   773   772  # 1972
767   775   777   777   776   779   787   790   791   792   802   799  # 1973
792   780   790   799   810   814   828   862   874   892   872   869  # 1974
870   859   857   870   867   856   854   862   861   855   846   847  # 1975
845   838   828   823   814   812                                      # 1976
```

```
# Portland, Oregon monthly average gasoline price (cents/gal)
# January 1973 through June 1982, n=114
#
36.9   36.9   36.9   36.9   37.1   38.7   38.9   38.9   38.4   39.3   40.4   43.4  # 1973
44.6   45.9   49.6   51.7   53.3   54.2   54.7   54.4   54.9   53.2   52.3   52.4  # 1974
52.8   52.6   52.1   52.5   53.5   54.8   56.7   57.4   57.6   57.6   57.3   57.4  # 1975
57.0   56.4   55.3   54.5   55.5   56.9   58.4   58.7   59.4   59.3   59.3   59.3  # 1976
59.6   60.3   60.9   61.3   61.7   62.1   62.4   62.3   62.4   62.4   62.4   62.6  # 1977
63.1   63.5   63.5   63.6   64.9   65.9   67.0   67.9   68.6   68.9   68.9   69.2  # 1978
69.5   70.4   72.3   75.1   80.9   86.1   92.2   94.3   98.5   98.9   99.6  101.1  # 1979
106.3 113.4 116.8 118.1 119.8 120.8 120.8 120.8 120.8 120.8 120.8 120.5  # 1980
122.5 129.8 131.3 131.9 132.7 134.0 135.1 135.0 135.0 133.6 132.7 132.5  # 1981
131.3 128.5 121.9 119.1 121.7 126.3                                       # 1982
```

```
# Iowa non-farm income per capita
# First quarter 1948 through third quarter 1975, n=111
#
 601   604   620   626   641   642   645   655   682   678   692   707
 736   753   763   775   775   783   794   813   823   826   829   831
 830   838   854   872   882   903   919   937   927   962   975   995
1001  1013  1021  1028  1027  1048  1070  1095  1113  1143  1154  1173
1178  1183  1205  1208  1209  1223  1238  1245  1258  1278  1294  1314
1323  1336  1355  1377  1416  1430  1455  1480  1514  1545  1589  1634
1669  1715  1760  1812  1809  1828  1871  1892  1946  1983  2013  2045
2069  2107  2144  2183  2231  2304  2343  2377  2393  2461  2494  2532
2565  2631  2682  2782  2849  2930  3029  3102  3181  3282  3391  3483
3568  3657  3705
```

```
# Monthly U.S. air passenger miles
# January 1960 through December 1977 , n=216
#
 2.42  2.14  2.28  2.50  2.44  2.72  2.71  2.74  2.55  2.49  2.13  2.28 # 1960
 2.35  1.82  2.40  2.46  2.38  2.83  2.68  2.81  2.54  2.54  2.37  2.54 # 1961
 2.62  2.34  2.68  2.75  2.66  2.96  2.66  2.93  2.70  2.65  2.46  2.59 # 1962
 2.75  2.45  2.85  2.99  2.89  3.43  3.25  3.59  3.12  3.16  2.86  3.22 # 1963
 3.24  2.95  3.32  3.29  3.32  3.91  3.80  4.02  3.53  3.61  3.22  3.67 # 1964
 3.75  3.25  3.70  3.98  3.88  4.47  4.60  4.90  4.20  4.20  3.80  4.50 # 1965
 4.40  4.00  4.70  5.10  4.90  5.70  3.90  4.20  5.10  5.00  4.70  5.50 # 1966
 5.30  4.60  5.90  5.50  5.40  6.70  6.80  7.40  6.00  5.80  5.50  6.40 # 1967
 6.20  5.70  6.40  6.70  6.30  7.80  7.60  8.60  6.60  6.50  6.00  7.60 # 1968
 7.00  6.00  7.10  7.40  7.20  8.40  8.50  9.40  7.10  7.00  6.60  8.00 # 1969
10.45  8.81 10.61  9.97 10.69 12.40 13.38 14.31 10.90  9.98  9.20 10.94 # 1970
10.53  9.06 10.17 11.17 10.84 12.09 13.66 14.06 11.14 11.10 10.00 11.98 # 1971
11.74 10.27 12.05 12.27 12.03 13.95 15.10 15.65 12.47 12.29 11.52 13.08 # 1972
12.50 11.05 12.94 13.24 13.16 14.95 16.00 16.98 13.15 12.88 11.99 13.13 # 1973
12.99 11.69 13.78 13.70 13.57 15.12 15.55 16.73 12.68 12.65 11.18 13.27 # 1974
12.64 11.01 13.30 12.19 12.91 14.90 16.10 17.30 12.90 13.36 12.26 13.93 # 1975
13.94 12.75 14.19 14.67 14.66 16.21 17.72 18.15 14.19 14.33 12.99 15.19 # 1976
15.09 12.94 15.46 15.39 15.34 17.02 18.85 19.49 15.61 16.16 14.84 17.04 # 1977
```

```
# Portland, Oregon average monthly bus ridership (/100)
# January 1973 through June 1982, n=114
#
 648   646   639   654   630   622   617   613   661   695   690   707 # 1973
 817   839   810   789   760   724   704   691   745   803   780   761 # 1974
 857   907   873   910   900   880   867   854   928  1064  1103  1026 # 1975
1102  1080  1034  1083  1078  1020   984   952  1033  1114  1160  1058 # 1976
1209  1200  1130  1182  1152  1116  1098  1044  1142  1222  1234  1155 # 1977
1286  1281  1224  1280  1228  1181  1156  1124  1152  1205  1260  1188 # 1978
1212  1269  1246  1299  1284  1345  1341  1308  1448  1454  1467  1431 # 1979
1510  1558  1536  1523  1492  1437  1365  1310  1441  1450  1424  1360 # 1980
1429  1440  1414  1424  1408  1337  1258  1214  1326  1417  1417  1329 # 1981
1461  1425  1419  1432  1394  1327                                     # 1982
```

```
# Weekday bus ridership, Iowa City, Iowa (monthly averages)
# September, 1971 through December, 1982, n=148
#
3603  4448  4734  4353  5438  5954  4838  4532  3599  3248  3230  2790
4738  5043  5367  5262  5065  5824  4612  4823  3564  3651  3625  3170
5022  5018  5326  5412  6345  6698  5605  5260  3789  4023  3775  3384
5303  5506  5645  5168  6173  6776  5957  5777  3833  4200  3891  3595
5862  5719  5957  5226  6467  6383  5589  5296  3943  4188  3673  3475
5392  5583  5738  4944  6552  6451  5354  5081  3874  3973  3751  3621
5328  5427  5419  5983  6568  8346  6800  6308  4857  4663  4499  3878
6031  6325  6764  6929  7557  8182  6892  6860  5181  5462  5328  4331
7279  7772  8262  7499  8081 10121  8745  8045  5906  5274  5283  4956
7879  8289  8422  7832  7857 10008  8057  7791  6147  5438  5494  5253
8154  8604  8832  7815 10179 11460  9641  9243  6824  6149  6161  5962
9068  9538  9988  7916
```

REFERENCES AND BIBLIOGRAPHY

Abraham, B. (1981). "Missing observations in time series." *Comm. Statist.*, A, **10**, pp. 1643–1653.

Abraham, B., and Box, G. E. P. (1978). "Deterministic and forecast adaptive time dependent models." *Applied Statistics*, **27**, pp. 120–130.

Abraham, B., and Box, G. E. P. (1979). "Bayesian analysis of some outlier problems in time series." *Biometrika*, **66**, pp. 229–236.

Abraham, B., and Ledolter, J. (1981). "Parsimony and its importance in time series." *Technometrics*, **23**, pp. 411–414.

Abraham, B., and Ledolter, J. (1983). *Statistical Methods for Forecasting*. New York: Wiley.

Abraham, B., and Ledolter, J. (1984). "A note on inverse autocorrelations." *Biometrika*, **71**, pp. 609–614.

Akaike, H. (1973). "Maximum likelihood identification of Gaussian auto-regressive moving average models." *Biometrika*, **60**, pp. 255–266.

Akaike, H. (1974). "A new look at the statistical model identification." *IEEE Trans. on Auto. Control*. **19**, pp. 716–723.

Anderson, T. W. (1971). *The Statistical Analysis of Time Series*. New York: Wiley.

Ansley, C. F. (1979). "An algorithm for the exact likelihood of a mixed autoregressive moving average process." *Biometrika*, **66**, pp. 59–65.

Ansley, C. F., and Newbold, P. (1979). "On the finite sample distribution of the residual autocorrelations in autoregressive–moving average models." *Biometrika*, **66**, pp. 547–553.

Ansley, C. F., and Newbold, P. (1981). "On the bias in estimates of forecast mean square error." *J. Amer. Stat. Assoc.*, **76**, pp. 569–578.

Ansley, C. F., Spivey, W. A., and Wrobleski, W. J. (1977). "On the structure of moving average processes." *J. Econometrics*, **6**, pp. 121–134.

Barham, S. Y., and Dunstan, F. D. J. (1982). "Missing values in time series." In *Time Series Analysis Theory and Practice 2*, ed. O. D. Anderson. Amsterdam: North-Holland.

Bartlett, M. S. (1946). "On the theoretical specification of sampling properties of autocorrelated time series." *J. Roy. Stat. Soc.*, B, **8**, pp. 27–41.

Bell, W . R., and Hillmer, S. C. (1983). "Modeling time series with calender variation." *J. Amer. Stat. Assoc.*, **78**, pp. 526–534.

Bhansali, R. J. (1981). "Effects of not knowing the order of an autoregressive process on the mean square error of prediction—I." *J. Amer. Stat. Assoc.*, **76**, pp. 588–597.

Birch, J. B., and Martin, R. D. (1981). "Confidence intervals for robust estimates of the first order autoregressive parameter." *J. Amer. Stat. Assoc.*, **2,** pp. 205–216.

Bloomfield, P. (1976). *Fourier Analysis of Time Series: An Introduction*. New York: Wiley.

Box, G. E. P., and Cox, D. R. (1964). "An analysis of transformations." *J. Roy. Stat. Soc.*, B, **26,** pp. 211–243.

Box, G. E. P., and Jenkins, G. M. (1976). *Time Series Analysis: Forecasting and Control*. Rev. ed. San Francisco: Holden Day.

Box, G. E. P., and Pierce, D. A. (1970). "Distributions of residual autocorrelations in autoregressive–integrated moving average models." *J. Amer. Stat. Assoc.*, **65,** pp. 1509–1526.

Brown, R. G. (1962). *Smoothing, Forecasting and Prediction of Discrete Time Series*. Englewood Cliffs, N.J.: Prentice-Hall.

Brubacher, S. R., and Wilson, G. T. (1976). "Interpolating time series with applications to the estimation of holiday effects on electricity demand." *Applied Statistics*, **25,** pp. 107–116.

Chan, K. H., Hayya, J. C., and Ord, J. K. (1977). "A note on trend removal methods: the case of polynomial regression versus variate differencing." *Econometrica*, **45,** pp. 737–744.

Chernick, M. R., Downing, D. J., and Pike, D. H. (1982). "Detecting outliers in time series data." *J. Amer. Stat. Assoc.*, **77,** pp. 743–747.

Chipman, J. S. (1979). "Efficiency of least squares estimation of linear trend when residuals are autocorrelated." *Econometrica*, **47,** pp. 115–128.

Cleveland, W. S. (1972). "The inverse autocorrelations of a time series and their applications." *Technometrics*, **14,** no. 2, pp. 277–298.

Cleveland, W. S., and Devlin, S. J. (1982). "Calendar effects in monthly time series: modeling and adjustment." *J. Amer. Stat. Assoc.*, **77,** pp. 520–528.

Cleveland, W. P., and Tiao, G. C. (1976). "Decomposition of seasonal time series: a model for the Census X-11 program." *J. Amer. Stat. Assoc.*, **71,** pp. 581–587.

Clinger, W., and Van Ness, J. W. (1976). "On unequally spaced time points in time series analysis." *Ann. Statist.*, **4,** pp. 736–745.

Cochrane, D., and Orcutt, G. H. (1949a). "Applications of least squares regression to relationships containing autocorrelated error terms." *J. Amer. Stat. Assoc.*, **44,** pp. 32–61.

Cochrane, D., and Orcutt, G. H. (1949b). "A sampling study of the merits of autoregressive and reduced form transformations in regression analysis." *J. Amer. Stat. Assoc.*, **44,** pp. 356–372.

Cryer, J. D., and Ledolter, J. (1981). "Small sample properties of the maximum likelihood estimator in the first order moving average model." *Biometrika*, **68,** pp. 691–694.

Damsleth, E., and Spjotvoll, E. (1982). "Estimation of trigonometric components in time series." *J. Amer. Stat. Assoc.*, **77,** pp. 381–387.

Davies, N., and Newbold, P. (1979). "Some power studies of a portmanteau test of time series models specification." *Biometrika*, **66,** pp. 153–155.

Davies, N., and Newbold, P. (1980). "Forecasting with misspecified models." *Applied Statistics*, **29,** pp. 87–92.

Davies, N., and Petrucelli, J. D. (1984). "On the use of the general partial autocorrelation function for order determination in ARMA(p, q) processes." *J. Amer. Stat. Assoc.*, **79,** pp. 374–377.

Davies, N., Spedding, T., and Watson, W. (1980). "Autoregressive moving average processes with non-normal residuals." *J. Time Series Anal.*, **1,** pp. 103–110.

Davies, N., Triggs, C. M., and Newbold, P. (1977). "Significance levels of the Box–Pierce portmanteau statistic in finite samples." *Biometrika*, **64,** pp. 517–522.

Davis, W. W. (1977). "Robust interval estimation of the innovation variance of an ARMA model." *Ann. Statist.*, **5,** pp. 700–708.

Denby, L., and Martin, R. D. (1979). "Robust estimation of the first-order autoregressive parameter." *J. Amer. Stat. Assoc.*, **74,** pp. 140–146.

Dent, W. T. (1977). "Computation of the exact likelihood for an ARIMA process." *J. Statist. Comp. and Simul.*, **5,** pp. 193–206.

Draper, N. R., and Smith, H. (1981). *Applied Regression Analysis*. 2nd ed. New York: Wiley.

Dunsmuir, W. (1981). "Estimation for stationary time series when data are irregularly spaced or missing." In *Applied Time Series Analysis II*, ed. D. Findley. New York: Academic Press.

Dunsmuir, W., and Robinson, P. M. (1981a). "Asymptotic theory for time series containing missing and amplitude modulated observations." *Sankhya*, A, **43,** pp. 260–281.

Dunsmuir, W., and Robinson, P. M. (1981b). "Estimation of time series models in the presence of missing data." *J. Amer. Stat. Assoc.*, **76,** pp. 560–568.

Dunstan, F. D. J. (1982). "Time series analysis in the detection of breast cancer." In *Time Series Analysis: Theory and Practice 1*, ed. O. D. Anderson. Amsterdam: North-Holland.

Durbin, J. (1960). "The fitting of time series models." *Rev. Int. Inst. Statist.*, **28,** pp. 233–244.

Durbin, J. (1970). "Testing for serial correlation in least-squares regression when some of the regressors are lagged dependent variables." *Econometrika*, **38,** pp. 410–421.

Durbin, J., and Watson, G. S. (1950). "Testing for serial correlation in least-squares regression: I." *Biometrika*, **37,** pp. 409–428.

Durbin, J., and Watson, G. S. (1951). "Testing for serial correlation in least-squares regression: II." *Biometrika*, **38,** pp. 159–178.

Durbin, J., and Watson, G. S. (1971). "Testing for serial correlation in least-squares regression: III." *Biometrika*, **58,** pp. 1–19.

Fox, A. J. (1972). "Outliers in time series." *J. Roy. Stat. Soc.*, B, **34,** pp. 340–363.

Franke, J., Hardle, W., and Martin, D,. (1984). *Robust and Nonlinear Times Analysis*. New York: Springer Verlag.

Fuller, W. A. (1976). *Introduction to Statistical Time Series*. New York: Wiley.

Fuller, W. A., and Dickey, D. A. (1979). "Distribution of the estimators for autoregressive time series with a unit root." *J. Amer. Stat. Assoc.*, **74,** pp. 427–431.

Fuller, W. A., and Hasza, D. P. (1981). "Properties of predictors for autoregressive time series." *J. Amer. Stat. Assoc.*, **76,** pp. 155–161.

Gaver, D. P., and Lewis, P. A. W. (1980). "First order autoregressive gamma sequences and point processes." *Adv. Appl. Prob.*, **12,** pp. 727–745.

Gersch, W., and Kitagawa, G. (1983). "The prediction of time series with trends and seasonalities." *J. Business Econ. Statist.* **1,** pp. 253–264.

Goldberg, S. I. (1958). *Introduction to Difference Equations*. New York: Science Editions.

Gonedes, N. J., and Roberts, H. V. (1977). "Differencing of random walks and near random walks." *J. Econometrics*, **6**, pp. 289–308.

Granger, C. W. J. (1978). "New classes of time series models." *The Statistician*, **27**, pp. 237–253.

Granger, C. W. J., and Newbold, P. (1976). "Forecasting transformed series." *J. Roy. Stat. Soc.*, B, **38**, pp. 189–203.

Gray, H. L., Kelley, G., and McIntire, D. (1978). "A new approach to ARMA modelling." *Comm. Statist.*, B, **7**, pp. 1–78.

Grenander, U., and Rosenblatt, M. (1957). *Statistical Analysis of Stationary Time Series*. New York: Wiley.

Hannan, E. J. (1970). *Multiple Time Series*. New York: Wiley.

Hannan, E. J., and Quinn, B. G. (1979). "The determination of the order of an autoregression." *J. Roy. Stat. Soc.*, B, **41**, pp. 190–195.

Hannan, E. J., and Rissanen, J. (1982). "Recursive estimation of mixed autoregressive–moving average order." *Biometrika*, **69**, pp. 81–94.

Harrison, P. J., and Stevens, C. F. (1976). "Bayesian forecasting." *J. Roy. Stat. Soc.*, B, **38**, pp. 205–247.

Harvey, A. C. (1981a). *The Econometric Analysis of Time Series*. Oxford: Phillip Allen.

Harvey, A. C. (1981b). "Finite sample prediction and overdifferencing." *J. Time Series Anal.*, **2**, pp. 221–232.

Harvey, A. C. (1981c). *Time Series Models*. New York: Halsted Press.

Harvey, A. C. (1984). "A unified view of statistical forecasting procedures." *J. Forecasting*, **3**, pp. 245–276.

Harvey, A. C., and Philips, G. D. A. (1979). "Maximum likelihood estimation of regression models with autoregressive–moving average disturbances." *Biometrika*, **66**, pp. 49–58.

Harvey, A. C., and Todd, P. H. J. (1983). "Forecasting economic time series with structural and Box–Jenkins models: a case study." *J. Business Econ. Statist.*, **1**, pp. 299–307.

Harvey, A. C., and Tomenson, J. (1981). "A note on testing for gaps in seasonal moving average models." *J. Roy. Stat. Soc.*, B, **43**, pp. 240–243.

Hasza, D. P. (1980). "The asymptotic distribution of the sample autocorrelation for an integrated ARMA process." *J. Amer. Stat. Assoc.*, **75**, pp. 349–352.

Hasza, D. P., and Fuller, W. A. (1982). "Testing for nonstationary parameter specifications in seasonal time series models." *Ann. Statist.*, **10**, pp. 1209–1216.

Henrici, P. (1974). *Applied and Computational Analysis*. Vol. 1. New York: Wiley.

Hillmer, S. C. (1982). "Forecasting with trading day variation." *J. Forecasting*, **1**, pp. 385–395.

Hillmer, S. C., and Tiao, G. C. (1982). "An ARIMA based approach to seasonal adjustment." *J. Amer. Stat. Assoc.*, **77**, pp. 63–70.

Hoff, J. C. (1983). *A Practical Guide to Box–Jenkins Forecasting*. Belmont, Calif.: Lifetime Learning.

Hopwood, W. S., McKeown, J. C., and Newbold, P. (1984). "Time series forecasting models involving power transformations." *J. Forecasting*, **3**, pp. 57–61.

Jenkins, G. M. (1982). "Some practical aspects of forecasting in organizations." *J. Forecasting*, **1**, pp. 3–21.

Jones, R. H. (1971). "Spectrum estimation with missing observations." *Ann. Inst. Stat. Math.*, **23,** pp. 387–398.

Jones, R. H. (1975). "Fitting autoregressions." *J. Amer. Stat. Assoc.*, **70,** pp. 590–592.

Jones, R. H. (1980). "Maximum likelihood fitting of ARMA models to time series with missing observations." *Technometrics*, **22,** pp. 389–395.

Jones, R. H., and Tryon, P. V. (1983). "Estimating time from atomic clocks." *J. Research NBS*, **88,** pp. 17–24.

Kalman, R. E. (1960). "A new approach to linear filtering and prediction problems." *Trans. ASME J. Basic Engineering*, **82,** pp. 34–45.

Kedem, B., and Slud, E. (1981). "On goodness of fit of time series models: an application of higher order crossings." *Biometrika*, **68,** pp. 551–556.

Kedem, B., and Slud, E. (1982). "Time series discrimination by higher order crossings." *Ann. Statist.*, **10,** pp. 786–794.

Keenan, D. M. (1982). "A time series analysis of binary data." *J. Amer. Stat. Assoc.*, **77,** pp. 816–821.

Kleinbaum, D. G., and Kupper, L. L. (1978). *Applied Regression Analysis and Other Multivariate Methods*, Boston: Duxbury Press.

Ledolter, J., and Abraham, B. (1981). "Parsimony and its importance in time series forecasting." *Technometrics*, **23,** pp. 411–414.

Levinson, N. (1947). "The Wiener RMS error criterion in filter design and prediction." *J. Math. Physics*, **25,** pp. 261–278.

Liu, L. M. (1980). "Analysis of time series with calendar effects." *Management Science*, **26,** pp. 106–112.

Ljung, G. M., and Box, G. E. P. (1978). "On a measure of lack of fit in time series models." *Biometrika*, **65,** pp. 67–72.

Ljung, G. M., and Box, G. E. P. (1979). "The likelihood function of stationary autoregressive–moving average models." *Biometrika*, **66,** pp. 265–270.

Lusk, E. J., and Wright, H. (1982). "Non-Gaussian series with non-zero means: practical implications for time series analysis." *Stat. Prob. Letters*, **1,** pp. 2–6.

McLeod, A. I. (1975). "The derivation of the theoretical autocorrelation function of autoregressive–moving average time series." *Appl. Statist.*, B, **24,** pp. 255–256. Correction in B, **26,** p. 194.

McLeod, A. I. (1978). "On the distribution of residual autocorrelations in Box–Jenkins models." *J. Roy. Stat. Soc.*, A, **40,** pp. 296–302.

Macpherson, B. D., and Fuller, W. A. (1983). "Consistency of the least squares estimator of the first order moving average parameter." *Ann. Statist.*, **11,** pp. 326–329.

Mallows, C. L. (1967). "Linear processes are nearly Gaussian." *J. Appl. Prob.*, **4,** pp. 313–329.

Maravall, A. (1983). "An application of nonlinear time series forecasting." *J. Business Econ. Statist.*, **1,** pp. 66–74.

Marshall, R. J. (1980). "Autocorrelation estimation of time series with randomly missing observations." *Biometrika*, **67,** pp. 567–570.

Martin, R. D. (1980). "Robust estimation of autoregressive models." In *Directions in Time Series*, ed., D. R. Brillinger and G. C. Tiao. Hayward, Calif.: Institute of Mathematical Statistics.

Martin, R. D. (1981). "Robust methods for time series." In *Applied Time Series Analysis II*, ed. D. Findley. New York: Academic Press.

Martin, R. D., Samarov, A., and Vandaele, W. (1983). "Robust methods for ARIMA models." In *Proceedings of ASA-CENSUS-NBER Conference on Applied Time Series Analysis of Economic Data*, ed. A. Zellner. Washington, D.C.: U.S. Dept. of Commerce, Bureau of the Census.

Masry, E. (1983). "Non-parametric covariance estimation from irregularly-spaced data." *Adv. Appl. Prob.*, **15,** pp. 113–132.

Meinhold, R. J., and Singpurwalla, N. D. (1983). "Understanding the Kalman filter." *Amer. Statistician*, **37,** pp. 123–127.

Montgomery, D. C., and Johnson, L. A. (1976). *Forecasting and Time Series Analysis*. New York: McGraw-Hill.

Muth, J. F. (1960). "Optimal properties of exponentially weighted forecasts of time series with permanent and transitory components." *J. Amer. Stat. Assoc.*, **55,** pp. 299–306.

Neftci, S. N. (1982). "Specification of economic time series models using Akaike's criterion." *J. Amer. Stat. Assoc.*, **77,** pp. 537–540.

Nelson, C. R. (1973). *Applied Time Series Analysis for Managerial Forecasting*. San Francisco: Holden Day.

Nelson, C. R., and Kang, H. (1981). "Spurious periodicity in inappropriately detrended time series." *Econometrica*, **49,** pp. 741–751.

Nelson, C. R., and Kang, H. (1983). "Pitfalls in the use of time as an explanatory variable in regression." *J. Business Econ. Statist.*, **2,** pp. 73–82.

Newbold, P. (1974). "The exact likelihood function for a mixed autoregressive–moving average process." *Biometrika*, **61,** pp. 423–426.

Newbold, P., and Granger, C. W. J. (1974). "Experiences with forecasting univariate time-series and the combination of forecasts." *J. Roy. Stat. Soc.*, A, **137,** pp. 131–165.

Newton, H. J., and Pagano, M. (1983). "The finite memory prediction of covariance stationary time series." *Siam J. Sci. Stat. Comp.*, **4,** pp. 330–339.

Nicholls, D. F. (1977). "A comparison of estimation methods for vector linear time series models." *Biometrika*, **64,** pp. 85–90.

Niemi, H. (1983). "On the effect of a nonstationary noise on ARMA models." *Scan. J. Stat.*, **10,** pp. 11–18.

O'Donovan, T. M. (1983). *Short Term Forecasting: An Introduction to the Box–Jenkins Approach*. New York: Wiley.

Pandit, S. M., and Wu, S. M. (1983). *Time Series and Systems Analysis with Applications*. New York: Wiley.

Pankratz, A. (1983). *Forecasting with Univariate Box–Jenkins Models: Concepts and Cases*. New York: Wiley.

Parzen, E. (1974). "Recent advances in time series modeling." *IEEE Trans. Auto Control*, **19,** pp. 723–730.

Parzen, E. (1981). "Time series model identification and prediction variance horizon." In *Applied Time Series Analysis II*, ed. D. Findley, New York: Academic Press.

Parzen, E. (1982). "ARARMA models for time series analysis and forecasting." *J. Forecasting*, **1,** pp. 67–82.

Parzen, E., ed. (1984). *Time Series Analysis of Irregularly Observed Data*. New York: Springer Verlag.

Petruccelli, J. D., and Davies, N. (1984). "Some restrictions on the use of corner method hypothesis tests." *Comm. Statist.—Theo. Meth.*, **13**, pp. 543–551.

Plosser, C. I., and Schwert, G. W. (1977). "Estimation of a noninvertible moving average process: the case of overdifferencing." *J. Econometrics*, **6**, pp. 199–224.

Priestly, M. B. (1978). "Nonlinear models in time series analysis." *The Statistician*, **27**, pp. 159–176.

Quenouille, M. H. (1949). "Approximate tests of correlation in time-series." *J. Roy. Stat. Soc.*, B, **11**, pp. 68–84.

Rao, T. S. (1981). "On the theory of bilinear time series models." *J. Roy. Stat. Soc.*, B, **43**, pp. 244–255.

Robinson, P. M. (1977). "Estimation of a time series from unequally spaced data." *Stoch. Proc. Appl.*, **6**, pp. 9–24.

Robinson, P. M. (1980). "Estimation and forecasting for time series containing censored and missing observations." In *Time Series Analysis*, ed. O. D. Anderson. Amsterdam: North Holland.

Pötscher, B. M. (1983). "Order estimation in ARMA-models by Lagrangian multiplier tests." *Ann. Statist.*, **11**, pp. 872–885.

Roy, R. (1977). "On the asymptotic behavior of the sample autocovariance function for an integrated moving average process." *Biometrika*, **64**, pp. 419–421.

Royston, J. P. (1983). "A simple method for evaluating the Shapiro–Francia W test of non-normality." *The Statistician*, **32**, pp. 297–300.

Sargan, J. D., and Bhargava, A. (1983). "Maximum likelihood estimation of regression models with first-order moving average errors when the root lies on the unit circle." *Econometrica*, **40**, pp. 617–636.

Shibata, R. (1976). "Selection of the order of an autoregressive model by Akaike's criterion." *Biometrika*, **63**, pp. 117–126.

Shumway, R. H., and Stoffer, D. S. (1982). "An approach to time series smoothing and forecasting using the EM algorithm." *J. Time Ser. Anal.*, **4**, pp. 253–264.

Slutsky, E. (1927). "The summation of random causes as the source of cyclic processes" (in Russian). *Problems of Economic Conditions*, **3**. English translation in *Econometrica*, **5** (1937), pp. 105–146.

Tiao, G. C., and Tsay, R. S. (1981). "Identification of nonstationary and stationary ARMA models." *Proc. Bus. Econ. Statist. Sec., Amer. Statist. Assoc.*, pp. 308–312.

Tiao, G. C., and Tsay, R. S. (1983a). "Consistency properties of the least squares estimates of autoregressive parameters in ARMA models." *Ann. Statist.*, **11**, pp. 856–871.

Tiao, G. C. and Tsay, R. S. (1983b). "Multiple time series modeling and extended sample cross correlations," *J. Business Econ. Statist.*, **1**, pp. 43–56.

Tryon, P. V., and Jones, R. H. (1983). "Estimation of parameters in models for cesium beam atomic clocks." *J. Res. NBS*, **88**, pp. 3–16.

Tsay, R. S., and Tiao, G. C. (1982). "Consistent estimates of autoregressive parameters and extended sample autocorrelation function for stationary and nonstationary ARMA models." Technical report no. 683. Department of Statistics, University of Wisconsin-Madison.

Tufte, E. R. (1983). *The Visual Display of Quantitative Information*. Cheshire, Conn.: Graphics Press.

Vandaele, W. (1983). *Applied Time Series and Box–Jenkins Models*. New York: Academic Press.

Walker, A. M. (1971). "On the estimation of a harmonic component in a time series with stationary independent residuals." *Biometrika*, **58**, pp. 21–36.

Wecker, W. E. (1981). "Asymmetric time series." *J. Amer. Stat. Assoc.*, **76**, pp. 16–21.

Whittle, P. (1953). "The analysis of multiple time series." *J. Roy. Stat. Soc.*, B, **15**, pp. 125–139.

Whittle, P. (1983). *Prediction and Regulation*. 2nd rev. ed. Minneapolis: University of Minnesota Press.

Wichern, D. W. (1973). "The behavior of the sample autocorrelation function for an integrated moving average process." *Biometrika*, **60**, pp. 235–239.

Wold, H. O. A. (1938). *A Study in the Analysis of Stationary Time Series*. (2nd ed., 1954.) Uppsala, Sweden: Almquist and Wiksell.

Woodward, W. A., and Gray, H. L. (1981). "On the relationship between the S array and the Box–Jenkins method of ARMA model identification." *J. Amer. Stat. Assoc.*, **76**, pp. 579–587.

Yule, G. U. (1926). "Why do we sometimes get nonsense-correlations between time series? A study in sampling and the nature of series." *J. Roy. Stat. Soc.*, **89**, pp. 1–69.

Yule, G. U. (1927). "On a method of investigating periodicities in disturbed series, with special reference to Wolfer's sunspot numbers." *Phil. Trans.*, **A226**, pp. 267–298.

ANSWERS TO SELECTED EXERCISES

CHAPTER 2

2.1 **a.** 16 **b.** 10.5 **c.** $5/(4\sqrt{11})$

2.2 0

2.3 **b.** $\gamma_k = \sigma_x^2$ for all k

2.5 **b.** $\text{Cov}(Z_t, Z_{t-k}) = \text{Cov}(X_t, X_{t-k})$ **c.** No; the mean is not constant.

2.9 **b.** Yes

2.11 Let γ_k be the autocovariance function for $\{X_t\}$. Then $\text{Corr}(Z_t, Z_{t-k}) = (\sigma_a^2 + \gamma_k)/(\sigma_a^2 + \gamma_0)$

CHAPTER 3

3.2 $2\sigma_a^2/n^2$

3.4 $2(2n - 1)\sigma_a^2/n^2$

CHAPTER 4

4.3 $\rho_1 = -\frac{10}{21}$ $\rho_2 = \frac{4}{21}$

4.4 AR(1)

4.5 ARMA(1, 1)

4.12 b. MA(2)

CHAPTER 5

5.1 A is stationary AR(2) and B is nonstationary IMA(1, 1). They have very similar π-weights and ψ-weights.

5.3 c. IMA(1, 1)

CHAPTER 6

6.1 MA(2) is not rejected.

6.3 ARMA(1, 1)

6.6 a. No **b.** No

6.8 I: a. Strongly positive **b.** Moderately positive **II: a.** Strongly negative **b.** Moderately positive

CHAPTER 7

7.1 $\hat{\phi}_1 = 1.11$ $\hat{\phi}_2 = -0.39$ $\hat{\sigma}_a^2 = 1.535$ and $\hat{\theta}_0 = 0.56$

7.2 $\hat{\rho}_1 = r_1 = -0.5$

7.3 204

CHAPTER 8

8.1 Yes; r_1 should be within ± 0.1 of zero.

8.3 r_1 and r_2 lie outside ± 2 standard deviations of zero, while r_3 is just inside. The modified Box–Pierce statistic is too big, though.

CHAPTER 9

9.1 10.1

9.2 a. $10.5 million and $11.55 million **c.** 7.73 to 13.27; 7.43 to 15.67 **d.** Updated value is 13.2

CHAPTER 10

10.1 a. $\psi_1 = -\theta_1$, $\psi_2 = -\theta_2$, $\psi_3 = 0$, $\psi_4 = 1$ **b.** 24.35, 20.75, 25, 40 **c.** 24 ± 2 or 22 to 26; 20.75 ± 2.2 or 18.55 to 22.95; 25 ± 2.3 or 22.7 to 27.3; 40 ± 2.3 or 37.7 to 42.3

10.3 ARMA $(p, q) \times (P, Q)s$ with a constant term

10.5 a. ARIMA $(1, 0, 1) \times (0, 1, 0)4$ **b.** ARIMA $(0, 1, 1) \times (0, 1, 1)12$

INDEX